T0136425

IoT-enabled Convolutional Neural Networks: Techniques and Applications

RIVER PUBLISHERS SERIES IN AUTOMATION, CONTROL AND ROBOTICS

The "River Publishers Series in Automation, Control and Robotics" is a series of comprehensive academic and professional books which focus on the theory and applications of automation, control and robotics. The series focuses on topics ranging from the theory and use of control systems, automation engineering, robotics and intelligent machines.

Books published in the series include research monographs, edited volumes, handbooks and textbooks. The books provide professionals, researchers, educators, and advanced students in the field with an invaluable insight into the latest research and developments.

Topics covered in the series include, but are by no means restricted to the following:

- Robots and Intelligent Machines
- Robotics
- Control Systems
- Control Theory
- Automation Engineering

For a list of other books in this series, visit www.riverpublishers.com

IoT-enabled Convolutional Neural Networks: Techniques and Applications

Editors

Mohd Naved

Amity International Business School (AIBS), Amity University, Noida, India

V. Ajantha Devi

Research Head, AP3 Solutions, Chennai, TN, India

Loveleen Gaur

Amity International Business School (AIBS), Amity University, Noida, India

Ahmed A. Elngar

Faculty of Computer and Artificial Intelligence, Beni-Suef University, Egypt

NEW YORK AND LONDON

Published 2023 by River Publishers
River Publishers
Alsbjergvej 10, 9260 Gistrup, Denmark
www.riverpublishers.com

Distributed exclusively by Routledge
605 Third Avenue, New York, NY 10017, USA
4 Park Square, Milton Park, Abingdon, Oxon OX14 4RN

IoT-enabled Convolutional Neural Networks: Techniques and Applications / by Mohd Naved, V. Ajantha Devi, Loveleen Gaur, Ahmed A. Elngar.

Routledge is an imprint of the Taylor & Francis Group, an informa business

ISBN 978-87-7022-725-4 (print)
ISBN 978-10-0087-969-8 (online)
ISBN 978-1-003-39303-0 (ebook master)

Contents

Preface

Convolutional neural networks (CNNs), a type of artificial neural network that has been popular in computer vision, are gaining popularity in a variety of fields, including radiology. CNN uses several building blocks like as convolution layers, pooling layers, and fully connected layers to learn spatial hierarchies of information automatically and adaptively through backpropagation. A neural network is a hardware and/or software system modelled after the way neurons in the human brain work. Traditional neural networks aren't designed for image processing and must be fed images in smaller chunks. CNN's "neurons" are structured more like those in the frontal lobe, the area in humans and other animals responsible for processing visual inputs. Traditional neural networks' piecemeal image processing difficulty is avoided by arranging the layers of neurons in such a way that they span the whole visual field. A CNN employs a technology similar to a multilayer perceptron that is optimised for low processing requirements. An input layer, an output layer, and a hidden layer with several convolutional layers, pooling layers, fully connected layers, and normalising layers make up a CNN's layers. The removal of constraints and improvements in image processing efficiency result in a system that is significantly more effective and easier to train for image processing and natural language processing. This Book article explains the core concepts of CNN and how they are used to diverse jobs, as well as their problems and future directions.

Through this edited volume we have intended to provide a structured presentation of CNN enabled IoT applications in vision, speech, and natural language processing. This book discusses a variety of CNN techniques and applications, including but not limited to IoT enabled CNN for speech denoising, smart app for visually impaired people, disease detection, ECG signal analysis, weather monitoring, texture analysis, etc.

Unlike other books on the market, this book covers the tools, techniques, and challenges associated with the implementation of CNN algorithms, computation time, and the complexity associated with reasoning and modelling various types of data. We have included CNN's current research trends and future directions. This edited book contains numerous new scientific results, and also compiles existing knowledge and research in the field. We have made a great effort to cover everything with citations while maintaining a fluent exposition, all corrections and suggestions will be greatly appreciated. Throughout the chapters, contributors included proper references as well as added all the needed diagrams and data within the content to make it easier for the readers to understand the concepts.

The following are the abstract of specific chapters included in this edited book:

Chapter-1:

Convolutional Neural Networks (CNNs) have shown to be revolutionary in computer vision applications, frequently surpassing standard models and even humans in image recognition tasks. Large models place a premium on cutting-edge performance and frequently find it difficult to scale down. The bibliometric approach is also used to identify the most advanced CNN in IoT. The research looked at papers in four major databases using the keywords "Convolutional Neural Network" and "IoT" or "Internet of Things": IEEE Xplore, SpringerLink, ACM Digital Library, and ScienceDirect. The Scopus database's publications are used in the bibliometric study. The network citation analysis and publishing trends, in particular, provide insight into the current domains and development patterns for CNN in IoT. The bibliometric research includes the most important and prolific writers, organisations, nations, sources, and documents. China has the most documents published, followed by the United States. Nanjing University of Posts and Telecommunications' College of Telecommunications and Information Engineering has published the most documents in the domains of IoT and CNN. Wang Y has published the most documents in this domain in the category of writers. In the fields of IoT and CNN, IEEE Access has the most papers.

Chapter-2:

The internet of things (IoT) has shown to be beneficial for the interconnection of computing devices embedded in items, allowing objects to transmit and receive data via the internet for day-to-day activities. It uses unique identities to connect computers equipment with mechanical, electronic, and other items such as animals or humans, allowing data to be transferred over a network without the need for human or computer involvement. Artificial intelligence (AI), deep learning, and machine learning are all being used to make data collection easier and more dynamic for IoT technologies. However, there is no categorical assessment of IoT research, particularly in a well-known research database, to identify the main fields where IoT is widely used and where it faces greater problems. As a result, this chapter discusses how CNN-enabled IoT techniques can be used in a variety of industries. The advantages and disadvantages of a CNN-enabled IoT-based system are highlighted. It summarises the study by field of application and identifies any gaps in the research. More partnerships between researchers and users/experts in areas where IoT applications are employed to overcome restrictions, improve IoT technology, and provide insight into the security of future work in cyberspace are recommended in this chapter.

Chapter-3:

Speech-controlled smart devices have recently become popular in Internet-of-Things (IoT) applications. Reverberation and noise are widely known for reducing the efficiency of human-machine interaction in indoor applications. As a result, speech augmentation has emerged as a significant front-end strategy for improving performance, garnering increased attention in recent years. This chapter focuses on single-channel speech augmentation algorithms based on deep learning (DL) for both denoising and dereverberation, with single and multiple speakers considered. Due to their parameter effectiveness and state-of-the-art performance, convolutional neural network (CNN) based models are provided for this difficult speech augmentation job. Following a description of one-stage and multi-stage CNN-based models, a series of experiments are carried out to demonstrate the benefits and drawbacks of using them to extract one desired speaker and multiple desired speakers. This research shows that CNN-based models can achieve great performance when only one desirable speaker is to be extracted, but that their performance degrades dramatically when numerous desired speakers are to be removed. Finally, future research topics are detailed after some potential ways for increasing the performance of extracting numerous desirable speakers are discussed.

Chapter-4:

In today's extravagant period, one of the symptoms of passionate newline in sight is the ability to recognise feelings, an element of human understanding that has been argued to be much more important than scientific and verbal intelligences. Newline workers have lost their interest or concentration in work activity, as well as their focus or newline performance in the working environment, as a result of gradual enrichment in IoT technology for smart newline environment level, technology disruptions, and degradation of performance in the industries. Furthermore, despite the rapid rise of IoT, in the newline field of modern intelligent services, current IoT-based systems notably lack cognitive newline intelligence, implying that they are unable to meet the needs of industrial services.

Newline Deep learning has become one of the most widely used approaches in a variety of machine learning applications and studies. While it is mostly used for content-based newline image retrieval, it can still be improved by using it in a variety of computer vision newline applications. According to the findings of a rigorous theoretical and practical analysis, a newline urgent need to address this issue by developing an emotional intelligent approach, Machine newline learning (deep learning, CIoT), which will mentor and counsel workers by monitoring their newline behaviour in the workplace. Newline The goal of this research was to develop a CNN model and controller area protocol-based emotional intelligence system (EIS) to automatically categorise expressions in the Facial Expression Recognition newline (FER2013) and kaggel picture databases.

Chapter-5:

According to the World Health Organization (WHO), millions of visually impaired people confront significant challenges in travelling independently around the world. They are always in need of assistance from folks who have normal vision. For visually impaired people, finding their way to their intended destination in a new environment is a huge difficulty. This report was written to aid these people in overcoming their difficulties in migrating to a new location on their own. To that end, we presented a way for visually impaired people to recognise the situation and scene elements automatically in real-time using a Convolutional Neural Network (CNN). Raspberry Pi, Ultrasonic Sensors, a camera, breadboards, Jumper wires, a buzzer, and an earphone are all part of the proposed system.

With the use of a Raspberry Pi and jumper wires, breadboards are utilised to connect the sensors. The sensors detect barriers and potholes, while the camera acts as a virtual eye for visually challenged persons, recognising these impediments from any direction (i.e., front, left and right). This system has a vital feature: when the blind receives the scene object, the system automatically calculates how far away he is from the obstacles, and a voice message informs and leads him through earphone. The CNN produced excellent results of 99.56 percent accuracy and a loss validation of 0.0201 percent, according to the collected testing findings.

Chapter-6:

Communication systems based on machine-to-machine contact are defined using Internet of Things (IoT)-based technologies. When the Internet of Things (IoT) is combined with a Convolution Neural Network (CNN), a system that can communicate with its environment using human speech can be created. Natural language processing (NLP) can work in conjunction with IoT-based deep learning systems to help with automation development. The Internet of Things (IoT) can connect a network of specialised devices and use deep learning to extract information such as sensor features, radio frequency features, and speech features. Speech-based recognition systems for home automation systems can be developed using IoT and NLP. To transmit particular orders to IoT devices, smart home applications can be integrated with voice command-based IoT devices. Furthermore, NLP-based IoT devices can assist impaired people in carrying out their regular tasks.

These devices can track their health and make voice-activated security changes. In addition, NLP-enabled IoT devices can aid in the automation of environmental data collecting, which includes geographic activities. However, language understanding, accent change, and voice change are all difficulties of NLP-based IoT application. These issues limit the efficiency and speed with which NLP-based IoT devices can be used. Deep learning technology combined with a large vocabulary library has opened up a plethora of possibilities for IoT speech and command recognition system training. CNN gadgets with IoT connectivity for voice recognition are a gift to society.

Chapter-7:

In therapeutic therapy, classification of ECG signals is crucial. Traditional methods have reached their limits of effectiveness, yet there is always room for improvement. The main purpose of this study is to automatically categorise and detect myocardial infarction using ECG signals. Deep Learning methods such as Convolutional Neural Network (CNN) and Long Short Term Memory (LSTM) algorithms were used in the proposed model Enhanced Deep Neural Network (EDN). On huge matrices of data, vector operations like matrix multiplication and gradient descent were performed in parallel with GPU support. In EDN, parallelism reduces the amount of time it takes for a procedure to run. The suggested model EDN has a greater accuracy (88.89%) than prior state-of-the-art techniques for the PTB database.

The EDN is 10 times faster than the LSTM due to the speed with which it converges during training. According to the algorithms' confusion matrices, the EDN achieved an accuracy of 87.80%. The recommended model displays performance improvisation based on metrics such as Precision, Recall, F1 measure, and Cohen Kappa Coefficient. These improvements to EDN's performance would help to save human lives.

Chapter-8:

In deep learning, image annotation is a difficult task. It is difficult for machines to recognise objects without the annotation. The major purpose of this chapter is to improve an automatic annotation system concept that comprises a pre-trained semantic segmentation model and the MATLAB Image Labeler Tool. In this chapter, we use pixel-wise semantic segmentation to automatically annotate Synthetic Aperture Radar photos of oil spills, which is possibly the most common task in computer vision. Due to their excellent feature representation, deep learning-based convolutional neural networks are currently redriving tremendous breakthroughs in semantic segmentation. To construct an automation algorithm, this proposed method uses a pre-trained DeepLabv3plus and Resnet18 as a backbone model. Each pixel in an input image is assigned to a separate category by DeepLabv3plus.

Image Labeler is being used to construct an automated algorithm for labelling oil spill photographs automatically. The article's originality stems from the usage of pre-trained deeplabv3+ as the backbone for image annotation using the Image Labeler function to increase the system's generalisation capacity. Over the oil spill SAR dataset, broad studies of proposed semantic image segmentation divisions utilising Resnet18 as backbone are conducted, and the findings are accurate when compared to Xception and Mobilenet backbone models.

Chapter-9:

Weather is a dynamic, chaotic, and multidimensional phenomenon that occurs in real time. Because of these qualities, forecasting and monitoring the weather is challenging. Wireless devices are critical components not only for businesses' development control, but also in day-to-day living for monitoring building security and movement streams, as well as estimating common metrics. Temperature, dampness, and weight are all components to be evaluated for this wander in atmosphere monitoring, thus sensors have been given the task of accomplishing everything considered. For purchasers and present-day applications, data gathering structures are well-known.

The proposed shape includes three sensors that process unusual factors as described above, as well as a rain fall recognisable evidence and storm bearing tempo estimation environment tool that is added to the stored data and compared to the previous 60 years of weather data to predict future weather using convolutional Neural Networks. Meteorologists used a range of methods for forecasting weather in the past, ranging from simple temperature, rain, air pollution, and moisture readings to complex computerised mathematical models. Convolutional Neural Networks (CNNs) are a powerful data modelling technique that uses deep learning to capture and represent complicated input/output interactions. Convolutional neural networks' real strength and advantage is their ability to simulate both linear and nonlinear relationships directly from data.

The quality and performance of these algorithms are assessed using an experimental approach in MATLAB 2013a. When compared to the conventional method, the use of convolutional neural networks delivered the most accurate weather prediction. The modelling findings, for the most part, reveal that reasonable forecast accuracy was achieved.

Chapter-10:

Efficient E-Learning (EL) environments for online learners are the responsibility of higher educational systems. Learners are engaged in instructional activities in an effective learning environment. The chapter is divided into three sections; the first section covered the Convolutional neural network (CNN). CNN includes a variety of models, but for the sake of this study, we used three models that we discovered to be the most effective in measuring students' SEt (SEt) in EL tasks. Because they have simple network architectures and exhibit efficiency in conditions and categories, we used All Convolutional network (All-CNN), Network-in-Network (NiN-CNN), and Very Deep Convolutional network (VD-CNN). These classifications are based on the circumstances of learners' facial expressions in an online setting.

The application of Machine Learning (ML) and Artificial Intelligence (AI) in EL is covered in the second part of the chapter. This chapter aims to illustrate the benefits of machine learning and artificial intelligence (AI) in E-Learning (EL) in general, as well as how the King Khalid University (KKU) EL Deanship is utilising AI and ML in its operations. In addition, academics have concentrated on the future of machine learning and artificial intelligence in any academic programme. The third section of the chapter delves into the role of the Internet of Things (IoT) in the education sector, outlining the benefits, types of security risks, and deployment obstacles.

The results for three CNN models are referred to for their advantages and challenges for online learners, while the results for ML and AI are based on qualitative analysis done through EL tools and techniques applied in KKU as an example, but the same modus operandi can be implemented by any institution in its EL platform. KKU uses Learning Management Systems (LMS) to provide online learning practises and Blackboard (BB) to share online learning resources, thus the researchers used these technologies to illustrate the findings. IoT has transformed the learning environment for both students and educators, according to the findings.

Chapter-11:

Texture analysis is important in computer vision since it is required for both the characterization and segmentation of image regions. It's used in a variety of technological fields, including food processing, materials characterization, remote sensing, and medical picture analysis, among others. Deep learning has redefined the state of the art in image recognition, and by extension, texture analysis and quantification, over the previous decade. Transfer learning with convolutional neural networks is driving much of the current developments for a variety of reasons.

Deep learning with convolutional neural networks will be discussed in this chapter, as well as a comparison of textures using different neural network topologies and traditional methodologies. Three case studies will be taken into consideration. In the first, Voronoi simulations of material textures will be examined, as well as convolutional neural networks' capacity to distinguish between different textures. In material science, these textures are crucial in the creation of geometallurgical and quantitative structure property connection models.

In the second case study, transfer learning will be used to demonstrate how froth imaging sensors in the mineral processing sectors may be considerably improved. Breakthroughs in this field could have a direct impact on flotation plant advanced real-time control. Textures linked with stochastic signals imitating stock price data will be studied in the final case study, and it will be demonstrated that these methods may be used to monitor minor changes in stock price data or any other signal more broadly.

Chapter-12:

By allowing data to be collected through various IoT-based devices and sensors, the Internet of Things (IoT) has revolutionised healthcare systems. These devices generate data in a variety of formats, such as text, images, and videos. As a result, getting accurate and useable data from the massive amounts of data generated by the Internet of Things is a difficult task. The diagnosis of various diseases utilising IoT data has lately arisen as an issue that requires sophisticated and effective approaches. It is difficult to make a good diagnosis due to the wide range of disease symptoms and signs. Currently available methods rely on either handmade features or a traditional machine learning model. As a result, the application of IoT-based enabled Convolutional Neural Network (CNN) in healthcare diagnostics is discussed in this chapter. The advantages and disadvantages of IoT-enabled CNN are highlighted. The chapter offers an intelligent IoT-based enabled CNN for determining a patient's health state. The CNN was utilised to diagnose data captured using IoT-based sensors for the ailment. As a result, the system makes use of the dataset's and CNN's features, ensuring high accuracy and reliability. The recommended system illustrates real-time health monitoring and evaluates the system's performance in terms of several metrics such as accuracy, recall, precision, and F1-score for a case study on healthcare dataset classifications.

List of Figures

List of Tables

List of Contributors

Abul Hassan, Muhammad, *Department of Computing and Technology, Abasyn University Peshawar, Pakistan; E-mail: abulhassan900@gmail.com*

Adebisi, O. A., *Department of Computer Engineering, Ladoke Akintola University of Technology, Nigeria; E-mail: oadebisi44@pgschool.lautech.edu.ng*

Adigun, M. O., *Department of Computer Science, University of Zululand, South Africa; E-mail: adigunm@unizulu.ac.za*

Ahmad, Iziz, *Department of Computing and Technology, Abasyn University Peshawar, Pakistan; E-mail: izazahmad445@gmail.com*

Ajagbe, S. A., *Department of Computer Engineering, Ladoke Akintola University of Technology LAUTECH, Nigeria; Department of Computer Science, University of Zululand, South Africa;*
E-mail: saajagbe@pgschool.lautech.edu.ng

Alahmari, F., *Department of Information Systems, College of Computer Science, King Khalid University, KSA; E-mail: fahad@kku.edu.sa*

Aldrich, C., *Western Australian School of Mines: Minerals, Energy and Chemical Engineering, Curtin University, Australia;*
E-mail: chris.aldrich@curtin.edu.au

Alqahtani, H., *Department of Information Systems, College of Computer Science, King Khalid University, KSA; E-mail: hsqahtani@kku.edu.sa*

Aravindh, S. R., *Research & Development Team, Crapersoft, India*

Awotunde, J. B., *Department of Computer Science, University of Ilorin, Nigeria; E-mail: awotunde.jb@unilorin.edu.ng*

Ayo, F. E., *Department of Computer Science, McPherson University, Nigeria; E-mail: ayofe@mcu.edu.ng*

Banupriya, C. V., *Department of Computer Application, Hindusthan College of Arts and Science, Coimbatore, Tamilnadu;*
E-mail: Banupriya.venkat@gmail.com

Bhoi, A. K., *Department of Computer Science and Engineering, Sikkim Manipal Institute of Technology (SMIT), Sikkim Manipal University (SMU), India; E-mail: akashkrbhoi@gmail.com*

Bojaraj, L., *Assistant Professor, KGiSL Institute of Technology, India*

Choudhury, Tanupriya, *School of Computer Science, University of Petroleum and Energy Studies, Dehradun, India; E-mail: tanupriya1986@gmail.com*

Devi, Ajantha, *Research Head, AP3 Solution Chennai, TN, India; E-mail: ap3solutionsresearch@gmail.com*

Farhatullah, *School of Automation, China University of Geosciences, Wuhan 430074, China; E-mail: farhatkhan8398@gmail.com*

Ganesan, A., *Research & Development Team, Crapersoft, India*

Jain, R., *Bhagwan Parshuram Institute of Technology, India*

Jimoh, R. G., *Department of Computer Science, University of Ilorin, Nigeria; E-mail: jimoh_rasheed@unilorin.edu.ng*

Junaid, Hazrat, *Department of Computer Science and Information Technology, University of Malakand, Pakistan; E-mail: abidj3692@gmail.com*

Kalimuthu, S., *Tech Lead, HCL Technologies, Coimbatore, Tamil Nadu, India; Research & Development Team, Crapersoft, India; E-mail: sivanantham.k@hcl.com*

Karthikeyan, M. P., *Department of BCA, School of CS & IT, Jain (Deemed-to-be) University, India; E-mail: karthi.karthis@gmail.com*

Karunya, N., *Department of Computer Science, Sri Ramakrishna College of Arts and Science, India; E-mail: karunyakarun@gmail.com*

Katal, A., *School of Computer Science, University of Petroleum and Energy Studies, Dehradun, India; E-mail: avita207@gmail.com*

Kavitha, A., *Department of Computer Science, Kongunadu Arts and Science College, India*

Ke, Y., *Key Laboratory of Sound and Vibration Research, Institute of Acoustics, Chinese Academy of Sciences, China; University of Chinese Academy of Sciences, China*

Kumar, Ashish, *School of Computer Science Engineering and Technology, Bennett University, Greater Noida, India*

Li, X., *Key Laboratory of Sound and Vibration Research, Institute of Acoustics, Chinese Academy of Sciences, China; University of Chinese Academy of Sciences, China*

Liu, X., *Western Australian School of Mines: Minerals, Energy and Chemical Engineering, Curtin University, Australia*

Luo, X., *Key Laboratory of Sound and Vibration Research, Institute of Acoustics, Chinese Academy of Sciences, China; University of Chinese Academy of Sciences, China*

Mahoob, Muhammad Awais, *Department of Computer Science, University of Engineering & Technology Texila, Pakistan; E-mail: awaisntu@gmail.com*

Mahdi, Hussain Falih, *School of Computer Science, University of Petroleum and Energy Studies, Dehradun, India; E-mail: hussain.mahdi@ieee.org*

Mamgai, R., *Bharati Vidyapeeth's College of Engineering, India*

Manimekalai, K., *Department of Computer Applications, Sri GVG Visalakshi College for Women, India; E-mail: gvgmanimekalai@gmail.com*

Naim, A., *Department of Information Systems, College of Computer Science, King Khalid University, KSA; E-mail: arshi@kku.edu.sa*

Oguns, Y. J., *Department of Computer Studies, The Polytechnic Ibadan, Nigeria; E-mail: oguns.yetunde@polyibadan.edu.ng*

Oladosu, J. B., *Department of Computer Engineering, Ladoke Akintola University of Technology LAUTECH, Nigeria;*
E-mail: jboladosu@lautech.edu.ng

Pandiammal, R., *Department of Computer Science, PPG College of Arts and Science, India; E-mail: panbhavya@gmail.com*

Sathish, S., *Research & Development Team, Crapersoft, India*

Shanmugam, S., *Iquants Engineering, India*

Sohail, Muhammad, *Department of Computer Science, Bahria University Islamabad Campus, Pakistan; E-mail: sohailbahria02@gmail.com*

Sudha, V., *Department of Computer Science, Sri Ramakrishna College of Arts and Science, India; E-mail: sudhavaiyapuri@gmail.com*

Vijendran, Anna Saro, *Sri Ramakrishna College of Arts and Science, India; E-mail: annasarovijendran@srcas.ac.in*

Vijetha, M., *Department of Computer Science, Sri Ramakrishna College of Arts and Science, India; E-mail: Vijethachandran1983@gmail.com*

Zheng, C., *Key Laboratory of Sound and Vibration Research, Institute of Acoustics, Chinese Academy of Sciences, China; University of Chinese Academy of Sciences, China; E-mail: cszheng@mail.ioa.ac.cn*

List of Abbreviations

AAMS	Axial accelerometer magnetometer sensor
AC	Academic connectivity
ACS	Acute coronary syndromes
ADAM	Adaptive momentum estimation
AE	Autoencoders
AI	Artificial intelligence
AIA	Automatic image annotation
ANN	Artificial neural network
ARIMA	Autoregressive integrated moving average
ASICs	Application-specific integrated circuit
ATCP	Applications, techniques, challenges, and prospects
AUC	Area under the ROC curve
B2C	Business-to-consumer
BB	Blackboard
B-CNN	Bilinear convolutional neural network
BDC	Big data collection
BLSTM	Bidirectional LSTM
BN	Batch normalization
BoVW	Bag of visual words
BPNN	Backpropagation in neural network
CAD	Computer-aided detection
CAD	Coronary artery disease
CAN	Controller area network
CASA	Computational auditory scene analysis
CBT	Cognitive behavioral therapy
CC	Central control
CDS	Connected devices
CHD	Coronary heart disease
CIoT	Cognitive Internet of Things
cIRM	Complex ideal ratio mask
CLO	Course learning outcome

CMGGAN	Conditional multi-generator generative adversarial network
CMS	Compressed magnitude/complex spectrum
CNNs	Convolutional neural networks
COLAB	Google Open Source Research Laboratory
Conv-GLU	Convolutional gated linear unit
CRNs	Convolutional recurrent networks
CVD	Cardiovascular disease
DAWN	Deep adaptive wavelet network
DBN	Deep belief network
DC	Data cloud
DCN	Dense convolutional network
DCNN	Deeper convolution neural network
DDoS	Distributed denial of service
DDTL	Distant domain transfer learning
Deconv-GLU	Deconvolutional gated linear unit
Deep-TEN	Deep texture encoding network
DL	Deep learning
DNN	Deep neural network
DNS	Deep noise suppression
DoS	Denial of service
DSS	Decision support system
ECG	Electrocardiography
ECG	Electrocardiogram
EDN	Enhanced deep neural network
EER	Extensive educational resource
EIS	Emotional intelligence system
EL	E-learning
ELD	EL deanship
ELM	Extreme learning machine
ESTOI	Extended short-time objective intelligibility
ExL	Experiential learning
F2F	Face-to-face
FACS	Facial Action Coding System
FASON	First- and second-order fusion network
FCN	Fully connected neural network
FANN	Fast artificial neural network
FFANN	Fast feedforward artificial neural network
FFT	Fast fourier transform

FG2CNN	Fuzzy genetic grey-wolf based CNN Model
FGOL	Future game plan for online learning
FN	False negative
FNN	Feedforward neural network
FP	False positive
FPGAs	Field programmable gate array
FTDI	Future technology devices international
G2NN	Genetic grey based neural network
GCRN	Gated convolutional recurrent networks
GCT-Net	Gated convolutional TCM Net
GFN	Guided filter network
GLCM	Gray level co-occurrence matrix
GLU	Gated linear unit
GLZLM	Gray level zone length matrix
Gnd	Ground
GPS	Global positioning system
GPUs	Graphics processing unit
GSM	Global system for mobile communications
GUI	Graphical user interface
HCI	Human–computer interface
HDC	Hybrid dilated convolution
HiCH	Hierarchical computing design
HOG	Histogram of oriented gradients
IC	Image creation
ICMD	Intelligent crop monitoring device
ICSP	
IDE	Integrated development environment
IHD	Ischemic heart disease
IIoT	Industrial Internet of Things
ILSVRC	ImageNet Large Scale Visual Recognition Challenge
IoMT	Internet of Medical Things
IoT	Internet of Things
IoU	Intersection over union
KKU	King Khalid University
KNN	K-nearest neighbor
L&T	Learning and teaching
LBP	Local binary pattern
LBP-TOP	Local binary patterns in three orthogonal planes

LDA	Linear discriminant analysis
LDP	Local directional pattern
LDP-TOP	Local directional patterns in three orthogonal planes
LDR	Light dependent resistor
LGBM	Light gradient-boosting machine
LIoT	Large IoT
LMS	Learning management service
LPS	Log power spectrum
LSTM	Long short-term memory
LTP	Local ternary pattern
MAP	Mean average precision
MAR	Multivariate adaptive regression
MI	Myocardial infarction
ML	Machine learning
MLA	Mineral liberation analyzer
MLOTS	ML's role in online training strategy
MLP	Multi-layer perceptron
MMSE	Minimum mean square error
MNIST	Modified national institute of standards and technology
MOOC	Massively online open course
MSE	Mean square error
MWSVM	Multiple-weight SVM
NFC	Near field communication
NGLDM	Neighborhood gray level dependence matrix
NiN	Network-in-network
NLP	Natural language processing
NNs	Neural networks
NPV	Negative predictive value
NST	Neural style transfer
NSTEMI	Non-ST segment elevation myocardial infarction
OCR	Optical character recognition
OnT	Online tutoring
OOB	Out-of-bag
PCA	Principal component analysis
PESQ	Perceptual evaluation of speech quality
PID	Proportional integral derivative
PPV	Positive predictive value

PReLU	Parametric ReLU
PTB	Pulmonary tuberculosis
PWM	Pulsewidth modulation
QEMSCAN	Quantitative evaluation of minerals by scanning electron microscopy
QoS	Quality of services
RANSAC	Random sample consensus
RBM	Restricted Boltzmann machine
RCNN	Residual conventional neural network
ReLU	Rectified linear unit
ResNet	Residual neural network
RF	Random forest
RFID	Radio-frequency identification
RI	Real ideal
rIBM	Real ideal binary mask
RIR	Room impulse response
rIRM	Ratio mask
RNN	Recurrent neural network
ROC	Receiver operating characteristic
ROI	Region of interest
RTT	Research available tech tool
SAR	Synthetic aperture radar
SDR	Signal-to-distortion ratio
SGDM	Stochastic gradient descent with momentum
SGLDM	Spatial gray level dependence matrix
SIR	Signal-to-interference ratio
SLAM	Simultaneous localization and mapping
SNN	Shallow neural network
SSE	Sum of squared errors
SSP	Statistical signal processing
S-TCM	Squeezed TCM
S-TCM	Squeezed temporal convolutional modules
STEMI	ST segment elevation myocardial infarction
STFT	Short-time Fourier transform
SVM	Support vector machine
TCNN	Texture convolutional neural network
T-F	Time–frequency
TL	Transfer learning
TMS	Target magnitude/complex spectrum

TN	True negative
TP	True positive
TSCN	Two-stage complex network
t-SNE	t-Distributed stochastic neighbor embedding
UA	Unstable angina
UCI	University of California Irvine machine learning repository
UIDs	Unique identifiers
VGG	Visual geometry group
VIN	voltage input
VIP	Visually impaired persons
VISOB	
WBAN	Wireless body area network
WBS	Wearable body sensor
WDBC	Wisconsin Diagnostic Breast Cancer
WHO	World Health Organization
WLAN	Wireless local area network
WPE	Weighted prediction error
WSNs	Wireless sensor networks

1

Convolutional Neural Networks in Internet of Things: A Bibliometric Study

Avita Katal[1,*], Hussain Falih Mahdi[2], and Tanupriya Choudhury[3]

[1]School of Computer Science, University of Petroleum and Energy Studies, Dehradun, India
[2]Department of Computer and Software Engineering, University of Diyala, Baquba, Iraq
[3]School of Computer Science, University of Petroleum and Energy Studies, Dehradun, India
E-mail: avita207@gmail.com; hussain.mahdi@ieee.org; tanupriya1986@gmail.com
*Corresponding Author

Abstract

Convolutional neural networks (CNNs) have shown to be ground breaking in computer vision applications, routinely outperforming standard models and even humans in image identification challenges. CNNs are frequently tested on computer vision tasks, although their effects are far-reaching. When deploying a CNN on resource-constrained IoT devices, there are two options: scale a big model down or utilize a tiny model developed particularly for resource-constrained settings. Small designs usually sacrifice accuracy for computational expense by using depth-wise convolutions rather than conventional convolutions as in big networks. Large models prioritize cutting-edge performance and frequently struggle to scale down enough. The goal of this study is to thoroughly examine and analyze the research landscape on CNN in Internet of Things. The bibliometric analysis is also done for detecting the state-of-the-art CNN in IoT. The approach is capable of describing trends of publishing within a specific time or body of literature by employing quantitative analysis and statistics. The study looked at publications with the keywords "convolutional neural network" and "IoT" or "Internet of

Things" in four major databases: IEEE Xplore, SpringerLink, ACM Digital Library, and ScienceDirect. The publications of the Scopus database are considered for the bibliometric analysis. A total of 1286 documents from 2015 to 2021 are used for the bibliometric analysis. The publishing patterns and network citation analysis, in particular, shed insight on the present domains and development patterns for CNN in IoT. The most significant and prolific authors, organizations, nations, sources, and documents are included in bibliometric study. The co-occurrence of all keywords is also examined in order to identify the most promising study topics in the field of CNN in IoT. The results of the study revealed that the number of papers published each year has grown rapidly on an average of 100–150 papers each year. China has published the maximum number of documents followed by United States. College of Telecommunications and Information Engineering, Nanjing University of Posts and Telecommunications, Nanjing, China has published the maximum documents in the domain of IoT and CNN. In the authors' category, Wang Y. has published the maximum number of documents in this domain. *IEEE Access* has maximum papers in the domain of IoT and CNN. This research will assist academicians and practitioners in gaining a thorough grasp of the current state and developments in this sector.

Keywords: Internet of Things, convolutional neural network, bibliometric analysis, VOSviewer.

1.1 Introduction

Internet of Things (IoT) has sparked a lot of attention in recent years. For example, an artificial-intelligence-based IoT robot might be utilized in a number of surveillance applications. Convolutional neural networks (CNNs) perform exceptionally well in a variety of machine learning applications. As IoT devices integrated with sensors pervade every part of contemporary life, requirement to run CNN analysis, a computationally intensive function, on devices that are resource constrained is becoming a requirement. CNNs are widely utilized in image and video processing applications. Despite their exceptional performance, CNNs are computationally demanding. There have been several suggestions for speeding CNNs with application-specific integrated circuit (ASICs), field programmable gate array (FPGAs), and graphics processing unit (GPUs). Due to their great performance and ease of use, GPUs are the primary platform for accelerating CNNs. The advancement of IoT enables the connection of smart devices over the Internet. Advances in IoT technology hold enormous potential for higher quality, more efficiency, and intelligent operations. Research on smart IoT operation solutions is gaining popularity as IoT technology advances. Recent study indicates that IoT has even more potential uses in information-intensive industrial areas.

Other expectations, including autonomous installation, automation system, and optimal functioning, are developing in the IoT service system, in contrast to inter-object communication support. Despite the fact that it enables Internet connectivity and automation features through pre-setting, it is challenging to sustain continuous functioning and value generation in the application area. To overcome these issues, user monitoring and action are required. The growth and rising popularity of various computational technologies (machine learning and deep learning) allows a wide variety of previously unsolved smart applications and difficulties to be considered again. The smart IoT service system is defined as a system that receives input from its surrounding, utilizes the data gathered to detect the situation, and interacts with the user environments using service regulations and specialized knowledge.

Machine-learning-based subsystems are rapidly being used in IoT edge devices, necessitating resource-efficient designs and implementations, particularly in battery-constrained settings. Since CNNs are non-exact, estimate computations can be utilized to reduce the required runtime and power usage on resource-constrained IoT edge devices without substantially affecting prediction performance. A CNN is a sort of ML technique that can tackle a wide range of issues, including visual processing and robust feature extraction for recognition and detection.

This chapter is structured as follows. Section 1.2 describes the related work, Section 1.3 demonstrates the research objectives, and Section 1.4 explains the overview of the selected studies for review followed with Section 1.5 covering the overview of bibliometric analysis. Section 1.6 explains the methodology followed for bibliometric analysis. Section 1.7 demonstrates the limitations and future work followed by conclusion in Section 1.8.

1.2 Related Work

Chen and Deng (2020) analyzed the research progress in the field of CNNs using the bibliometric method. A simple statistical analysis and co-citation network was used to examine textual extracts from CNNs. Experiments demonstrate that CNNs are used in a variety of computer vision applications, including fault and image identification diagnostics, seismic detection, location, and automatic detection of fractures and signals, image analysis, and pattern classification, whereas Nakhodchi and Dehghantanha (2020) presented a detailed overview of the use of deep learning in cybersecurity studies and gap bridging. During 2010 and 2018, around 740 papers from the ISI Web of Science database were evaluated. The number of articles as well as the number of citations are addressed utilizing bibliometric analysis. The authors also provided analyses based on nations and continents, study topics, authors, institutions, phrases, and keywords.

Sakhnini *et al.* (2021) offered a bibliometric review of scholarly publications focusing on the security concerns of IoT-aided smart grids. All articles and journals were subjected to a bibliometric analysis, with the results organized by chronology, author, and important topics. Moreover, the researchers review the numerous cyber-threats confronting the smart grid, many security techniques presented in the research, and the research needed in the area of intelligent grid security. Gamboa-Rosales *et al.* (2020) presented a bibliometric analysis presented a bibliometric analysis of the decision-making using Internet of Things and machine learning. According to their study, Hanumayamma Innovations and Technologies, Inc., Beijing University of Posts and Telecommunications, Vellore Institute of Technology, Vellore, and others are among the most productive enterprises. In this regard, the most prolific nations are, in order, India, the United States of America, China, the United Kingdom, Canada, Spain, France, and Germany. The relationship between the most prolific organizations and nations demonstrates the diversity and high quality of the articles covered by this study topic.

1.3 Research Questions

The current research examines the bibliometrics of Internet of Things enabled convolutional neural networks. The following research questions have been formulated to achieve the stated goal:

RQ1: In terms of publishing output, what are the general publishing patterns from 2015 to October 2021?

RQ2: Which countries have the maximum number of documents and citations in the domain of IoT-enabled CNN?

RQ3: Which organizations have the maximum number of documents in the domain of IoT-enabled CNN?

RQ4: Who has published the maximum number of articles in the area of IoT-enabled CNN?

RQ5: Which journals, proceedings, or book chapters have the maximum number of articles for the topic IoT-enabled CNN?

RQ6: Which document has the maximum number of citations?

RQ7: Which keywords occur the maximum number of times and what are the hot research topics?

1.4 Literature Review

Field monitoring is a critical function in the IoT, in which several IoT nodes collect data and transmit data to a ground unit or the cloud for computation, reasoning, and evaluation. Whenever the observations are high-dimensional, this transfer gets costly. IoT networks with limited bandwidth and low-power gadgets are difficult to manage such regular high broadcasts.

There are several fields in which the combination of both IoT and CNN technologies has been applied. Some of them are listed below.

Shin *et al.* (2019) presented an IoT solution outfitted with a smart surveillance robot that employs machine learning to bypass the constraints of traditional CCTV. The customer is provided an app that enables customers to directly control the robot. CNN-based ML is advantageous in the setting of image context analysis, which is provided for the exact identification and classification of pictures or pattern, and with which a high recognition performance may be predicted.

Njima *et al.* (2019) created a localized architecture that moves the online forecasting difficulty to an offline pre-processing phase, based on CNN. Its goal is to precisely pinpoint a sensor node by calculating its geographical area. Received Signal Strength Indicator (RSSI) fingerprints are used to create 3D radio pictures. The model findings verify the use of various settings, optimization methodologies, and model architectures.

Song *et al.* (2021) proposed FDA3, a powerful federated defense technique that can aggregate defensive knowledge against hostile samples from several sources. Their suggested cloud-based architecture, inspired by federated learning, allows IIoT devices to share security capabilities against various assaults.

Apart from the above-mentioned domains of defense and malware classification, there are various other domains like healthcare (Niu *et al.*, 2021), image classification (Li *et al.*, n.d.), agriculture (Shylaja *et al.*, 2021), etc., in which the integration of IoT and CNN plays a vital role.

Niu *et al.* (2021) developed an approach for medical image categorization based on Distance Domain Transfer Learning (DDTL). They developed a DDTL model for COVID-19 diagnosis using unlabeled Office-31,

Catech-256, and chest X-ray image large datasets as original data and a small batch of COVID-19 lung CT image data as target domain. CNN, Alexnet, and Resnet were chosen as non-transfer supervised foundation systems.

Li *et al.* (n.d.) constructed a deep-learning-based IoT image identification system that uses CNN to build image processing algorithms and PCA and LDA to identify picture features.

Shylaja *et al.* (2021) proposed a new machine-learning-enabled smart IoT media to help the agriculture sector. They developed an Intelligent Crop Monitoring Device (ICMD) to analyze crops in the crop land around the clock, seven days a week. This type of monitoring gadget improves agricultural productivity and service quality, as well as associated products.

1.5 Overview of Bibliometric Analysis

A statistical examination of published scientific articles, books, or book chapters is described as bibliometric analysis, and it is an effective approach of quantifying the influence of publication in the research world (de Moya-Anegón *et al.*, 2007). The number of times a piece of research has been mentioned by other writers may be used to determine its academic influence. A bibliometric analysis or citation classics research design is a popular method for determining an item's effect.

A bibliographic analysis is a statistical examination of books, papers, or other printed materials. In recent years, bibliometric analysis has been used to demonstrate the state, features, history, and growing patterns of knowledge in a range of professional disciplines. This can help interested academicians who lack proficiency in some disciplines get a complete understanding. From a micro to macro standpoint, bibliometrics may express a substantial amount of academic research. When the scope of review is extensive, bibliometric analysis should be performed on bigger datasets (Donthu *et al.*, 2021). Bibliometric techniques may be used to clearly examine the performance of several disciplines. As a result, the VOSviewer clusters will be employed in this effort to conduct a thorough and systematic evaluation of IoT-enabled CNN research. VOSviewer is a software for creating and viewing bibliometric maps (Van Eck, N. J., & Waltman, L. , 2009). The bibliometric research group can use the software for free. VOSviewer may be used to generate co-citation maps for authors or journals, as well as keyword co-occurrence maps.

Scholars employ bibliometric analysis to achieve a range of objectives, including detecting developing alterations in article and journal quality, cooperation patterns, and study characteristics, as well as studying the cognitive architecture of a certain field in the recent research. The information extracted through bibliometric analysis is generally very large, but its explanations are commonly reliant on both impartial and interpretive assessment founded via educated methods and practices. In another terms, bibliometric analysis can be utilized to systematically make sense of huge volumes of unorganized information in order to grasp and map the cumulative scientific information and developmental intricacies. As a consequence, well-conducted bibliometric experiments can build a firm base for progressing a profession in novel and constructive methods, which facilitates and enables researchers to (1) obtain a one-stop summary, (2) recognize weak areas, (3) develop new research concepts, and (4) position their intentional commitments to the field.

As a result of the preceding discussion, it can be concluded that bibliometric studies are important for presenting trends in the study's spheres and that they are conducted on a regular basis to report on the top authors, publication venues, top organizations and countries, citation landscape, and other related information, as well as research trends in the domain under study. As a consequence, the goal of this research is to give a full analysis of IoT-enabled CNN through bibliometric analysis of the relevant literature.

1.6 Methodology for Bibliometric Analysis

1.6.1 Database Collection

Web of Science (https://www.webofknowledge.com/), Scopus (https://www.scopus.com/), Google Scholar (https://scholar.google.com/), and Scimago (https://www.scimagojr.com/) are all well-known databases that contain a diverse spectrum of academic articles. We searched the bibliometric literature on IoT-enabled CNN to locate the best digital library. Clarivate Analytics Web of Science and Scopus are the most often used databases for analyzing and acquiring citation data from the literature. For the bibliometric survey, this study relied on Scopus, the most well-known database.

1.6.2 Methods for Data Extraction

Around 1286 papers were collected from the Scopus database, together with their publishing metadata. The data was exported in CSV format from Scopus and entered into the VOSviewer for bibliometric analysis.

The following are the queries used for searching Scopus for documents:

Search Parameters: TITLE-ABS-KEY (internet AND of AND things AND convolutional AND neural AND networks) AND (LIMIT-TO DOCTYPE, "cp") OR LIMIT-TO (DOCTYPE, "ar") OR LIMIT-TO (DOCTYPE, "ch") AND (EXCLUDE (PUBYEAR, 2022)) AND (LIMIT-TO (PUBSTAGE, "final")) AND (LIMIT-TO (LANGUAGE, "English")).

The inquiry is limited to articles, book chapters, and conference papers. Review papers are not included since they summarize carefully chosen scientific knowledge (Ellegaard & Wallin, 2015). This information is often transmitted in the literature, along with a complete bibliography for the subject. Bibliometric research, on the other hand, concentrates on statistical data and is almost never used in combination with a field bibliography.

The keywords are important since they provide context for the study topic. As a result, the keywords "blockchain" and "healthcare" are used to get documents from the Scopus database. The documents are sorted by year, language, and document type. From 2015 to October 2021, papers published in journals, conferences, and book chapters in English were used for this investigation.

1.6.3 Year-Wise Publications

The yearly pattern of publication activity allows for an examination of the stage of growth, knowledge gathering, and maturity of IoT-enabled CNN. Because there are relatively few publications before 2015, only papers

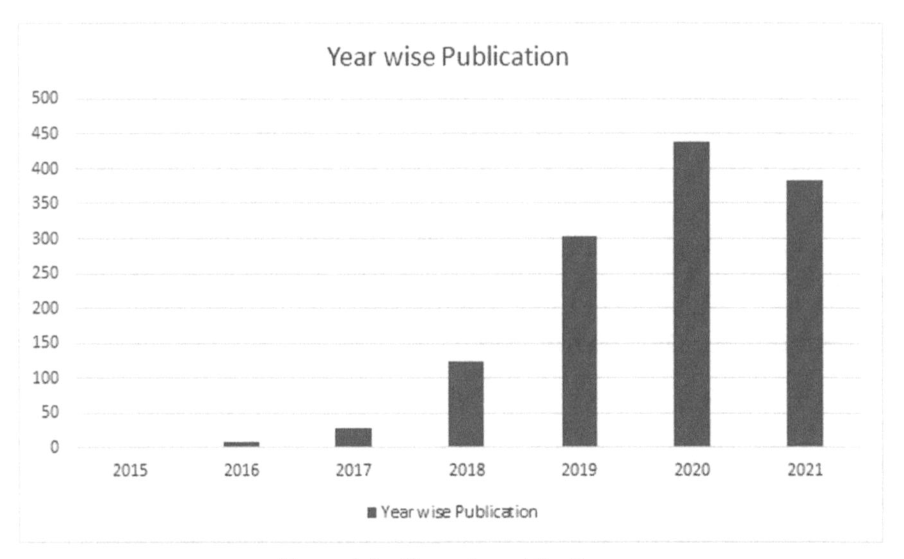

Figure 1.1 Year-wise publications.

from 2015 to 2021 are evaluated. After 2015, the number of publications concerning IoT-enabled CNN increased from 7 in 2016 to 28 in 2017, 123 in 2018, 303 in 2019, 439 in 2020, and 384 in 2021 through October 31st, 2021. Figure 1.1 depicts a graph of publications in the field of IoT-enabled CNN.

1.6.4 Network Analysis of Citations

1.6.4.1 Citation analysis of countries

The citation criteria were satisfied by 25 nations out of 171, with a minimum of 5 papers per nation and a minimum of 100 citations per nation. China has the maximum number of documents as 473 with total citations of 2614, followed by the United States with 233 documents with total citations of 2115 followed by India with 202 documents with total citations of 769. Citation network analysis in respect of countries is shown in Figure 1.2.

Table 1.1 shows the top 25 countries with maximum documents.

1.6.4.2 Citation analysis of organizations

With a minimum of 2 documents per organization and a minimum of 50 citations per organization as a threshold, 18 organizations out of 2797

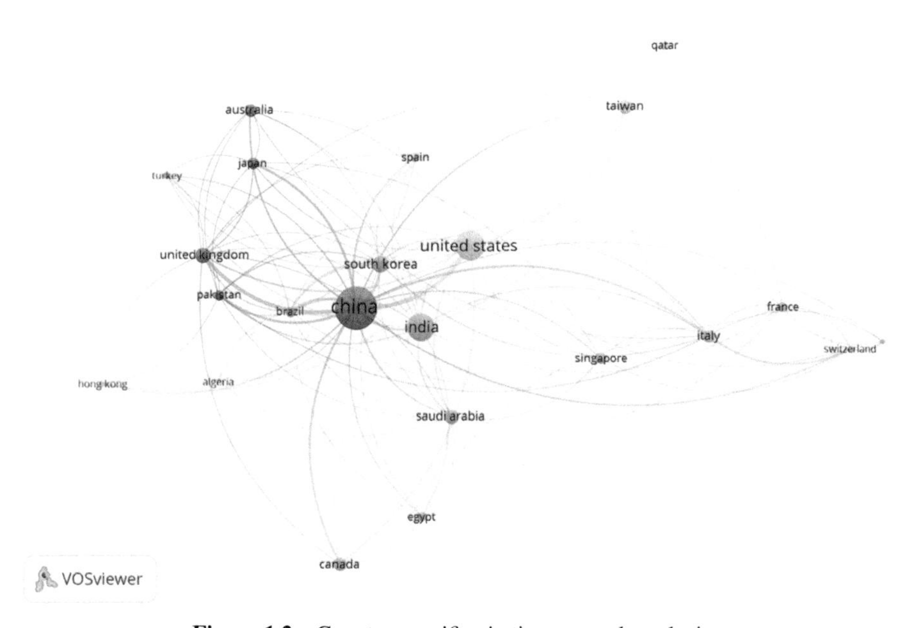

Figure 1.2 Country-specific citation network analysis.

Table 1.1 Top 25 countries in terms of documents

Country	Documents	Citations	Total link strength
China	473	2614	143
United States	233	2115	70
India	202	769	38
South Korea	71	448	43
United Kingdom	65	1082	43
Saudi Arabia	53	300	26
Taiwan	52	327	18
Canada	45	407	17
Australia	44	356	12
Italy	44	576	37
Japan	37	408	20
Singapore	32	232	10
Brazil	29	291	38
Pakistan	29	434	27
Egypt	26	175	9
France	25	192	8
Spain	23	433	20
Bangladesh	19	136	0
Qatar	19	106	2
Turkey	13	135	14
Hong Kong	11	122	9
Switzerland	11	318	19
Austria	9	209	7
Belgium	9	183	0
Algeria	6	138	10

satisfied the requirement. College of Telecommunications and Information Engineering, Nanjing University of Posts and Telecommunications, Nanjing, China, has six documents. Citation network analysis in terms of organizations is represented in Figure 1.3.

Table 1.2 shows the top 18 organizations in terms of number of documents published.

1.6.4.3 Citation analysis of authors

Using a minimum of 5 documents by the author and a minimum number of 100 citations, 28 meet the thresholds out of 3874 authors. Wang Y. has a maximum number of documents that is 39 with total citations of 409 followed by Li J. having 25 documents with 250 citations. Citation network analysis in terms of authors is represented in Figure 1.4.

Figure 1.3 Citation network analysis in terms of organizations.

Table 1.2 Top organizations

Organization	Documents	Citations	Total link strength
College of Telecommunications and Information Engineering, Nanjing University of Posts and Telecommunications, Nanjing 210003, China	6	78	0
Department of Electrical Engineering and Computer Sciences, University of California, Berkeley, Berkeley, CA, United States	5	113	0
Department of Information Systems and Cyber Security, University of Texas at San Antonio, San Antonio, Tx 78249, United States	4	54	0
Department of Software, Sejong University, Seoul 143-747, South Korea	4	149	5
Department of Software Engineering, College of Computer and Information Sciences, King Saud University, Riyadh 11543, Saudi Arabia	3	59	0
Institute of Microelectronics, Tsinghua University, Beijing 100084, China	3	56	2
Intelligent Media Laboratory, Digital Contents Research Institute, Sejong University, Seoul 143-747, South Korea	3	149	3

(Continued)

Table 1.2 Continued.

Organization	Documents	Citations	Total link strength
School of Electrical and Electronics Engineering, Nanyang Technological University, Singapore	3	89	0
Department of Computer Engineering, College of Computer and Information Sciences, King Saud University, Riyadh 11543, Saudi Arabia	2	59	0
Department of Computer Science, Guelma University, Guelma 24000, Algeria	2	138	0
Department of Electrical Engineering and Computer Science, Syracuse University, Syracuse, NY, United States	2	92	0
Department of Electrical Engineering, City University of New York, City College, NY, United States	2	92	0
Department of Electrical Engineering, University of Southern California, Los Angeles, CA 90089, United States	2	55	0
Department of Electrical Engineering, University of Southern California, Los Angeles, CA, United States	2	92	0
Graduate Program in Applied Informatics, Universidade De Fortaleza, Fortaleza 60811-905, Brazil	2	74	2
Integrated Systems Laboratory, Eth Zurich, Zürich 8092, Switzerland	2	113	2
School of Computer Science, South China Normal University, Guangzhou, China	2	54	0
University of Washington, Seattle, WA, United States	2	191	0

Table 1.3 shows the top authors based on the number of documents published.

1.6.4.4 Source citation analysis

The source citation analysis is performed using a threshold of 3 documents per source and a minimum of 50 citations per source. Only 18 of the 556 sources fulfill the criteria. *IEEE Access* has the maximum number of documents, that is, 97 and total citations of 937 followed by *IEEE Internet of Things Journal* with 84 documents and total citations of 684. Citation network analysis in terms of sources is shown in Figure 1.5.

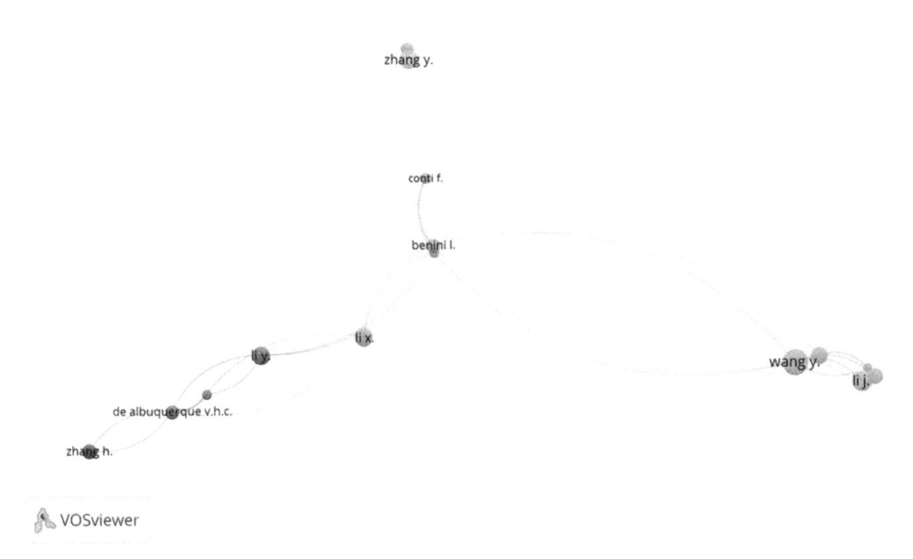

Figure 1.4 Citation network analysis in terms of authors.

Table 1.3 Top authors based on the number of documents published

Author	Documents	Citations	Total link strength
Wang Y.	39	409	19
Li J.	25	250	4
Li X.	21	149	7
Zhang Y.	21	135	5
Li Y.	20	204	4
Yang J.	18	219	12
Zhao Y.	16	368	8
Zhang H.	13	102	3
Zhou Y.	13	149	0
Benini L.	12	315	19
De Albuquerque V.H.C.	12	188	11
Li Z.	12	177	0
Wang C.	11	104	3
Chen Z.	10	243	1

(Continued)

Table 1.3 Continued.

Author	Documents	Citations	Total link strength
Li W.	10	127	0
Yuan B.	8	165	0
Conti F.	7	202	10
Zhang Q.	7	153	1
Muhammad K.	6	171	9
Ren A.	6	164	0
Rossi D.	6	243	18
Ding C.	5	119	0
Gui G.	5	122	10
Kumar N.	5	131	0
Li L.	5	174	0
Qiu Q.	5	153	0
Spanos C.J.	5	113	0
Zou H.	5	113	0

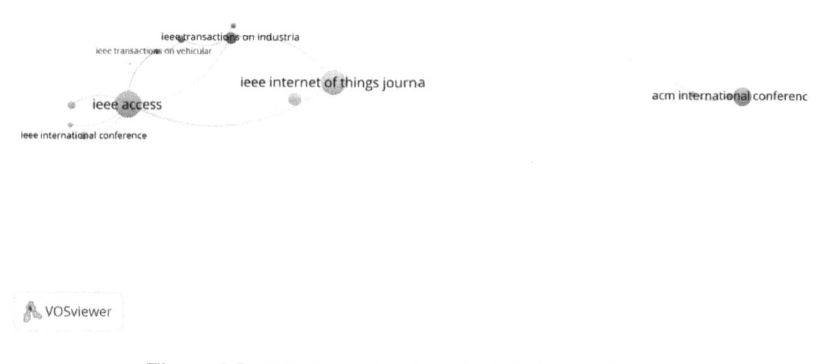

Figure 1.5 Citation network analysis in terms of sources.

Table 1.4 shows the top sources on the basis of the published documents.

1.6.4.5 Citation analysis of documents

A document's citation analysis is achieved by taking into account a threshold of 70 citations per document. Only 20 papers out of 1348 passed the test. *Deep Sense: A Unified Deep Learning Framework for Time-Series Mobile Sensing Data Processing* (Yao *et al.*, 2017) has the maximum citations, that is, 218. Citation network analysis in terms of documents is shown in Figure 1.6.

Table 1.4 Top sources that publish most documents

Source	Documents	Citations	Total link strength
IEEE Access	97	937	16
IEEE Internet of Things Journal	84	684	11
ACM International Conference Proceeding Series	47	57	1
Sensors (Switzerland)	22	236	3
IEEE Transactions on Industrial Informatics	19	503	6
2020 IEEE International Conference on Informatics, IoT, And Enabling Technologies, ICIOT 2020	17	59	0
Procedia Computer Science	15	161	0
Wireless Communications and Mobile Computing	10	93	0
Computer Communications	8	82	2
Future Generation Computer Systems	8	62	3
IEEE International Conference on Communications	8	75	1
IEEE Sensors Journal	8	55	0
IEEE Transactions on Computer-Aided Design of Integrated Circuits and Systems	7	220	3
IEEE Transactions on Circuits and Systems I: Regular Papers	5	203	3
Security and Communication Networks	5	50	0
Computer Networks	4	102	1
IEEE Journal of Biomedical and Health Informatics	4	198	2
IEEE Transactions on Vehicular Technology	4	113	4

Table 1.5 shows the author names of the documents with maximum citations.

1.6.5 Co-occurrence Analysis for KEYWORDS/HOT RESEARCH AREAS

Since one keyword frequency assessment may immediately and successfully define fields of study and fundamental material of a certain issue, this article created some keyword co-occurrence networks, thus marking the areas of IoT-enabled CNN research.

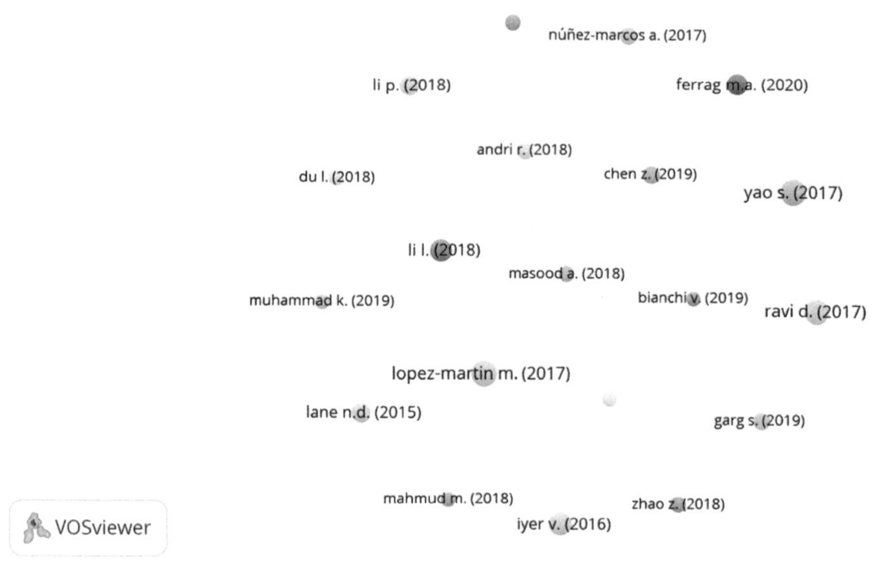

Figure 1.6 Document-based citation network analysis.

Table 1.5 Top documents based on the citations

Document	Citations	Links
Yao (2017)	218	0
Lopez-Martin (2017)	205	0
Ravi (2017)	197	0
Iyer (2016)	174	0
Li (2018)	165	0
Ferrag (2020)	137	0
Li (2018)	126	0
Lane (2015)	122	0
Andri (2018)	96	0
Chen (2019)	95	0
Garg (2019)	93	0
Núũez-Marcos (2017)	91	0
Zheng (2019)	87	0

(Continued)

Table 1.5 Continued.

Document	Citations	Links
Du (2018)	86	0
Zhao (2018)	82	0
Masood (2018)	79	0
Mahmud (2018)	76	0
Bianchi (2019)	75	0
Muhammad (2019)	75	0
Conti (2017)	70	0

1.6.5.1 Co-occurrence for all keywords

When analyzing co-occurrences, many keywords are considered. To be considered, the keywords must have at least five occurrences. Only 687 of the 9624 keywords tested positive. The word Internet of Things appears 1071 times with a total link strength of 11,141. Figure 1.7 depicts a network analysis of co-occurrences for all terms. The following are the parameter settings: The kind of research is co-occurrence; the method of counting is full counting; and all keywords are the focus of the study. The top 10 keywords

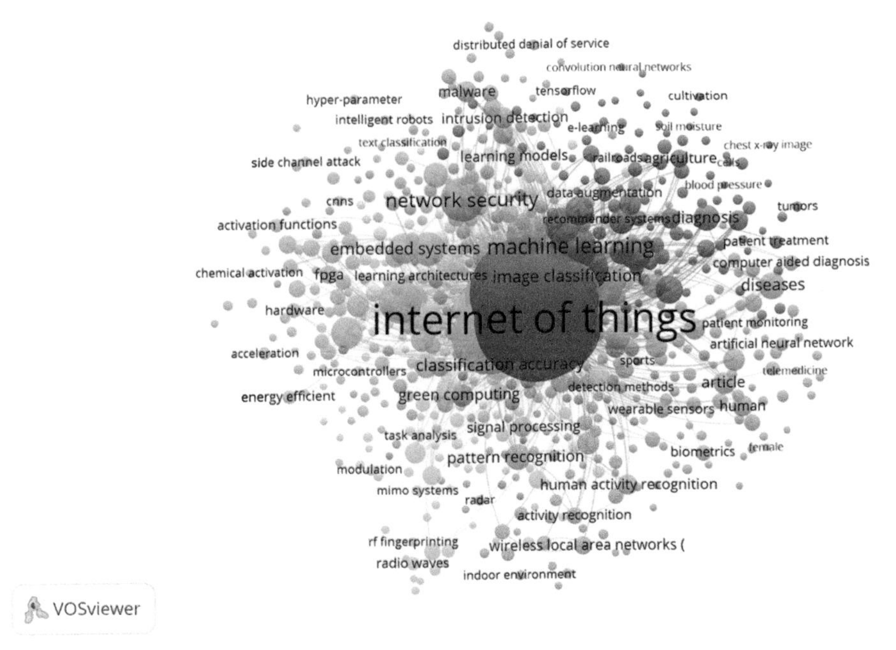

Figure 1.7 Network analysis of co-occurrences for all keywords.

Table 1.6 Top 10 keywords of blockchain in healthcare research on the basis of the frequency

Keyword	Occurrences	Total link strength
Internet of Things	1071	11,141
Convolutional neural networks	796	8088
Deep learning	626	6781
Convolutional neural network	553	5812
Convolution	541	5997
Neural networks	376	4049
Internet of Things (IoT)	336	3765
Deep neural networks	309	3494
Learning systems	239	2895

of IoT-enabled CNN research on the basis of the frequency are shown in Table 1.6.

The density visualization map based on occurrences-weights, as shown in Figure 1.8, reveals the fascinating study themes.

The size of the circle represents the weight of the keyword link in the overlay visualization, and the gradient color from blue to yellow shows the average citation score of a term.

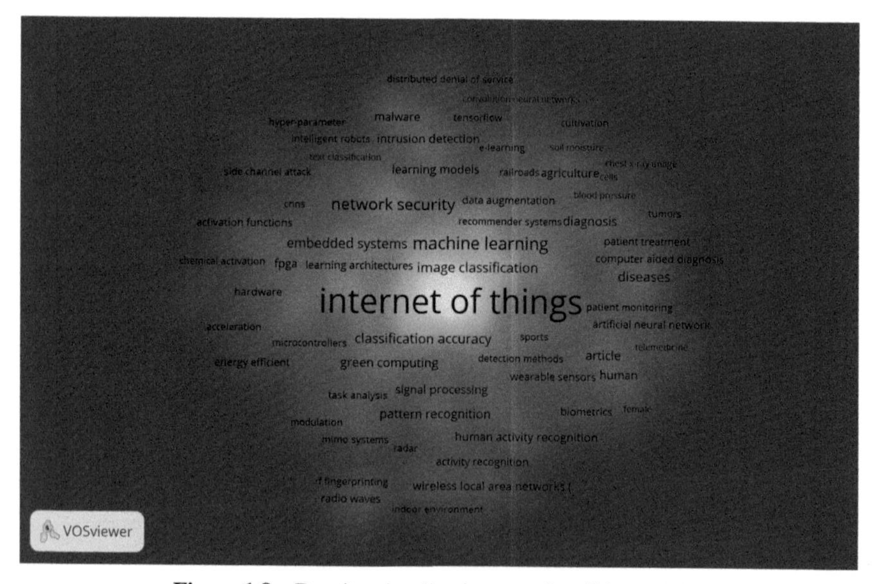

Figure 1.8 Density visualization map for all keywords.

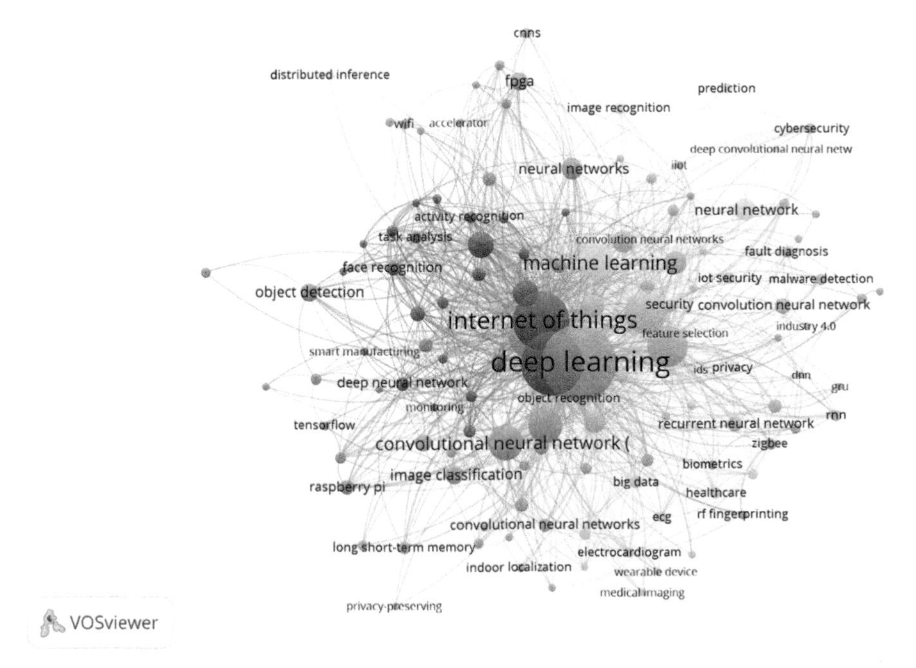

Figure 1.9 Network analysis of co-occurrences for author keywords.

1.6.5.2 Co-occurrence for author keywords

Considering the minimum occurrences of a particular keyword 5 as a threshold, out of 3273 keywords, 130 met the threshold. Network analysis of co-occurrences in terms of author keywords is shown in Figure 1.9.

The top 10 author keywords of IoT-enabled CNN research on the basis of the frequency are shown in Table 1.7.

The density visualization map depending on author keyword occurrences-weights revealed the hot study subjects, as illustrated in Figure 1.10.

1.6.5.3 Co-occurrence for index keywords

Considering the minimum occurrences of a particular keyword 5 as a threshold, out of 7367 keywords, 595 met the criteria. Network analysis of co-occurrences for index keywords is shown in Figure 1.11.

The top 10 index keywords of IoT-enabled CNN research on the basis of the occurrences are shown in Table 1.8.

As shown in Figure 1.12, the density visualization map based on index keyword occurrences-weights demonstrates the hot research topics.

Table 1.7 Top 10 author keywords of IoT-enabled CNN research on the basis of the frequency

Keyword	Occurrences	Total link strength
Deep learning	385	752
Convolutional neural network	265	444
Internet Of Things	224	493
IoT	123	255
Convolutional neural networks	121	219
CNN	112	216
Machine learning	95	218
Convolutional neural network (CNN)	85	136
Internet Of Things (IoT)	85	168
Edge computing	46	120

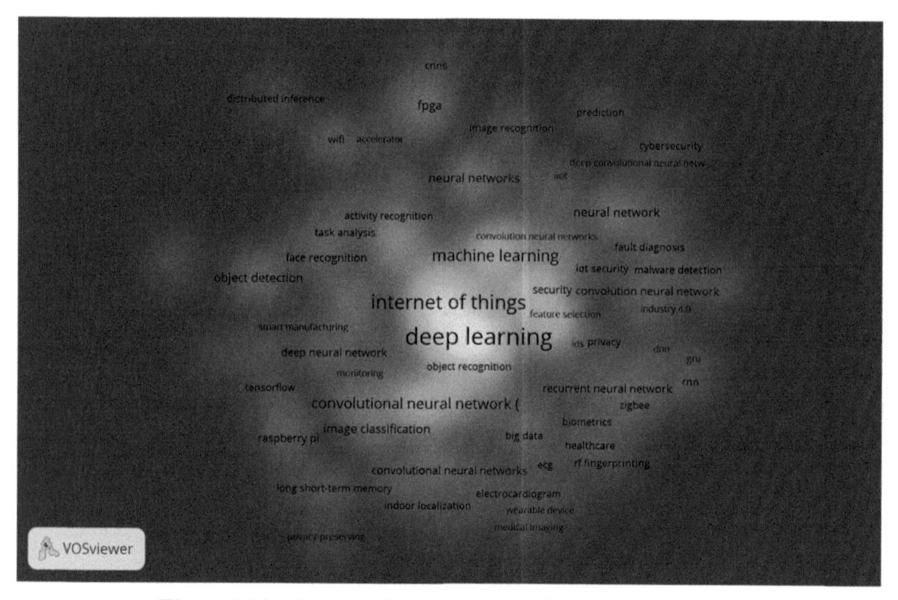

Figure 1.10 Density visualization map for author keywords.

1.7 Limitations and Future Work

- *Computational Complexity and Resource Availability:* Without a doubt, the benefits of deep neural networks (DNNs) and IoT complement each other. However, many scientific researchers are attempting to answer the difficulty of properly combining DNNs with IoT. Cloud computing provides intrinsic benefits such as virtualization, large-scale integration,

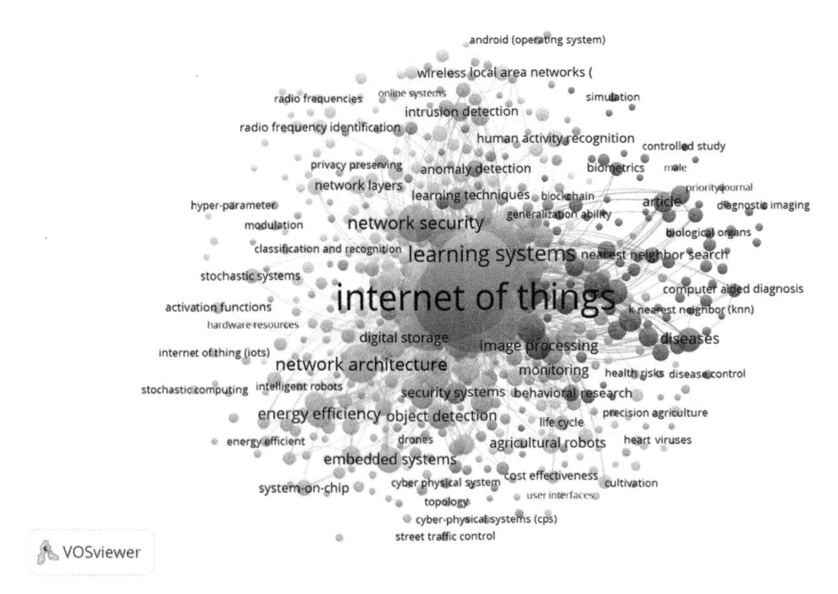

Figure 1.11 Network analysis of co-occurrences for index keywords.

Table 1.8 Top 10 index keywords of IoT-enabled CNN research on the basis of the frequency

Keyword	Occurrences	Total link strength
Internet of Things	1013	9431
Convolutional neural networks	729	6383
Convolution	538	5331
Deep learning	530	5359
Convolutional neural network	377	3858
Neural networks	355	3552
Deep neural networks	305	3095
Internet of Things (IoT)	293	2942
Learning systems	238	2524
Classification (of information)	144	1521

high dependability, high scalability, and cheap cost. Because of the huge DNN models and computational complexity, computing the inference results on devices with minimal resources is problematic. The obvious answer is to install the DNN model on a server in a cloud data center. However, with so many computing jobs being done in the cloud, the data that must be transported is huge in size and quantity, putting significant strain on the network bandwidth and computational power of the cloud computing infrastructure.

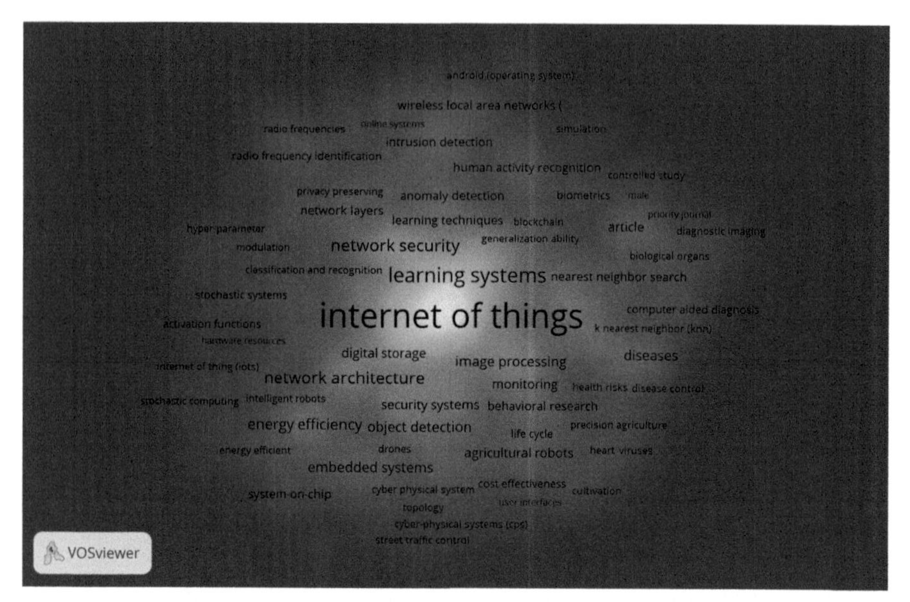

Figure 1.12 Density visualization map for index keywords.

- *Network Overhead:* Furthermore, practically, all IoT applications demand ultra-low power consumption, limited storage space use, and real-time data processing, particularly those that are sensitive to delay or are highly interactive. The large volume of data transfer has significantly increased the strain on the backbone network, resulting in massive expansion and maintenance expenditures for service providers. Traditional approaches for DNN computations rely on massive clouds to meet the exorbitant resource needs of DNNs. However, utilizing this strategy may result in significant delays and lost energy.
- *Data Privacy:* In practical use, IoT applications have critical real-time needs as well as user data privacy concerns. Deploying DNNs at edge computing nodes rather than faraway cloud servers is more efficient and safer.
- *Sensor Deployment*: Deep learning methods' performance is dependent on data sources. Even if the model's architecture is correctly constructed, the deep model is ineligible to play a role if there is insufficient clean data. As a result, determining how to install data gathering gadgets is a significant research subject. The number of sensors employed and how well they are dispersed has an impact on the reliability of the

information collected. The information included in the data is truly the key to resolving issues. For the whole work-flow of an IoT program, a data collection module must be built. For example, Li *et al.* (2016) purposefully constructed a photo collecting module DeepCham to increase the model's identification accuracy. In fact, it incorporated the concept of crowdsourcing within the data collecting module. In the creation of practical DL-based IoT applications, a cost-effective, dependable, and credible paradigm for the collection of data is needed.

- *Performance Degradation:* Training a deep network necessitates time-consuming procedures. As we all know, the depths of a deep learning network affect its ability to extract crucial characteristics. The gradient vanishment problem, on the other hand, develops when models go deeper, degrading performance. To that purpose, Hinton and Salakhutdinov (2006) suggest stacking RBMs as a method for pre-training models. Furthermore, replacing the sigmoid function with the ReLU function assists to alleviate the gradient vanishment problem. Another big issue that may arise when learning deep models is generalizing. The major answer is to add new data or to lower the number of variables. To decrease the number of parameters, one effective way is to use convolutional kernels, and applying the dropout (Srivastava *et al.*, 2014) is another option. Furthermore, a big breakthrough in CNN has been realized in recent years, and the number of levels in CNN algorithms has grown from 5 to over 200. Methods covered in these standard CNN (such as developing smaller convolutional kernels or batch normalization) could be useful when using the deep learning approach methods to find challenges in the wireless communication field.
- *Resource-Constrained Embedded System*: Deep learning is a powerful approach for analyzing vast volumes of data, but it requires a lot of equipment. It is still challenging to develop a detailed model of a resource-constrained embedded system. Till date, two types of research have been performed in an effort to solve the challenge. Considering end devices (such as a smart phone) to be data collectors. All data is sent to competent servers for analysis. Nevertheless, during this method, data exposure, network breakdown, and other difficulties may occur. Another alternative is to reduce complexity of the network while losing some efficiency so that some training sessions may be performed on end devices.

1.8 Conclusion

Convolutional neural networks (CNNs) have shown to be ground breaking in computer vision applications, regularly outperforming standard models and even humans in image identification challenges. CNNs are widely evaluated on computer vision applications, despite their far-reaching impacts. When implementing a CNN model in a resource-constrained context, like IoT devices or cell phones, striking a compromise among predictive performance and processing cost is critical to ensuring the model functions appropriately. The Internet of Things (IoT) has a huge influence on society and the economy as a revolutionary network system combined with computer, control, and communication technology. IoT and CNN integration has been widely used in medical, intelligent transportation, smart homes, and other industries. CNN would likely play a role in drawing important conclusions from this vast volume of data, hence assisting in the development of smarter IoT. Convolutional neural networks (CNNs) excel in a wide range of machine learning and deep learning applications. This chapter presented a visual and bibliometric analysis of current IoT-enabled CNN concepts. This bibliometric study looked at 1286 legitimate IoT CNN documents from the Scopus database from 2015 to October 2021. Since 2016, the number of publications on blockchain technology has exploded. The VOSviewer-identified research hotspots have supplied thorough information on the literature. Theoretical knowledge and hot research subjects on blockchain in healthcare are largely dispersed across: (1) Internet of Things; (2) convolutional neural networks; and (3) deep learning based on the high-frequency keywords apart from the blockchain and healthcare.

References

Chen, H., & Deng, Z. (2020). Bibliometric analysis of the application of convolutional neural network in computer vision. *IEEE Access*, *8*, 155417–155428. https://doi.org/10.1109/ACCESS.2020.3019336

Donthu, N., Kumar, S., Mukherjee, D., Pandey, N., & Lim, W. M. (2021). How to conduct a bibliometric analysis: An overview and guidelines. *Journal of Business Research*, *133*, 285–296. https://doi.org/10.1016/J.JBUSRES.2021.04.070

Ellegaard, O., & Wallin, J. A. (2015). The bibliometric analysis of scholarly production: How great is the impact? *Scientometrics*, *105*(3), 1809–1831. https://doi.org/10.1007/S11192-015-1645-Z

Gamboa-Rosales, N. K., Castorena-Robles, A., Casas-Valadez, M. A., Cobo, M. J., Castaneda-Miranda, R., & Lopez-Robles, J. R. (2020). Decision making using Internet of Things and machine learning: A bibliometric approach to tracking main research themes. *2020 International Conference on Data Analytics for Business and Industry: Way Towards a Sustainable Economy, ICDABI 2020*. https://doi.org/10.1109/ICDABI51230.2020.9325656

Hinton, G. E., & Salakhutdinov, R. R. (2006). Reducing the dimensionality of data with neural networks. *Science*, *313*(5786), 504–507. https://doi.org/10.1126/SCIENCE.1127647

Li, D., Salonidis, T., Desai, N. v., & Chuah, M. C. (2016). DeepCham: Collaborative edge-mediated adaptive deep learning for mobile object recognition. *Proceedings - 1st IEEE/ACM Symposium on Edge Computing, SEC 2016*, 64–76. https://doi.org/10.1109/SEC.2016.38

Li, J., Li, X., & Ning, Y. (n.d.). *Deep Learning Based Image Recognition for 5G Smart IoT Applications*.

de Moya-Anegón, F., Chinchilla-Rodríguez, Z., Vargas-Quesada, B., Corera-Álvarez, E., Muñoz-Fernández, F., González-Molina, A., & Herrero-Solana, V. (2007). Coverage analysis of Scopus: A journal metric approach. *Scientometrics*, *73*(1), 53–78. https://doi.org/10.1007/S11192-007-1681-4

Nakhodchi, S., & Dehghantanha, A. (2020). A bibliometric analysis on the application of deep learning in cybersecurity. *Security of Cyber-Physical Systems*, 203–221. https://doi.org/10.1007/978-3-030-45541-5_11

Niu, S., Liu, M., Liu, Y., Wang, J., & Song, H. (2021). Distant domain transfer learning for medical imaging. *IEEE Journal of Biomedical and Health Informatics*, *25*(10), 3784–3793. https://doi.org/10.1109/JBHI.2021.3051470

Njima, W., Njima, W., Zayani, R., Terre, M., & Bouallegue, R. (2019). Deep CNN for indoor localization in IoT-sensor systems. *Sensors*, *19*(14), 3127. https://doi.org/10.3390/S19143127

Sakhnini, J., Karimipour, H., Dehghantanha, A., Parizi, R. M., & Srivastava, G. (2021). Security aspects of Internet of Things aided smart grids: A bibliometric survey. *Internet of Things*, *14*, 100111. https://doi.org/10.1016/J.IOT.2019.100111

Shin, P. W., Kim, B., & Hwang, S. (2019). An IoT platform with monitoring robot applying CNN-based context-aware learning. *Sensors*, *19*(11), 2525. https://doi.org/10.3390/S19112525

Shylaja, S. L., Fairooz, S., Venkatesh, J., Sunitha, D., Rao, R. P., & Prabhu, M. R. (2021). IoT based crop monitoring scheme using smart device with machine learning methodology. *Journal of Physics: Conference Series, 2027*(1). https://doi.org/10.1088/1742-6596/2027/1/012019

Song, Y., Liu, T., Wei, T., Wang, X., Tao, Z., & Chen, M. (2021). FDA3: Federated defense against adversarial attacks for cloud-based IIoT applications. *IEEE Transactions on Industrial Informatics*, *17*(11), 7830–7838. https://doi.org/10.1109/TII.2020.3005969

Srivastava, N., Hinton, G., Krizhevsky, A., Sutskever, I., & Salakhutdinov, R. (2014). Dropout: A simple way to prevent neural networks from overfitting. *Journal of Machine Learning Research, 15*(56), 1929–1958. http://jmlr.org/papers/v15/srivastava14a.html

Van Eck, N. J., & Waltman, L. (2009). Software survey: VOSviewer, a computer program for bibliometric mapping. *Scientometrics, 84*(2), 523–538. https://doi.org/10.1007/S11192-009-0146-3

Yao, S., Hu, S., Zhao, Y., Zhang, A., & Abdelzaher, T. (2017). DeepSense: A unified deep learning framework for time-series mobile sensing data processing. *26th International World Wide Web Conference, WWW 2017*, 351–360. https://doi.org/10.1145/3038912.3052577

2

Internet of Things Enabled Convolutional Neural Networks: Applications, Techniques, Challenges, and Prospects

Sunday Adeola Ajagbe[1,*], Matthew O. Adigun[2], Joseph B. Awotunde[3], John B. Oladosu[4], and Yetunde J. Oguns[5]

[1,4]Department of Computer Engineering, Ladoke Akintola University of Technology LAUTECH, Nigeria
[1,2]Department of Computer Science, University of Zululand, South Africa
[3]Department of Computer Science, University of Ilorin, Nigeria
[5]Department of Computer Studies, The Polytechnic Ibadan, Nigeria
E-mail: saajagbe@pgschool.lautech.edu.ng; adigunm@unizulu.ac.za; awotunde.jb@unilorin.edu.ng; jboladosu@lautech.edu.ng; oguns.yetunde@polyibadan.edu.ng
*Corresponding Author

Abstract

The Internet of Things (IoT) has been proven useful for the interconnection of computing devices embedded in objects to enable objects to send and receive data through the Internet for day-to-day activities. It connects computing devices with mechanical, electronics, and other objects such as animals or a human with the help of unique identifiers that have the ability for data transmission through a network without the intervention of a computer or human being. It helps to work smartly and gain control over activities, it is popular in homes automation, and it is essential to business as it provides real-time monitoring into how businesses work. The IoT setup also provides supply chain and logistics operations with information on machine performance and ensure fast delivery of information. The use of artificial intelligence

(AI), deep learning, and machine learning for IoT technology cannot be undermined especially to ensure that easier and more dynamic data collection is possible. However, there is no categorical appraisal of IoT research that used convolutional neural network (CNN) techniques for its implementation, to identify the prominent field where IoT and CNN are highly embraced in technological advancement and where IoT-enabled CNN technologies are facing challenges. Therefore, this chapter reviews the applicability of IoT-enabled CNN applications and techniques in various fields. The challenges and prospects of an IoT-enabled CNN system, the prominent field of research that embraces IoT-enabled CNN, were discussed. It summarizes the research according to the field of application and identified areas of limitations. This chapter recommends more collaborations between researchers and the users and experts in the areas where such IoT applications are used to overcome limitations, enhance IoT technologies and provide an insight into the security of future work in cyberspace; these include improvement in the IoT-based mobility and scalability, management of quality of services (QoS), among others.

Keywords: Artificial intelligence, convolutional neural networks, deep learning (DL), IoT-applications, IoT-cybersecurity, machine learning (ML), industrial IoT.

2.1 Introduction

The Internet of Things (IoT) is a network of networked computing devices, mechanical and digital technologies, goods, animals, and people that use unique identifiers (UIDs) as a tool to enable data transfer without human or computer contact. It has a number of benefits for industries and professionals. It uses a huge number of sensors to generate massive amounts of data that may be used for a variety of applications and purposes. It can also be used in a variety of smart devices that are connected to the Internet, such as household appliances, sensors, phones, automobiles, and computer systems (since they are Internet-enabled). It has connectivity, interaction, and data interchange with a wide range of applications, from home automation and wearable technology to large-scale infrastructure development [1]. The IoT enables data integration and effective interchange between the individual in need and the service provider [2, 3]. It entails quality potentials to process data in the cloud environment to ensure the user's advantage.

IoT has been proven useful for the interconnection of computing devices embedded in objects to enable objects to send and receive data through the Internet for day-to-day activities. It connects computing devices with mechanical, electronics, and other objects such as animals or a human with the help of unique identifiers that have the ability for data transmission through a network without the intervention of a computer or human being. The application of CNN has been making the use of IoT more effective in the cloud computing field, as it ensures the data for IoT transmission error-free and provides ease of use of dataset. A person who has had a cardiac monitor implanted and a farm animal that has had a biochip transponder implanted, an automobile with in-built sensors to inform the driver on various issues, or any device that can be allocated an IP address and can send data over a network are all examples of things in the IoT. The IoT device connects to an IoT gateway or another cutting-edge device to share sensor data, which is either forwarded to the cloud or analyzed locally. Since useful decision-making relies on data, IoT may use artificial intelligence (AI) to make data collection processes much easier and dynamic.

In Figure 2.1, a typical IoT communication setup is depicted, with an application domain (interacting with services, which, in turn, interact with the data and the operator). The active satellite interacts with the data and network domain, while the network domain will, in turn, communicate with the dish and IoT device. The IoT device that makes use of the bridge through the gateway also communicates with the user of the system. The interaction and communication in this setup are enabled throughout the Internet and it is not possible without an Internet facility. The cloud environment provides infrastructures to support all these devices and ensures data transfer and other related interactions. Researchers are improving on these innovations, ensuring ease of use and improving the efficiency of the framework.

The advent of COVID-19 pandemic brings technical solutions where vital in keeping many cities operational, and the long-term effects of engaging technologies on metropolitan environments may remain even after COVID-19 has passed. Online food and pharmacy shopping is made easier with the help of robots, drones, and contactless payments, particularly for the old and weak [4]. Manufacturers and enterprises can use IoT-enabled devices to control and automate their services so that it helps remote workers in services delivery [5, 6]. Due to the high-speed, low-cost Internet that 5G cells provide for urban areas, online health consultations, online learning, and cultural services are now possible. The epidemic has spread to about 216 countries

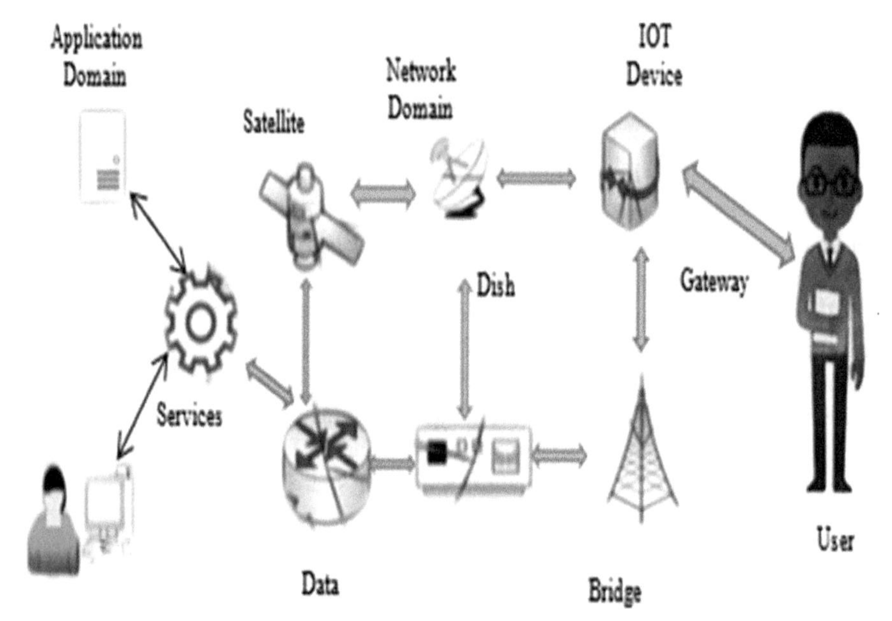

Figure 2.1 IoT communication setup.

throughout the globe, making it the most deadly disease of the twenty-first century. Despite the availability of contemporary, advanced medical therapy, the disease is spreading through outbreaks. In the healthcare profession, both ML and DL research works have significantly made progress in a variety of areas (especially in data and image analysis), including giving help for eventual medical diagnosis [7].

Several solutions have been provided by IoT applications, and some services are delivered remotely with the introductions of IoT. Dian *et al.* [8] conduct a thorough review of the most recent and significant studies in the field of wearable IoT and cloud computing for data analysis and decision making. The study categorized the wearables into four primary clusters: (1) medical, (2) sports, and daily activities, (3) tracking and location, and (4) safety. The essential distinctions between the algorithms in each cluster are classified and studied, as well as the research difficulties and open concerns in each cluster for further investigations. However, the tools that enable IoT to analyze and transfer the data and make a decision were not discussed. Also, the application of IoT goes beyond the four classes that the study identified in the study [9]. The existing cloud computing approach is inefficient for analyzing massive amounts of data for AI or CNN applications in a short amount of time and meeting the needs of users. Therefore, there is a need to further break the frontier of research in the areas of IoT-enabled CNN, to study and analyze the application, techniques, challenges, and prospects of the duo.

2.1.1 Contribution of the Chapter

In the chapter, we expose the novelties of IoT-enabled based on CNN architecture, viz-a-viz the applications, techniques, challenges, and prospects (ATCP) were discussed. This chapter also discusses the prominent field where AI/CNN and IoT are embraced and where they face challenges and present the applicability of CNN-enabled IoT techniques in various fields intending to explore the prospect of the research area and recommend the area of improvement.

2.1.2 Chapter Organization

The organization of this chapter is as follows. The introduction of IoT- and CNN-enabled devices is entailed in Section 2.1. Section 2.2 describes the overview of artificial intelligence (AI) applications and IoT in various fields, Section 2.3 describes the CNN architecture, with seven distinct CNN explorations, and Section 2.4 is the applicability of CNN techniques in the IoT environment and reveals the challenges of applicability of CNN-enabled IoT in various fields. Finally, Section 2.5 concludes the chapter and highlights the directions of further studies.

2.2 Application of Artificial Intelligence in IoT

The use of CNN applications as artificial intelligence (AI), deep learning, or machine learning for IoT technology cannot be undermined especially to ensure easy implementation of data-driven technologies and make sure that data collection or acquisition is possible. In the comprehensive survey conducted by [10], the study was aimed at identifying the applications of AI in combating the challenges posed by the COVID-19 outbreak. As a result, it encompasses all AI methodologies used in this field to ensure holistic survey. The goal of this work is to create an intelligent computing algorithm for forecasting COVID-19 outbreaks. The Facebook prophet algorithm forecasts 90-day future values, including the peak date of confirmed

COVID-19 cases, for six of the world's worst-affected countries, and six of India's high-incidence states. The model also reveals five important change points in the growth curve of verified Indian cases, indicating the impact of the government's initiatives on the infection's rate of spread. The model's goodness-of-fit is 85% for six countries and six Indian states [11].

Although DL and ML are branches of AI, there are networks under DL and ML for successful implementations of projects in each case. The CNN has a recurrent neural network, generative adversarial network, auto-encoder, deep belief network, restricted Boltzmann machine, and human activity recognition. ANNs are examples of supervised and unsupervised networks in DL [12]. Each DL model has its merits and demerits, and these are evident in the performance evaluation of the proposed model. In the same vain, the conventional ML algorithms delineate the superiority of the DL model over other models. This section serves as an impetus for advanced research in the field of AI-IoT applications for better performance, in addition to revealing the potential of AI in IoT applications. Table 2.1 is a review of AI-enabled IoT studies in different fields of applications; it reveals and describes AI and IoT studies in different fields of applications as well as their references. Many novel studies were identified in the areas of health, natural language processing, technology, data mining, and security, among others.

2.3 Convolutional Neural Networks and its Architecture

A CNN is a type of neural network that has excelled in a number of competitions involving image processing and computer vision. Image classification and segmentation, natural language processing (NLP), processing of video, object detection, and speech recognition are just a few of CNN's fascinating application areas. The high learning tendencies of CNN is traced to the use of numerous feature extraction stages which can automatically learn the representations from data, and the transmission of these can be easier when the architecture interface with the receiving end with an Internet-enabled device (IoT). The availability of a large amount of dataset as well as hardware developments have spurred CNN research, and fascinating CNN architectures have lately been revealed [25, 26]. The activation and loss functions of various types, parameter optimization, regularization, and architectural design improvements are all among many things that can be done like a few of the unique ways to improve CNNs that have been studied. On the other hand, architectural advancements have resulted in a massive rise in CNN's representational capacity. The spatial explorations and channel information,

Table 2.1 Review of AI-enabled IoT studies in different fields of applications

AI-IoT studies and their descriptions.	Field of application	Reference
AI-IoT survey study.	Health	[13]
Deployment of AI-IoT into the chatbot agents works and ensures natural language processing.	Natural language processing	[14]
AI-IoT was employed to ensure data mining and management to control congestion in the network.	Data mining	[15]
4IR, or "Industry 4.0," is a concept in which industrial gadgets and machines are connected to the Internet and interact to make choices using AI (M2M communication).	Artificial intelligence	[16]
The frameworks for centralized and distributed IoT-enabled AI technology were deployed. For various network designs, key technological challenges such as random access and spectrum sharing are examined.	Technology	[17]
For successful data management, a high-level scheme architecture was created and applied at the edge-level micro services, together with an AI technique.	Technology	[18]
The research proposes a hybrid AI/ML detection model as a solution for combating and mitigating IoT cyber risks in cloud computing settings, both at the host and network levels.	Technology	[19]
The work reviewed novel technologies applications on AI-IoT and sum it up to AI-IoT.	Technology	[20]
The study created a CNN model to enable the IoT platform's contextually aware services, as well as tests to test the CNN model's accuracy using a collection of photos acquired by the robot. The study's experimental results showed that the learning accuracy was over 98%, indicating that the study improved image context recognition learning. The paper's contribution was the creation of image-and-context-aware learning and intelligence using a CNN model enhancement of the proposed IoT platform, as well as the realization of an IoT platform with an active CCTV robot.	Technology	[21]
The techniques and concepts of IoT and AI were explored, and the possibility of applying blockchain for providing security was also discussed. The study's main focus was on the use of integrated technologies to improve data models, gain better	Data modeling and intelligence prediction	[22]

(Continued)

Table 2.1 Continued.

AI-IoT studies and their descriptions.	Field of application	Reference
insights, and discover new things, global verification, and innovative audit systems. Academics and industry professionals can exchange their thoughts and new research in the convergence of these technologies in this book, which is meant for both practitioners and scholars. Contributors provide their technical assessment and compare it to current technologies. There are also theoretical explanations and experimental case studies relating to real-time scenarios. IT workers, researchers, and academics working on fourth-generation technology will benefit from this study.		
To combat the COVID-19 pandemic, three research directions were proposed using AI-based methods: To improve diagnosis, deep convolutional neural networks (DCNN) with transfer learning were used to classify chest X-ray images. Patient pandemic risk prediction based on patient features, comorbidities, first symptoms, and vital signs for disease prognosis; and deep neural networks are being used to forecast disease transmission and case fatality rates. Additionally, some of the issues of open datasets and research opportunities were discussed.	Health	[7]
The number of AI/ML algorithms at the edge is growing in tandem with the number of low-cost sensors allowing IoT remote sensing using AI/ML algorithms at the edge. For establishing a distributed remote sensing network and interfacing with the cloud, raspberry PI, and other boards with appropriate sensor devices are readily available. The paper describes the components and preliminary findings of establishing a small distributed remote sensing network for recognizing and identifying a range of target kinds using NVIDIA Jetson Nano edge devices with low-cost acoustic and image sensors, as well as IoT Greengrass. Additionally, cloud services were used to enable auditing and monitoring, resulting in a secure and dependable operational service environment.	Security	[23]
Comprehensive analysis of IoT security requirements and difficulties were presented, as well as a discussion of the unique role of DL and a review of state-of-the-art research work in IoT contexts employing DL methodologies. A comparative examination of DL algorithms such as RNN, LSTM, DBN, and AE was also performed.	Technology	[24]

Table 2.1 Continued.

AI-IoT studies and their descriptions.	Field of application	Reference
Based on the notions of applications of CNN, the TCNN, an intelligent model for IoT-IDS that combines CNN with general convolution, was proposed. The TCNN is built using synthetic minority oversampling methods with nominal continuity to accommodate unbalanced datasets in the model. It was then used with useful feature engineering approaches such as attribute transformation and reduction. Using the Bot-IoT data repository, the provided model is compared to two classical ML methods, random forest, logistic regression, and LSTM on the CNN architecture.	Security	[25]

architecture depth and width, and multi-path information processing, in particular, have received a lot of attention in its networks taxonomy. The use of a layer block as a structural unit is also becoming more widespread [27, 28].

The CNNs are biologically inspired structures that are at the heart of computer vision's deep learning algorithms [29]. It is a feed-forward neutral network that uses many convolutional layers to extract features and fully connected layers with softmax to make inference [30]. In at least one of its layers, the system uses a mathematical linear operation called convolution instead of ordinary matrix multiplication, as indicated by the phrase "convolutional neural networks." As in a normal multi-layer neural network, at least one convolutional layer and at least one fully linked layer are present in a CNN. They are a common strategy in image recognition and computer vision frameworks, and they are verifiable and significant in the field of data science. Because the number of nodes in the input layer of CNN is dictated by inputting required data size, such as the resolution of an image, nodes number in the input layer is determined by inputting the required data size. The convolution computation names the convolutional layers. They are also thought to be the key to extracting the most distinguishing traits from the original dataset. Convolutional kernels are used to build feature maps, which are then activated by nonlinear activation functions including the sigmoid function, tan function, and rectified linear units function [31].

The CNNs exhibit remarkable performance among DL networks especially with huge data, meaning that they have more layers that are deeper and more interconnected. The CNN receives a two-dimensional picture or voice as the input signal. With the help of a chain of hidden layers, which includes convolutional layers and pooling layers, the CNN extracts hierarchical properties from the input data. The CNN will surely aid in the creation of smarter IoT-based solutions by deriving valuable insights from this vast volume of data. Exploring the possibilities of CNN for IoT data analytics becomes extremely important in this regard. AI and its various architectures spawned the CNN approach [12]. Linear regression, a machine learning application, is a concept for estimating the number of active cases, deaths, and recoveries. On the basis of the expected number of active cases, deaths, and recovery across India, the lockdown would be extended. A date-wise analysis of current data was performed to predict the number of active cases, recoveries, and death, and necessary parameters such as daily recovery, daily deaths, and the increasing rate of COVID-19 cases were included. To make expected findings more understandable, each analysis and forecast was graphically represented [32].

The most prominent DL network model is CNN. There are three layers in CNN: (1) convolutional, (2) pooling, and (3) fully connected layer. The convolutional layer is made up of a finite number of filters that combine with the input data to identify a large number of important features in the input image. The pooling layer uses a down-sampling approach to reduce the number of the resultant features, reducing the overall computational efforts. The system goes deeper by repeating the convolution-pooling sequences numerous times, depending on the data and the required accuracy. The method collects more high-dimensional features from the input dataset, followed by one or more fully connected layers for classification [33]. There are many C++, MATLAB, and Python-based frameworks to execute various tasks on the CNN architecture [34]. CNNs are also one of the most often used strategies in image recognition, object identification, and computer vision frameworks, and they are undeniably significant in data science. It employs deep learning algorithms, which take video/image input, give weights and bias to various parts of the image, separate them from one another, and transform the data into a numerical dataset. They try to make use of the spatial information contained within an image's pixels. As a result, they rely on discrete convolution [35].

Although the introduction of new generations of networks has posed substantial issues in terms of meeting the needs of many new applications, the majority of which necessitate advanced infrastructure to offer the necessary resources to assure high quality of service. Among biologically inspired artificial intelligence (AI) techniques, CNNs are the most extensively used algorithms. CNN's beginnings can be traced back to [25, 36] neurological

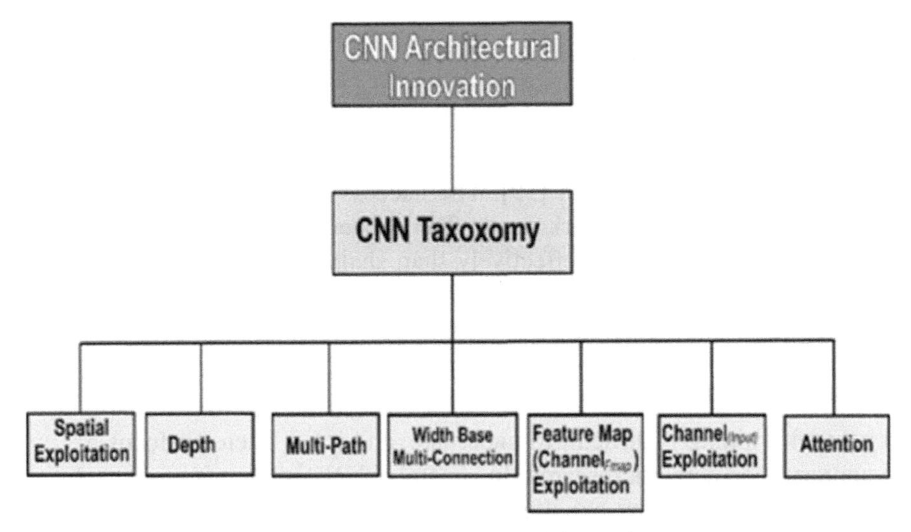

Figure 2.2 CNN architectural innovation.

experiments. As a result of a survey conducted in [25], which focuses on the intrinsic taxonomy found in recently reported CNN architectures and divides current CNN architecture improvements into seven groups. The spatial exploitation, depth, multi-path, width, feature map, channel exploitation, as well as attention are the seven categories of the CNN taxonomy. In addition, a basic grasp of CNN components, current problems, and CNN applications is offered, and the taxonomy of the survey is presented in Figure 2.2. In general, all these groups have influences on the architectural performance of the networks.

2.3.1 CNN Based on Spatial Exploration

The number of layers, the biases, the weights, and the processing units, stride, activation function, learning rate, and other parameters and hyper-parameters abound in CNNs for effective implementation [25]. The levels of differences in correlation can be examined by utilizing different filter sizes because convolutional operations analyze the neighborhood of input pixels. Filters of various sizes encompass various levels of granularity; typically, small size filters extract fine-grained information, while large size filters extract coarse-grained information. The spatial exploration for CNN brought about LeNet, VGG, GoogleNet, AlexNet, and ZfNet which were the early products of spatial exploration [37, 38]. This exploration does not consider IoT.

2.3.2 Depth of CNNs

The depth of CNN architecture refers to the deep on the convolutional layers of the networks, they are based on the premise that as the network's depth increases, and more nonlinear mappings, it helps in the improvement of goal function in the networks [39]. The success of supervised training has been attributed to the network depth. The deeper networks can express some classes of function more effectively than shallow designs, according to theoretical studies. This operates based on Highway Networks, ResNet, Inception-ResNet, and Inception-V3, V4 [40, 41].

2.3.3 Multi-Path of CNNs

Deep network training is difficult to undertake, and it has been the focus of contemporary deep network research. In general, CNNs excel at complicated tasks. Meanwhile, they may experience performance degradation, gradient disappearing, or explosion issues, which are caused by an increase in the depth rather than overfitting [42, 43]. The disappearing gradient problem leads to a higher test error as well as a bigger training error [44]. The concept of cross-layer connectivity was developed for deep network training. Multiple pathways can be used to connect one layer to another while skipping some intermediary layers, allowing specialized information to travel across the layers. The Highway Network, ResNet, and DenseNet are the major techniques of multi-path [45–47].

2.3.4 Width for Multi-Connection CNNs

The focus of [48] was primarily on leveraging the potential of depth, as well as the efficacy of multi-pass regulatory links in network regularization. But, according to [49], the network's width is equally crucial. The multi-layer perceptron obtained the advantage of mapping complex functions over perceptron by using numerous processing units in parallel within a layer. It shows that, like depth, width is an important component in developing learning principles. The authors in [50] demonstrated that NNs with ReLU activation functions must be wide enough to maintain universal approximation while still increasing in depth. The pyramidal Net, wide ResNet, ResNet, Inception, and Xception are all family of the exploration of CNN architecture [51].

2.3.5 Feature-Map Exploitation for CNN

The hierarchical learning and automatic feature extraction capabilities make it possible for the CNN architecture to become useful for MV tasks.

The performance of modules for classification and segmentation are heavily influenced by feature selection. The weights related to a kernel are called a mask and are tuned in CNN to choose features dynamically. In addition, various steps of feature extraction are performed, allowing for the extraction of a wide range of features. Some of the feature maps, on the other hand, play a little function in object discrimination [52, 53]. Large feature sets may produce a noise effect, resulting in overfitting of the network and most of the time causes the squeeze and excitation of the networks. Conversely, a low feature set may result in the underfitting of the networks.

2.3.6 The CNN Channels for Exploitation

The performance of image processing techniques, both conventional and DL systems, is heavily influenced by a picture representation. A good image representation is one that can define the key elements of an image with a small amount of code. Some conventional filters are used in MV tasks to extract different degrees of information from the same image type. The model then uses these various representations as an input to improve performance [54]. CNN is now a compelling feature learner that can extract discriminating features automatically depending on the challenge [55]. Channel exploration for CNN produces channel exploited CNN using transfer learning (TL) and [56] develops a novel CNN design called channel boosted CNN in 2018, based on the idea of increasing the input channel numbers to improve the network's representation capacity [56]. Channel boosting is accomplished by using auxiliary deep generative models to artificially create extra channels, which are subsequently exploited by deep discriminative models.

2.3.7 Attention Exploration for CNN

Different levels of abstraction play a key role in determining the neural networks (NNs) discrimination power. In addition to learning numerous hierarchies of abstractions, concentrating on context-relevant features is important for image identification and localization [25], [57]. The effect that makes the learning numerous hierarchies of abstractions, concentrating on context-relevant features is important for image identification and localization is also known as attention in the human visual system [58]. Because convolutional operations allow for weight sharing, distinct sets of features inside an image can be retrieved by sliding kernels with the same set of weights on the picture, making CNN parameters more efficient. The convolution operations can be classified based on the type and size of filters, the type of padding, and the direction of convolution. This operation can be executed in three steps: data collection and feature generation, preprocessing and feature selection, and data partition, learning, and analysis/evaluation [59, 60].

Figure 2.3 shows CNN system operational layout steps that reveal the activities in each step of the convolutional operation. It is divided into three distinct steps for use of the CNN architecture. They are as follows:

(i) Data collection and feature generation: All AI experiments are about data; so the first thing is to acquire data, and all the acquired data has features. Therefore, this is the first step in CNN operation layout.

(ii) Data preprocessing and feature selection: The operation to be carried out here include: missing data input, feature analysis, data normalization,

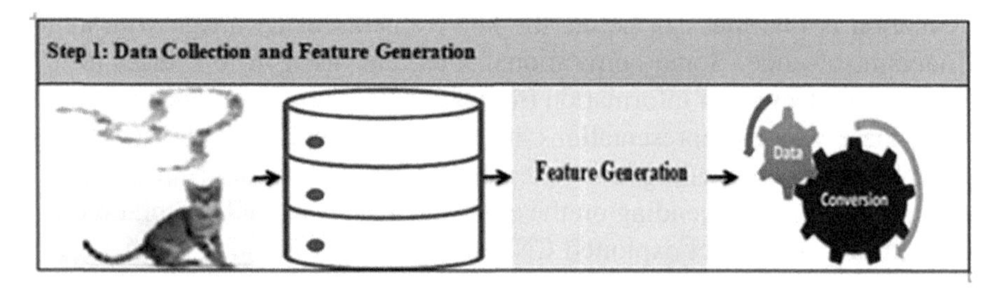

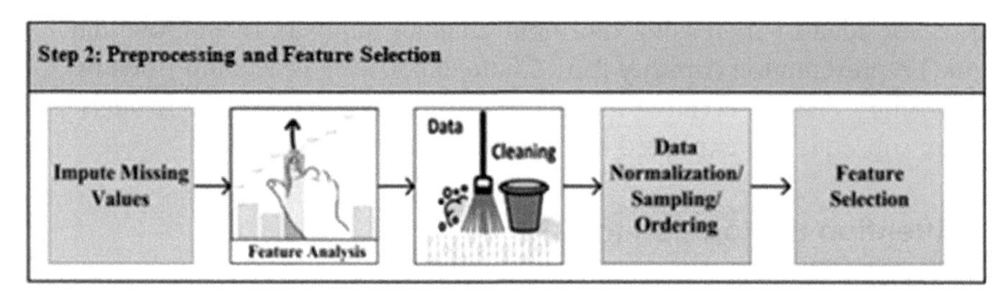

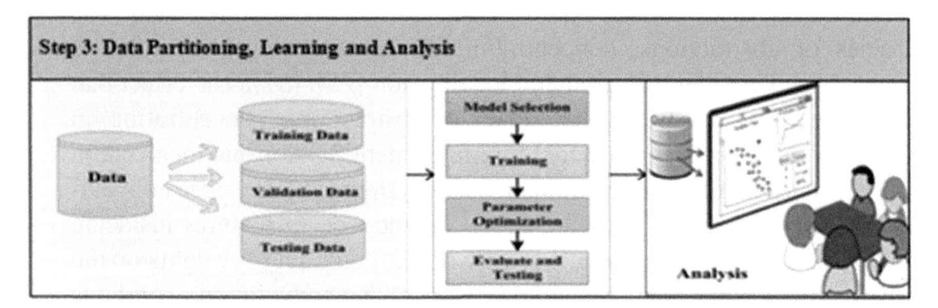

Figure 2.3 Typical CNN system operational layout steps.

and the selection of the relevant features to the project or experiment at hand.

(iii) Data partitioning, learning, and analysis: In some studies, this may be broken into two, depending on the implementation and the data involved. Then, data partitioning and learning would be one and analysis or evaluation would be the second one. The data would be partitioned into training, validation, and testing dataset; for better CNN operation and to avoid overfitting and underfitting of the models, attention is paid to the dataset and model for the implementation. The analysis or evaluation is done based on the accuracy of the model(s), recall, and other metrics depending on the particular operation. The metric for data classification is different from data segmentation operation.

2.4 The CNN Techniques in IoT Environment

The Internet of Things (IoT) is a network of smart objects like sensors, home appliances, phones, automobiles, and computers that are connected via the Internet [61]. Although the IoT is not a new concept, it is currently a big issue around the world. IoT refers to the interconnection of electronic devices to allow data to be exchanged between them for specific domain applications. This internetworking concept in IoT makes human life much easier than it was previously. A TCNN, which combines CNN with causal convolution, was devised and applied using the CNN technique in the construction of an effective and efficient DL-based intrusion detection system for the IoT [62]. CNNs are also one of the most often used strategies in image recognition, object identification, and computer vision frameworks, and they are undeniably significant in data science. CNN approaches employ deep learning algorithms, which take video/image input, assign weights and bias to various parts of the image, and then distinguish them from one another [63]. They try to make use of the spatial information contained within an image's pixels. As a result, they rely on discrete convolution [64].

"Can an existing optimized CNN model be used to automatically generate a competitive CNN for an IoT application whose objects of interest are a fraction of the categories that the original CNN was meant to classify, resulting in a proportionally scaled-down resource requirement?" was looked into. The notion and proposed method for the automatic synthesis of resource scalable CNNs from an existing optimized baseline CNN was referred to as resource scalability. The result shows that synthesized CNN has the learning power to handle the provided IoT application requirements while also providing competitive accuracy. The suggested method is quick, and unlike current CNN design practice, it does not necessitate iterative rounds of training trial and error [65]. With no prior knowledge, the authors in [66] proposed a mobile-based methodology for counting people in high- and low-packed public locations in Saudi Arabia under a variety of scenario conditions. The deeper convolution neural network (DCNN) was introduced such that it is based on a pre-trained CNN called VGG-16 with some modifications to the last layer of the CNN to improve the training model's efficiency [67]. The emphasis is on crowd counts as well as producing high-quality density maps. The proposed method improves efficiency while accepting photos of various sizes and scales as inputs. The model's backbone is made up of pure convolutional layers, which allow for the utilization of images of various sizes and resolutions. The counting system mobile application attempts to shorten users' wait times by displaying the crowd size of the place they are visiting.

Kimbugwe *et al.* [68] gives an in-depth look at how DL algorithms have been used to improve QoS in the IoT. According to the study, QoS in IoT-based systems is disrupted when the systems' security when IoT resources are not adequately managed. As a result, the goal of this study is to learn how DL has been used to improve QoS in IoT by preventive security and privacy breaches in IoT-based environments and assuring proper and efficient resource allocation and management. It selected the most commonly utilized DL algorithms for dealing with IoT QoS concerns and described the state-of-the-art of those techniques. For resource-constrained contexts, Lawrence and Zhang [69] developed IoTNet, a CNN-IoT enabled technique that achieves state-of-the-art performance within the domain of tiny efficient models. Instead of using depth-wise convolutions, IoTNet sacrifices accuracy for computational expense in a different way than prior approaches by factoring standard 3×3 convolutions into pairs of 1×3 and 3×1 standard convolutions. The study compares IoTNet against state-of-the-art efficiency focused on the algorithms and scaled-down big frameworks on datasets that are most representative of the complexity of challenges encountered in resource-constrained contexts.

The CNN techniques are doing pretty well in several areas, especially for data analysis and decision making. Many CNN techniques were applied to solve some problems in IoT environments, and the remarkable achievements of these are evident [70]. Contributions in the area of healthcare, agriculture, meteorology, intelligence learning, smart city, biometrics, eCommerce, and some of these novel works are discussed. Figure 2.4 is a typical application of CNN and IoT techniques setup in various fields.

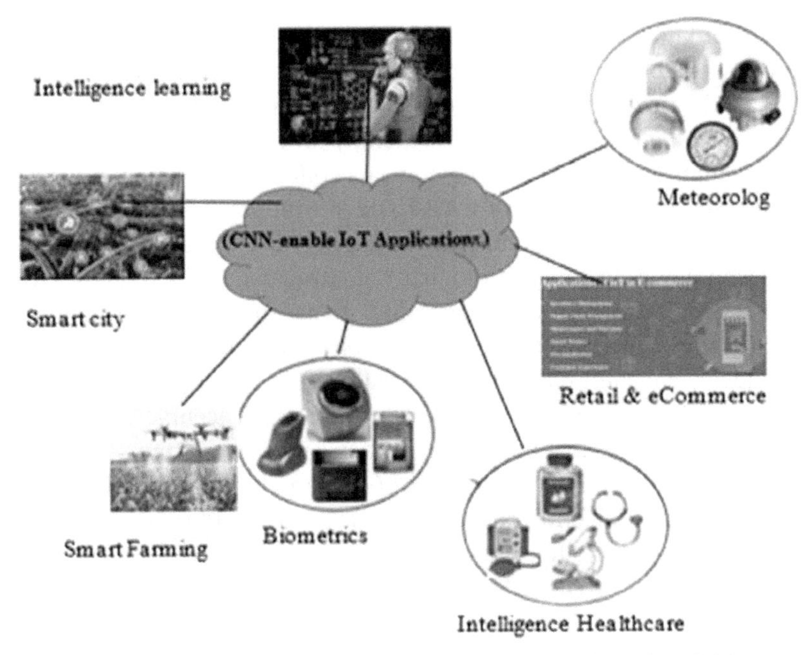

Figure 2.4 Applications of CNN and IoT techniques in various fields.

2.4.1 Intelligence Healthcare System

The healthcare system has been made an intelligence system, through the evolution of IoT-based applications. The IoT-based technology helped to make diagnosis, management, test, data access, and contact tracing (especially during a pandemic) among other healthcare easier for both the healthcare personnel and the patients [71]. IoT is converting traditional healthcare systems into more personalized ones, making it easier to diagnose, treat, and monitor patients. The current global pandemic caused by the novel highly contagious respiratory illness COVID-19 second variant is the most serious global public health disaster since the 1918 influenza pandemic [72]. Just like every other pandemic, since the outbreak of COVID-19, researchers have been working feverishly to leverage a wide range of technologies to battle the global threat, and IoT technology is one of the forerunners in this field. For the COVID-19 pandemic cases, IoT-enabled devices were connected and applications are used to reduce the risk of COVID-19 spreading to others by detecting the virus early, patients' monitoring, and follow-up on the prescribed protocols once the patients have recovered [73]. The study reported a setback in the areas of mobility.

With a better knowledge of the COVID-19 pandemic natures, the authors in [74] proposed architecture that was used for the diagnosis of probable COVID-19 pandemic cases in real time, and, most importantly, the architecture could be used to monitor and forecast therapy for confirmed patients. The suggested system is divided into five layers: data collection, isolation, preprocessing of dataset, and layer analysis. Also, the health physician application layer and cloud infrastructure were used to connect the layers. The system uses smart technologies to collect biological data from patients and sends it to a cloud-IoT server for analysis and processing. As a result, any abnormalities in patient data will be notified to the patient's physicians via the COVID-19 monitoring device and alert platform. The results of the four algorithms used revealed that LGBM outperformed them all with a 97% accuracy, and this was followed by XGBoost that has 90% accuracy, and random forest that has 76% accuracy. In the confusion matrix, the LGBM model had a recall rate of 96%, indicating a good detection rate; the study is similar to [75]. Kumar *et al.* [22] proposed a 24-hour intelligence system that is non-invasive for the blood level glucose monitoring system. The system is meant for the management of invasive blood monitoring challenges. The system provides accurate reading and generates alerts using IoT so that undesirable effects caused by severe variations in blood glucose levels can be avoided. The system was able to make a decision with an ML model that has been trained with data. The system is implemented utilizing FPGA and achieves optimum efficiency and throughput while consuming little energy [76].

2.4.2 Intelligence Learning System

The educational application should be more beneficial to art students and designers than to children. The understanding of object position and three-dimensional state is increasingly vital and crucial in painting and design. The primary and secondary relations of items can be found by separating several objects and finding the relation, especially in the field of picture segmentation. The location relationship of objects can be readily understood after removing the complex background. The majority of the assistance provided to children and students will be in the area of thinking construction. It can effectively assist students in forming a three-dimensional perspective in their thinking, such as in three-dimensional geometry questions in the college entrance exam [77, 78]. The knowledge-level assessment in e-learning systems using ML and user activity analysis was proposed. The project aimed at designing a futuristic intelligent and autonomous e-learning, where ML and user activity analysis serve as an automatic knowledge level evaluator. To alter the content presentation and have a more realistic evaluation of e-learning, it is necessary to measure their knowledge level. Many classification methods are used to forecast the learners' knowledge level, with the results being reported. In addition, the study provides a modern design for a dynamic learning environment that follows current e-learning trends. The experimental results show that when evaluating knowledge levels, an SVM model outperforms other algorithms with 98.6% of examples correctly classified and an MAR of 0.0069 [79].

2.4.3 Smart City

An intelligence-based city is referred to as a smart city; this involves but is not limited to a smart transportation system, intelligence vehicle, and traffic and highway management. Being an intelligence system, AI techniques play a pivotal role in the development of such systems and the interoperation strength lies in IoT since the connection is the cloud. The use of IoT also provides supply chain and logistics operations with information on machine performance. Payvar *et al.* [80] worked on vehicle classification; the study described an NN-based image classifier that has been trained to categorize vehicle photos into four categories. Binarization is used to optimize the neural network, and the resulting binarized network is placed on an IoT-class processor for execution. The binarization reduces CNN's memory footprint by about 95% while increasing performance by more than 6%. In addition, the paper showed that by employing the processor's proprietary instruction "pop-count," the performance of the binarized vehicle classifier can be improved by more than two times, making the CNN-based image classifier appropriate for even the smallest embedded processors [80]. The work recommended improvement in the area of scalability.

With the end objective of on-street vehicle identification, [73] proposes a pre-handled quicker regional convolution neural network (faster RCNN). The framework includes a faster RCNN preprocessing pipeline. The preprocessing technique is used to improve the faster RCNN's preparation and detection time. To recognize paths, a preprocessing path identification pipeline based on Sobel edge administrator and Hough transform is used. After that, the gallery organizes a rectangle district, which is a less interesting location (ROI). When compared to faster RCNN without preprocessing, the proposed technique enhances the preparation speed of faster RCNN [73]. The system will be helpful in transportation and security, especially in urban centers.

In the case of a pandemic like COVID-19 in metropolises and cities, [81] proposes an end-to-end IoT infrastructure to promote social distancing. The design includes the most common IoT use cases in relation to COVID-19. Using the IoT architecture, the novelty of the work was the presentation of short- and long-term strategies for managing the social distancing method, although none of the CNN techniques was used.

2.4.4 Agriculture

Gikunda and Jouandeau [82] propose a classification taxonomy for CNN applications in agriculture. Finally, the report provided a complete assessment of research on the use of cutting-edge CNNs in agricultural production systems. The make of a two-fold contribution. For starters, the benchmarking findings can help end-users of agricultural DL applications choose the right architecture to utilize. Second, the in-depth research describes the state-of-the-art CNN complexity and brings out probable future paths to better optimize the operating performance for agricultural software developers of DL tools. The study concluded by identifying real-time image classification, interactive image detection, and classification, as the areas of improvements in agricultural domains, despite amazing progress in the use of cutting-edge CNN in agriculture in general. Gikunda and Jouandeau [82] investigated agricultural challenges that use the primary state-of-the-art CNN architectures that have achieved the maximum accuracy in a multi-class classification problem in the ImageNet large-scale visual recognition challenge. Their analysis sheds light on the use of cutting-edge CNN models for smart farming and identifies computer-vision-related smart farming research and development issues.

The article demonstrated the value of cutting-edge CNN in IoT-based smart farming. A survey was also conducted on the use of the identified CNNs in agricultural services. The results show that in all of the agricultural sector scenarios studied, state-of-the-art CNN provided greater precision. When compared to other image-processing algorithms, it achieves superior accuracy in the majority of the problems for agricultural service. Despite outstanding performance in using state-of-the-art CNN in agriculture in general, future researchers may explore gray areas in relation to smart farms, such as real-time image classification, interactive object identification, and classification. The mobility and scalability problems were reported in real-time applications of these technologies.

2.4.5 Meteorology

With no prior knowledge, a mobile-based approach is suggested for counting people in low- and high-packed public locations in Saudi Arabia under diverse scene conditions. The suggested model is based on the VGG-16 pre-trained CNN model, with some tweaks to the last layer of the CNN to improve the training model's efficiency. The suggested method supports photos of arbitrary sizes/scales as inputs, in addition to improving efficiency. The applicability of the suggested method was assessed by implementing IoT architecture, which requires surveillance cameras to be connected to the Internet in order to gather live images of various public locations [83]. Although the prominent one in relation to IoT is computational cost, because of the high computational cost of these topologies, big CNNs are often impractically sluggish, especially for embedded IoT devices. The early detection, isolation of the sick person, and identifying probable contacts are important when a pandemic first breaks out in cities. IoT protocols, particularly Bluetooth low energy, as well as RFID, NFC, and Wi-Fi, are gaining a lot of traction as answers to these problems [84]. Integration, automation, improved communication, and self-monitoring IoT-based systems are increasing, and they help in the production of smart solutions (such as smart city, smart farming, technological process, meteorology, eHealth, and biometrics) that can be used in the analysis of the problem in different field and proffer solution. Figure 2.2 depicts the applications of IoT devices. The study reported a setback in the areas of mobility.

2.4.6 Biometrics Applications

Rattani *et al.* [85] created a CNN architecture for combining biometric data from several sources. CNN-based multibiometric fusion has the advantages of (1) being able to execute early, and late fusion, and (2) the fusion architecture itself being learnable during network training. Experiments on large-scale VISOB data show that multibiometric CNNs outperform the traditional fusion method. Based on CNN techniques, [86] created a localization framework that moves the online prediction complexity to an offline preprocessing step (CNN). The indoor localization problem is described as 3D radio-image-based region recognition, inspired by the exceptional performance of such networks in the image classification field. Its goal is to precisely locate a sensor node by determining its position region. The received signal strength indicator fingerprints are used to create 3D radio pictures. The simulation findings validate the parameters, optimization strategies, and model designs that were employed [87]. Based on a fingerprint, finger-vein, and facial recognition system with multiple biometrics, [88] suggested a hybrid system incorporating the effect of tree efficient models: CNN, random forest classifier, and softmax. In a traditional fingerprint system work, the segregate the foreground and background regions using K-means and DBSCAN algorithms, and image preprocessing was done, then the features are extracted using CNN and a dropout technique, and softmax is used as an identifier. The region of interest picture contrast enhancement utilizing the exposure fusion framework is fed into the CNN algorithms in a traditional finger vein system [89]. The results of these systems are combined to improve human identification. However, the interoperatabilities of the system and management of quality-of-service (QoS) were the issues observed.

2.4.7 E-Commerce and E-Business

In the context of business-to-consumer (B2C) in form of retails and e-commerce, Sohaib *et al.* [90] provide a proposed integrated framework of IoT and cloud computing for people with sensory, motor (restricted use of hands), and cognitive (learning and language disorders) impairments. It helps to work smartly and gain control over activities, it is popular in homes automation, and it is essential to business as it provides real-time monitoring into how businesses work. It also provides supply chain and logistics operations with information on machine performance. The work achieved a milestone by ensuring ease of use and improving security in an IoT environment. The framework uses state-of-the-art technology to ensure e-business activities and identified security challenges as an issue.

2.5 Challenges of Applicability of IoT-Enabled CNN in Various Fields

Embedded systems are utilized in a variety of settings to achieve a wide range of applications, in the area of security, traffic control, e-health, smart home, and smart cities. Digital systems regulate physical things in various applications, resulting in a constant interplay between the digital and physical worlds [91]. Meanwhile, the fundamental characteristics of these applications, particularly when embedded systems are taken into account, provide difficulties. Also, CNN is similar to black boxes in that they are difficult to analyze and explain, confirming they can be difficult occasionally [92, 93]. Analyzing the massive amount of data in the cloud environment takes a long time, and this is detrimental to the IoT applications and overall network performance of the system [94]. The authors in [95] carried out training on a CNN model with a noisy picture dataset, which may result in an increase in misclassification errors. The addition of a small quantity of random noise in the input image can fool the network, causing the original image to be distorted and slightly perturbed versions of the image to be classified differently [96].

Meanwhile, IoT in conjunction and other emerging technologies such as 5G networks, AI, and cloud computing has the potential to revolutionize healthcare, commerce, agriculture, education, security, and other sectors; the high cost of mass-scale deployment has made it difficult for some e-health operators and governments to do so [97]. Typically, CNNs have many benefits that override its drawback in many operations, and that is why they are mostly embraced in data science projects. Meanwhile, challenges of techniques of CNNs with this digitalization include computational cost data rate, coverage, energy usage, and practicality which are all factors to be considered along with IoT [98]. CNN is applicable in the creation of an IoT-based DL-based intrusion detection system that is both effective and efficient. It adopts the necessary principles of a TCNN, that combines CNN with causal convolution was designed and implemented. The CNN techniques are also one of the more prevalent strategies utilized in image recognition. They are notable in the field of data science for their object detection and computer vision frameworks. CNN employs DL algorithms, which take video or image input, assign weights and bias to various parts of the image, and then distinguish them from one another. They are based on discrete convolution and try to use the spatial information among the pixels of an image [99]. The categorical appraisal of IoT research that used CNN techniques for its implementation is scarce, especially the one that identifies the prominent areas of applications of IoT-enabled and CNN techniques were used. To improve this area of research, the IoT-enabled CNN applications challenging area were also revealed for possible improvements. Table 2.2 presents IoT-enabled CNN studies, techniques, fields of applications, as well as limitations of the studies in order to ensure the prospects of the applications.

There are numerous prospects in the combination of these prominent and promising computing fields CNN and IoT, but a number of modern challenges are associated with techniques. The deployment of technologies such as drones, robots, autonomous vehicles, and other IoT tools has brought the revolution to data capturing and transmission and data analysis for informed decision making has become easier with CNN techniques. However, these achievements are threatened through, computational cost, coverage, energy

Table 2.2 The IoT-enabled CNN studies in different field techniques and challenges

IoT-enabled CNN studies	Techniques	Fields of applications	Challenges	References
The IoTNet, a new model for resource-constrained situations which provides state-of-the-art performance in the realm of tiny efficient models, was proposed to manage computational complexity in image classification.	IoTNet	Technology	Augmentation strategies of CNN were not utilized	[100]
A mobile-based strategy called DCNN for counting individuals is proposed. The proposed mobile-friendly model is capable of detecting information quickly; people are counted everywhere and in various places.	VGG-16 (a pre-trained CNN)	Security and economic values	Non-uniform distribution of clutter, computa-tional cost, practicality with IoT	[101]
A CNN model for picture learning and intelligence that is context-aware with an active CCTV robot IoT platform was developed to detect anomalous conditions in various industrial settings, factories, logistic warehouses, smart farms, and public areas.	IoT-Net	Security	The model developed exhibited little operation time due to battery inefficiency	[21]
A CNN sleep apnea detection IoT-enabled method that is based on ECG and EDR signals was developed. The model could detect sleep apnea on a minute-by-minute basis or a window of 30 seconds.	IoT-sensors	Healthcare	Smaller models suitable for sleep apnea detection in a few seconds for the patient with minimal tuning were not used	[90]

Table 2.2 Continued.

IoT-enabled CNN studies	Techniques	Fields of applications	Challenges	References
A unique IoT-enabled depth-wise separable CNN with SVM was proposed for COVID-19 diagnosis and classification. The algorithm aims to identify all class labels of COVID-19 using CXR images.	DWS-CNN-IoT	Healthcare	The data rate, and energy consumption	[91]
A system for detecting data packet breaches in network security is introduced. The system follows a CNN-based approach for instruction detection.	CNN-NS	Security	Computational cost, coverage, energy consumption, and security	[92]
The cost-effective ways of improving healthcare living facilities by integrating remote health monitoring and Internet of medical things were proposed, and the system was used to monitor diabetic patients' vital signals in real time successfully.	IoMT-sensors	Healthcare	Use of multiple protocols and devices, resulting in high energy consumption, and data intrusion	[93]
Digital technology plays a vital role in supporting social, professional, and economic activities when people are confined to their homes. The Internet of Things has a track record of providing high-quality remote healthcare and automation services, allowing for social separation while sustaining population health and well-being.	IoT-sensor	Health	The lack of suitable safeguards and rules in handling personal medical data, and privacy, increases the vulnerabili-ties associated	[102]

(Continued.)

Table 2.2 Continued.

IoT-enabled CNN studies	Techniques	Fields of applications	Challenges	References
IoMT-based systems must be able to not only maintain acceptable performance in the face of such changes but also react appropriately if necessary. These IoMT systems also create security concerns since they frequently regulate conditions in which a system failure might be fatal.	IoT-sensor	Health	Security	[103]

consumption securing the security of the architecture devices and data. The lack of suitable safeguards and rules in handling personal medical data, as well as algorithms for assuring privacy, and security, improves the vulnerabilities associated with the CNN-IoT system.

2.6 Conclusion and Future Direction

The emergence of new generations of networks is assisting cloud computing environment and indeed IoT-based solutions. It connects computing devices with mechanical, electronics, and other objects such as animals or a human with the help of unique identifiers that have the ability for data transmission through a network without the intervention of a computer or human being. The chapter discusses the application of AI- and IoT-enabled in various fields of human endeavor, highlighting the novelty of these state-of-the-art technologies. We identified prominent researches in the area of health, technology, security, where IoT-enabled and CNN technologies were applied, and discussed their contributions to knowledge. The CNN and its architectures were reviewed and the research was summarized according to the field of application and the areas of limitations were identified with particular attention on healthcare, technology, humanity, security, infrastructure development, and agriculture with respect to techniques. Our contribution is three-fold. First, we identified the IoT-enabled CNN applications in the popular field. Second, we discussed some insights for selecting appropriate techniques to use for end-users of CNN tools in the IoT environment. Third, we identify CNN-IoT challenging areas for improvements. The three CNN operational layout steps' effective usages of a CNN architecture are discussed as follows: (1) data collection and feature generation, (2) data preprocessing and feature selection, and (3) data partitioning, learning, and analysis.

This chapter recommends more collaborations between researchers and the users and experts in the areas where such IoT applications are used to overcome limitations, enhance IoT technologies, and provide an insight into the security of future work in cyberspace. Going by the numerous challenges such as computational cost, data rate, coverage, energy consumption, security, and the applicability of CNN with IoT identified in various fields, the diction of research toward implementation of CNNs enables IoT to provide the solution that will peoples' needs in the area of healthcare, education, security, infrastructure development, agriculture, and so on, so as to address factors that are inimical to people to harness the esteem benefit in AI-IoT. Specifically, we suggest the following direction of future work based on our review: Since CNNs are primarily used for image processing, implementing state-of-the-art CNN architectures on sequential data necessitates the translation of 1D data into 2D data. The interoperation of the system and management of quality of service (QoS) were the issues observed in some studies, and the mobility and scalability problems were reported in real-time applications of IoT-enabled technologies to support the novelties. The use of DCNNs for sequential data is more popular because of their good feature extraction ability, and efficient calculations with a few numbers of parameters for effective data transmission in a cloud environment should be looked into.

References

[1] Y. Perwej, K. Haq, F. Parwej and M. M. Hassan, "The Internet of Things (IoT) and its application domains," *International Journal of Computer Applications*, vol. 182, no. 49, pp. 36–49, 2019.

[2] J. Oyeniyi, I. Ogundoyin, O. Oyediran and L. Omotosho, "Application of Internet of Things (IoT) to enhance the fight against Covid-19 pandemic," *International Journal of Multidisciplinary Sciences and Advanced Technology*, vol. 1, no. 3, pp. 38–42, 2020.

[3] A. Al-Fuqaha, M. Guizani, M. Mohammadi, M. Aledhari and M. Ayyash, "Internet of Things: A survey on enabling technologies, protocols and applications," *IEEE Communications Surveys & Tutorials*, vol. 17, no. 4, pp. 1–34, 2015.

[4] O. A. Adebisi, J. Ojo and T. Bello, "Computer-aided diagnosis system for classification of abnormalities in thyroid nodules ultrasound images using deep learning," *IOSR Journal of Computer Engineering (IOSR-JCE)*, vol. 22, no. 3, pp. 60–66, 2020.

[5] J. B. Awotunde, S. Ajagbe, M. Oladipupo, J. Awokola, O. Afolabi, M. Timothy and Y. Oguns, "An improved machine learnings diagnosis technique for COVID-19 pandemic using chest X-ray images," in *Proceedings of Applied Informatics: Communications in Computer and Information Science ICAI 2021*, H. Florez and M.F. Pollo-Cattaneo (eds), 2021.

[6] M. Achterberg, B. Prasse, L. Long Ma, S. Trajanovski, M. Kitsak and P. Mieghem, "Comparing the accuracy of several network-based COVID-19 prediction algorithms," *International Journal of Forecasting*, pp. 1–62, 2020.

[7] J. Somasekar, P. Kumar, A. Sharma and G. Ramesh, "Machine learning and image analysis applications in the fight against COVID-19 pandemic: Datasets, research directions, challenges and opportunities," in *Materials Today: Proceedings*, pp. 1–4, 2020.

[8] F. Dian, R. Vahidnia and A. Rahmati, "Wearables and the Internet of Things (IoT), applications, opportunities, and challenges: A survey," *IEEE Access*, 2020.

[9] R. Abdul, C. Chinmay and W. Celestine, "Exploratory data analysis, classification, comparative analysis, case severity detection, and Internet of Things in COVID-19 telemonitoring for smart hospitals," *Journal of Experimental & Theoretical Artificial Intelligence*, pp. 1–24, 2021.

[10] H. Mohammad and N. Tayarani, "Applications of artificial intelligence in battling against Covid-19: A literature review," *Chaos, Solitons and Fractals*, pp. 1–60, Aug. 21, 2020.

[11] D. Sujata, C. Chinmay, K. Sourav, K. Subhendu and F. Jaroslav, "BIFM: Big-data driven intelligent forecasting model for COVID-19," *IEEE Access*, pp. 1–13, 2021.

[12] T. J. Saleem and M. A. Chishti, "Deep learning for the Internet of Things: Potential benefits and use-cases," *Digital Communications and Networks*.

[13] O. Nadeem, M. Saeed, M. Tahir and R. Mumtaz, "A survey of artificial intelligence and Internet of Things (IoT) based approaches against Covid-19," in *Proceedings of the 2020 IEEE 17th International*

Conference on Smart Communities: Improving Quality of Life Using ICT, IoT and AI (HONET), 2020.
[14] R. Milton, D. Hay, S. Gray, B. Buyuklieva and A. Hudson-Smith, "Smart IoT and soft AI," *Living in the Internet of Things: Cybersecurity of the IoT - 2018*, vol. 2018, pp. 1–6, 2018.
[15] A. Osuwa, E. Ekhoragbon and L. Fat, "Application of artificial intelligence in Internet of Things," in *Proceedings of the 2017 9th International Conference on Computational Intelligence and Communication Networks (CICN)*, 2017.
[16] S. El-Gendy, "IoT based AI and its implementations in industries," in *Proceedings of the 2020 15th International Conference on Computer Engineering and Systems (ICCES)*, 2020.
[17] H. Song, J. Bai, Y. Yi, J. Wu and L. Liu, "Artificial intelligence enabled Internet of Things: Network architecture and spectrum access," *IEEE Computational Intelligence Magazine*, vol. 15, no. 1, pp. 44–51, 2020.
[18] O. Debauche, S. Mahmoudi, S. Mahmoudi, P. Manneback and F. Labeau, "A new edge architecture for AI-IoT services deployment," in *Proceedings of 17th International Conference on Mobile Systems and Pervasive Computing (MobiSPC)*, 2020.
[19] T. Zewdie and A. Girma, "IoT security and the role of AI/ML to combat emerging cyber threats in cloud computing environment," *Issues in Information Systems*, vol. 21, no. 4, pp. 253–263, 2020.
[20] T. Sung, P. Tsai, T. Gaber and C. Lee, "Artificial Intelligence of Things (AIoT) technologies and applications," in *Proceedings of the Wireless Communications and Mobile Computing*, pp. 1–2, 2021.
[21] M. Shin, W. Paik, B. Kim and S. Hwang, "An IoT platform with monitoring robot applying CNN-based context-aware learning," *Sensors*, vol. 19, no. 11, pp. 2525, 2019.
[22] R. Kumar, Y. Wang, T. Poongodi and A. Imoize, *Internet of Things, Artificial Intelligence and Blockchain Technology*. Berlin, Germany: Springer, 2021.
[23] K. Bennett and J. Robertson, "Remote sensing: Leveraging cloud IoT and AI/ML services," in *Proceedings of SPIE 11746, Artificial Intelligence and Machine Learning for Multi-Domain Operations Applications III, 117462L*, 2021.
[24] Y. Otoum and A. Nayak, "On securing IoT from deep learning perspective," in *Proceedings of the 2020 IEEE Symposium on Computers and Communications (ISCC)*, University of Ottawa, 2020.

[25] A. Khan, A. Sohail, U. Zahoora and A. S. Qureshi, "A survey of the recent architectures of deep convolutional neural networks," *Artificial Intelligence Review*, vol. 53, pp. 5455–5516, 2020. https://doi.org/10.1007/s10462-020-09825-6

[26] K. Zhang, Z. Zhang, Z. Li and Y. Qiao, "Joint face detection and alignment using multitask cascaded convolutional networks," *IEEE Signal Processing Letters*, vol. 23, pp. 1499–1503, 2016.

[27] S. Zagoruyko and N. Komodakis, "Wide residual networks," in *Proceedings British Machine Vision Conference*, pp. 87.1–87.12. https://doi.org/10.5244/C.30.872016

[28] T. Tong, G. Li, X. Liu and Q. Gao, "Image super-resolution using dense skip connections," in *Proceedings of the 2017 IEEE International Conference on Computer Vision (ICCV)*, pp. 4809–4817, 2017.

[29] A. Aljumah, "IoT-based intrusion detection system using convolution neural networks," *PeerJ Computer Science*, vol. 7, pp. e721, 2021.

[30] T. Bezdan and N. Džakula, "Convolutional neural networks layers and architectures," in *International Scientific Conference on Information Technology and Data Related Research*, Sinteza, 2019.

[31] B. Deng and W. Chen, "Deep learning backend for single and multisession I-vector speaker recognition," in *Proceedings of the Conference on Computer Vision and Pattern Recognition*, 2017.

[32] Y. Shen, T. Han, Q. Y. X. Yang, Y. Wang, F. Li and H. Wen, "CS-CNN: Enabling robust and efficient convolutional neural networks inference for Internet-of-Things applications," *IEEE Access Special Section on Multimedia Analysis for Internet-of-Things*, vol. 6, pp. 13439–13448, 2018.

[33] P. Wadhwa, A. Aishwarya, A. Tripathi, P. Singh, M. Diwakar and N. Kumar, "Predicting the time period of extension of lockdown due to increase in rate of COVID-19 cases in India using machine learning," in *Materials Today: Proceedings*, pp. 1–6, 2020.

[34] K. D. A. S. a. M. C. Sandeep Sony, "A systematic review of convolutional neural network-based structural condition assessment techniques," *Article in Engineering Structures*, pp. 1–53, 2021.

[35] S. Pouyanfar, S. Sadiq, Y. Yan, H. Tian, Y. Tao, M. Reyes, M. Shyu, S. Chen and S. Iyenger, "A survey on deep learning: algorithms, techniques, and applications," *ACM Computing Survey*, vol. 51, no. 5, 2018.

[36] D. H. Hubel and T. N. Wiesel, "Receptive fields, binocular interaction and functional architecture in the cat's visual cortex," *Journal of Physiology*, vol. 160, pp. 106–154, 1962. https://doi.org/10.1113/jphysiol.1962.sp006837
[37] M. Kafi, M. Maleki and N. Davoodian, "Functional histology of the ovarian follicles as determined by follicular fluid concentrations of steroids and IGF-1 in Camelus dromedarius," *Research in Veterinary Science*, vol. 99, pp. 37–40, 2015. doi: 10.1016/j.rvsc.2015.01.001
[38] H. C. C. Shin, H. R. Roth, M. Gao, L. Lu, Z. Xu, I. Nogues, J. Yao, D. Mollura, and R. M. Summers, "Deep convolutional neural networks for computer aided detection: CNN architectures, dataset characteristics and transfer learning," *IEEE Transactions on Medical Imaging*, vol. 35, pp. 1285–1298, 2016. doi: 10.1109/TMI.2016.2528162
[39] G. F. Montufar, R. Pascanu, K. Cho and Y. Bengio, "On the number of linear regions of deep neural networks," in *Proceedings of the Advances in Neural Information Processing Systems*, pp. 2924–2932, 2014.
[40] H. Wang and B. Raj, "On the origin of deep learning," pp. 1–72, 2017. doi: 10.1016/0014-5793(91)81229-2
[41] Q. Nguyen, M. Mukkamala and M. Hein, "Neural networks should be wide enough to learn disconnected decision regions," 2018, arXiv Prepr arXiv180300094.
[42] C. Dong, C. C. Loy, K. He and X. Tang, "Image super-resolution using deep convolutional networks," *IEEE Transactions on Pattern Analysis and Machine Intelligence*, vol. 38, pp. 295–307, 2016.
[43] Y. N. Dauphin, H. De Vries and Y. Bengio, "Equilibrated adaptive learning rates for non-convex optimization," *Advance in Neural Information Process Systems*, vol. 2015, pp. 1504–1512, 2015.
[44] Y. N. Dauphin, A. Fan, M. Auli and D. Grangier, "Language modeling with gated convolutional networks," in *Proceedings of the 34th International Conference on Machine Learning*, vol. 70, pp. 933–941, 2017.
[45] R. K. Srivastava, K. Greff and J. Schmidhuber, *Highway Networks*, 2015. doi: 10.1002/esp.3417
[46] G. Larsson, M. Maire and G. Shakhnarovich, "Fractalnet: Ultra-deep neural networks without residuals," pp. 1–11, 2016, arXiv Prepr arXiv160507648

[47] G. Huang, Z. Liu, L. Van Der Maaten and K.Q. Weinberger, "Densely connected convolutional networks," in *Proceedings of the 30th IEEE Conference on Computer Vision and Pattern Recognition, CVPR 2017*, pp. 2261–2269, Jan. 2017, doi: 10.1109/CVPR.2017.243

[48] K. He, X. Zhang, S. Ren and J. Sun, "Deep residual learning for image recognition," *Multimedia Tools and Applications*, vol. 77, pp. 10437–10453, 2015. doi: 10.1007/s11042-017-4440-4

[49] K. Kawaguchi, J. Huang and L. P. Kaelbling, "Effect of depth and width on local minima in deep learning," *Neural Computation*, vol. 31, pp. 1462–1498, 2019. doi: 10.1162/neco_a_01195

[50] Z. Lu, H. Pu, F. Wang, *et al.*, "The expressive power of neural networks: A view from the width," in *Proceedings of the Advances in Neural Information Processing Systems*, pp. 6231–6239, 2017.

[51] B. Hanin and M. Sellke, "Approximating continuous functions by ReLU nets of minimal width," 2017, arXiv Prepr arXiv171011278

[52] G. Huang, Y. Sun, Z. Liu, *et al.*, "Deep networks with stochastic depth," in *Proceedings of the European Conference on Computer Vision*, Springer, pp. 646–661, 2016.

[53] J. Hu, L. Shen and G. Sun, "Squeeze-and-excitation networks," in *Proceedings of the 2018 IEEE/CVF Conference on Computer Vision and Pattern Recognition*, IEEE, pp. 7132–7141, 2018.

[54] M. Oquab, L. Bottou, I. Laptev and J. Sivic, "Learning and transferring mid-level image representations using convolutional neural networks," in *Proceedings of the IEEE Computer Society Conference on Computer Vision and Pattern Recognition*, IEEE, pp. 1717–1724, 2014.

[55] J. Yang, W. Xiong, S. Li and C. Xu, "Learning structured and non-redundant representations with deep neural networks," *Pattern Recognition*, vol. 86, pp. 224–235, 2019.

[56] A. Khan, A. Sohail and A. Ali, "A new channel boosted convolutional neural network using transfer learning," 2018. arXiv Prepr arXiv180408528

[57] J. Kuen, X. Kong, G. Wang and Y.P. Tan, "DelugeNets: Deep networks with efficient and flexible cross-layer information inflows," in *Proceedings of the 2017 IEEE International Conference on Computer Vision Work, ICCVW 2017*, pp. 958–966, Jan. 2018, doi: 10.1109/ICCVW.2017.117

[58] X. Mao, C. Shen, and Y.-B. Yang, "Image restoration using very deep convolutional encoder-decoder networks with symmetric skip

connections," in *Proceedings of the Advances in Neural Information Processing Systems*, pp. 2802–2810, 2016.

[59] Y. LeCun, Y. Bengio and G. Hinton, "Deep learning," *Nature*, vol. 521, pp. 436–444, 2015. https://doi.org/10.1038/nature14539

[60] G. Huang, Y. Sun, Z. Liu *et al.*, "Deep networks with stochastic depth," in *Proceedings of the European Conference on Computer Vision, ECCV 2016*, Springer, pp. 646–661, 2016.

[61] S. Sajitha, "Vehicle detection using faster regional convolutional neural network," *International Journal of Advanced Research in Electrical, Electronics and Instrumentation Engineering*, pp. 2701–2704, 2019.

[62] C. Chinmay and N. Arij, "Intelligent internet of things and advanced machine learning techniques for COVID-19," *EAI Endorsed Transactions on Pervasive Health and Technology*, pp. 1–14, 2021.

[63] A. Derhab, A. Aldweesh, A. Emam and F. Khan, "Intrusion detection system for Internet of Things based on temporal convolution neural network and efficient feature engineering," in *Proceedings of the Wireless Communications and Mobile Computing*, vol. 6689134, pp. 16-23, 2020.

[64] S. A. Ajagbe, KA. Amuda, M. O. Oladipupo, O. Afe and K. Okesola, "Multi-classification of Alzheimer disease on magnetic resonance images (MRI) using deep convolution neural network approaches," *International Journal of Advanced Computer Research (IJACR)*, vol. 11, no. 53, pp. 51–60, 2021.

[65] A. Dhillon and G. Verma, "Convolutional neural network: A review of models, methodologies and applications to object detection," *Progress in Artificial Intelligence*, pp. 1–28, 2019.

[66] M. Motamedi, F. Portillo, M. Saffarpour, D. Fong and S. Ghiasi, "Resource-scalable CNN synthesis for IoT applications," pp. 1–7, 2018, https://arxiv.org/pdf/1901.00738.pdf.

[67] A. Maha, S. Jarrayaa and M. M. K. Alia, "CNN-based croed counting through IOT: Application for Saudi public places," in *Computers, Materials, & Continua*, 2019.

[68] N. Kimbugwe, T. Pei and M. Kyebambe, "Application of deep learning for quality of service enhancement in Internet of Things: A review," *Energies*, vol. 14, pp. 6384, 2021.

[69] T. Lawrence and L. Zhang, "IoTNet: An efficient and accurate convolutional neural network for IoT devices," *Sensors*, vol. 19, no. 24, pp. 5541, 2019.

[70] E. Adeniyi, R. Ogundokun and J. B. Awotunde, "IoMT-based wearable body sensors network healthcare monitoring system," in *IoT in Healthcare and Ambient Assisted Living*. Singapore : Springer, pp. 103–121.

[71] C. Chinmay, B. Gupta and S. Ghosh, "Identification of chronic wound status under tele-wound network through smartphone," *International Journal of Rough Sets and Data Analysis, Special issue on: Medical Image Mining for Computer-Aided Diagnosis*, vol. 2, no. 2, pp. 56–75, 2015.

[72] M. Nasajpour, S. Pouriyeh, R. Parizi, M. Dorodchi, M. Valero and H. Arabnia, "Internet of Things for current COVID-19 and future pandemics: An exploratory study," *Journal of Healthcare Informatics Research*, vol. 4, pp. 325–364, 2020.

[73] M. Otoom, N. Otoum, M. Alzubaidi, Y. Etoom and R. Banihani, "An IoT-based framework for early identification and monitoring of COVID-19 cases," *Biomedical Signal Processing and Control*, vol. 62, pp. 102–149, 2020.

[74] J. B. Awotunde, S. A. Ajagbe, I. R. Idowu and J. Ndunagu, "An enhanced cloud-IoMT-based and machine learning for effective COVID-19 diagnosis system," in *Intelligence of Things: AI-IoT Based Critical-Applications and Innovations.* Cham: Springer, 2021, pp. 55–76.

[75] C. R. J. Kumar, M. B. Arunsi, R. Jenova and M. A. Majid, "VLSI design of intelligent, self-monitored and managed, strip-free, non-invasive device for diabetes mellitus patients to improve glycemic control using IoT," in *Proceedings of 16th International Learning & Technology Conference 2019*, 2019.

[76] P. Viet, N. Le, N. Hoang and B. Thu, "DGCNN: A convolutional neural network over large-scale labeled graphs," *International Neural Network Society*, vol. 108, pp. 533–543, 2018.

[77] Y. Zhou and Y. Su, "Development of CNN and its application in education and learning," in *Proceedings of the 2019 9th International Conference on Social Science and Education Research (SSER 2019)*, 2019.

[78] N. Ghatashe, "Knowledge level assessment in e-learning systems using machine learning and user activity analysis," *International Journal of Advanced Computer Science and Applications, (IJACSA)*, vol. 6, no. 4, pp. 107–113, 2015.

[79] K. Y. Huang, C. H. Wu, Q. B. Hong, *et al.*, "Speech emotion recognition using deep neural network considering verbal and nonverbal speech sounds," in *Proceedings of the ICASSP, IEEE International Conference on Acoustics, Speech and Signal Processing*, 2019.
[80] S. Payvar, M. Khan, R. Stahl, D. Mueller-Gritschneder and J. Boutellier, "Neural network-based vehicle image classification for IoT devices," in *Proceedings of the 2019 IEEE International Workshop on Signal Processing Systems (SiPS)*, 2019.
[81] E. A. Lee, "CPS foundations," in *Proceedings of the 47th Design Automation Conference*, ACM, 2010, pp. 737–742.
[82] A. Gatouillat, Y. Badr, B. Massot and E. Sejdić, "Internet of Medical Things: A review of recent contributions dealing with cyber-physical systems in medicine," *IEEE Internet of Things Journal*, vol. 5, no. 5, pp. 3810–3822, 2018. doi:10.1109/JIOT.2018.2849014
[83] S. Siddiqui, M. Shakir, A. Khan and I. Dey, "Internet of Things (IoT) enabled architecture for social distancing during pandemic," *Frontier Communications Networking*, vol. 2, no. 614166, 2021.
[84] P. K. Gikunda and N. Jouandeau, "Modern CNNs for IoT based farms," in *Information and Communication Technology for Development for Africa. ICT4DA 2019. Communications in Computer and Information Science*, F. Mekuria, E. Nigussie and T. Tegegne (eds), vol. 1026. Cham: Springer, 2019.
[85] M. MaAlotibia, S. Jarrayaa, M. Ali and K. Moria, "CNN-based crowd counting through IoT: Application for Soudi public places," in *Proceedings of the 16th International Learning and Technology Conference 2019*, 2019.
[86] J. K. Ephraim Lo, "Internet of Things (IoT) discovery using deep neural networks," in *Proceedings of the WECV*, 2020.
[87] A. Cardenas, S. Amin, B. Sinopoli, A. Giani, A. Perrig and S. Sastry, "Challenges for securing cyber physical systems," in *Proceedings of the Workshop on Future Directions in Cyber-Physical Systems Security*, 2009.
[88] Y. Huang, Y. Cheng, A. Bapna, *et al.*, "GPipe: Efficient training of giant neural networks using pipeline parallelism," 2018, arXiv Prepr arXiv181106965.
[89] J. Kuen, X. Kong, G. Wang, Y. P. Tan, "DelugeNets: Deep networks with efficient and flexible cross-layer information inflows," in *Proceedings of the 2017 IEEE International Conference*

on Computer Vision Work, ICCVW 2017, pp. 958–966, Jan. 2018, doi: 10.1109/ICCVW.2017.117

[90] O. Sohaib, H. Lu and H. Hussain, "Internet of Things (IoT) in E-commerce: For people with disabilities," in *Proceedings of the 12th IEEE Conference on Industrial Electronics and Applications (ICIEA)*, 2017.

[91] S. A. Ajagbe, MO. Oladipupo and E. Balogun, "Crime belt monitoring via data visualization: A case study of folium," *International Journal of Information Security, Privacy and Digital Forensic*, vol. 4, no. 2, pp. 35–44, 2020.

[92] A. Rattani, N. Reddy and R. Derakhshani, "Multi-biometric convolutional neural networks for mobile user authentication," in *Proceedings of the 2018 IEEE International Symposium on Technologies for Homeland Security (HST)*, 2018.

[93] W. Njima, I. Ahriz, R. Zayani, M. Terre and R. Bouallegue, "Deep CNN for indoor localization in IoT-sensor systems," *Sensors (Basel, Switzerland)*, vol. 19, no. 14, pp. 3127, 2019.

[94] C. Szegedy, W. Zaremba, I. Sutskever, *et al.*, "Intriguing properties of neural networks," in *Proceedings of the 2nd International Conference on Learning Representations, ICLR 2014 - Conference Track Proceedings*, 2014.

[95] E. Cherrat, R. Alaoui and H. Bouzahir, "Convolutional neural networks approach for multimodal biometric identification system using the fusion of fingerprint, finger-vein and face images," *PeerJ Computer Science*, vol. 6, pp. e248, 2020.

[96] M. Alotibi, S. Jarraya, M. Ali and K. Moria, "CNN-based crowd counting through IoT: Application for Saudi public places," *Procedia Computer Science*, 2019.

[97] A. John, B. Cardiff and D. John, "A ID-CNN based deep learning technique for sleep apnea detection in IoT sensors," 2021.

[98] D. Le, V. Parvathy, D. Gupta, A. Khanna, J. Rodrigues and K. Shankar, "IoT enabled depthwise separable convolution neural network with deep support vector machine for COVID-19 diagnosis and classification," *International Journal of Machine Learning and Cybernetics*, pp. 23–34, 2021.

[99] Z. Gu, S. Nazir, C. Hong and S. Khan, "Convolution neural network-based higher accurate intrusion identification system for the network security and communication," *Security and Communication Networks*, vol. 2020, pp. 12–21, 2020.

[100] O. AlShorman, B. AlShorman, M. Al-khassaweneh and F. Alkahtani, "A review of internet of medical things (IoMT) – based remote health monitoring through wearable sensors: A case study fordiabetic patients," *Indonesian Journal of Electrical Engineering and Computer Science*, vol. 20, no. 1, pp. 414–422, 2020.

[101] A. Salehi, P. Baglat and G. Gupta, "Review on machine and deep learning models for the detection and prediction of Coronavirus," in *Materials Today: Proceedings*, pp. 1–6, 2020.

[102] D. N. Le, V. S. Parvathy, D. Gupta *et al.*, "IoT enabled depthwise separable convolution neural network with deep support vector machine for COVID-19 diagnosis and classification," *International Journal of Machine Learning and Cybernetics*, vol. 12, pp. 3235–3248, 2021. https://doi.org/10.1007/s13042-020-01248-7

[103] X. Mao, C. Shen and Y. B. Yang, "Image restoration using very deep convolutional encoder-decoder networks with symmetric skip connections," in *Proceedings of the Advances in Neural Information Processing Systems*, pp. 2802–2810, 2016.

[104] R. Vinayakumar, K. P. Soman and P. Poornachandrany, "Applying convolutional neural network for network intrusion detection," in *Proceedings of the 2017 International Conference on Advances in Computing, Communications and Informatics, ICACCI 2017*, 2017.

3

Convolutional Neural Network-Based Models for Speech Denoising and Dereverberation: Algorithms and Applications

Chengshi Zheng, Yuxuan Ke, Xiaoxue Luo, and Xiaodong Li

Key Laboratory of Sound and Vibration Research, Institute of Acoustics, Chinese Academy of Sciences, China
University of Chinese Academy of Sciences, China
E-mail: cszheng@mail.ioa.ac.cn

Abstract

Recently, speech controlled smart devices play an important role in Internet of Things (IoT) applications. Both reverberation and noise may significantly reduce the efficiency of the human–machine interaction for indoor applications. Therefore, speech enhancement becomes a critical front-end technique to improve the performance, which has attracted increasing attention in recent years. This chapter focuses on deep learning (DL) based monaural speech enhancement algorithms for both denoising and dereverberation, and both single and multiple speakers are considered to be extracted. More specially, convolutional neural network (CNN) based models are presented for this challenging speech enhancement task due to its parameter efficiency and state-of-the-art performance. After describing one-stage and multi-stage CNN-based models, numerous experiments are conducted to show the advantage and disadvantage when applying them to extract one desired speaker and multiple desired speakers. This study reveals that CNN-based models can achieve high performance when there is only one desired speaker to be extracted, while their performance may degrade a lot for multiple desired speakers. Some potential strategies are discussed on improving the performance of extracting multiple desired speakers and future research directions are outlined finally.

Keywords: Internet of Things, speech dereverberation, deep learning, multi-stage, nonlinear compression, multiple speakers.

3.1 Introduction

The Internet of Things (IoT) becomes an important part of our daily life, which aims at connecting a vast number of devices and sharing their data over the Internet. Among them, speech-controlled devices are quite general for IoT applications in smart home. For indoor applications, clean speech is frequently corrupted by both environmental noise and room's late reverberation, which may degrade speech quality/intelligibility and increase word error rate simultaneously, leading to affected user experience. Therefore, speech enhancement, including noise reduction and dereverberation, becomes a critical and hot front-end technique to improve the convenience of using speech-controlled IoT devices. We focus on the research of monaural speech enhancement algorithms in this chapter, as a majority of IoT devices only have one microphone for speech acquirement.

In the last half century, a vast number of noise reduction algorithms have already been proposed, which can be, at least roughly, divided into two kinds: conventional statistical signal processing (SSP) based algorithms and neural network (NN) based algorithms. Typical SSP based algorithms include spectral subtraction [1], Wiener filtering [2], minimum mean square error (MMSE) based short-time spectral amplitude estimator and MMSE-based log-spectral amplitude estimator [3, 4], subspace-based approaches [5], and so on. These conventional SSP-based algorithms can effectively suppress stationary and/or quasi-stationary noise, but their performance may degrade a lot, especially when handling the extremely non-stationary noise [6]. On the contrary, noise reduction algorithms with neural networks, such as shallow neural networks (SNNs) and deep neural networks (DNNs) or deep learning (DL) networks, can often achieve much better performance in more challenging scenarios, such as non-stationary noise environments and low signal-to-interference ratio (SIR) cases. In the early phase, the NN-based supervised speech enhancement algorithms were inspired by the theory of computational auditory scene analysis (CASA) [7, 8]. These methods can be named as time–frequency (T-F) mask mapping based methods, aiming to estimate real ideal ratio mask (rIRM), real ideal binary mask (rIBM), and complex ideal ratio mask (cIRM). After estimating the T-F mask, it was then multiplied with the noisy complex spectrum for noise reduction [9, 10]. In the late phase, more efficient methods that map the clean spectrum directly from the noisy spectrum were proposed, namely spectral mapping methods [11, 12]. The time-domain mapping-based algorithms were proposed in [13], which have been shown to be more effective in speech separation.

The environmental noise usually refers to additive noise, while the early reverberation is a typical kind of convolutional interference and the late reverberation is often assumed to be uncorrelated/independent with the early reflections, and, thus, the late reverberation can also be regarded as additive noise [14]. Accordingly, many SSP-based algorithms have already been proposed for dereverberation, such as spectral subtraction [1], weighted prediction error (WPE) [15, 16], and inverse filtering [17, 18]. Recently, more studies perform the dereverberation task in a supervised way [19, 20, 21, 22]. The same as NN-based denoising methods, mask mapping and spectral mapping are the two effective training schemes, where the spectral mapping includes log-power spectrum (LPS) mapping [23], target magnitude/complex spectrum (TMS) mapping [19], and compressed magnitude/complex spectrum (CMS) mapping [22]. The above-mentioned DL networks for noise reduction have already been extended for speech dereverberation, and they have shown better performance than traditional SSP-based dereverberation algorithms.

In most cases, noise reduction and dereverberation are considered separately. It is known to all that, in real-world acoustical environments, additive noise and reverberation often co-exist and both need to be suppressed to reduce their influence on speech quality and speech intelligibility. It has not been sufficiently addressed to handle both the noise and the reverberation, especially when multiple speakers need to be enhanced simultaneously. Recently, several NN-based monaural speech enhancement algorithms have been proposed, aiming to achieve joint removal of the noise and reverberation. For instance, in [26], a framework that jointly implements denoising and dereverberation was proposed, where the multi-target optimization is decomposed into four stages, and the denoising module and the dereverberation module are implemented sequentially. Moreover, a phase-aware mask and a complex ratio mask were proposed in [27] and [28], respectively, tackling the denoising task and the dereverberation task with a single-stage framework. An end-to-end WaveNet was proposed in [29] to suppress both the noise and reverberation.

Typical DL networks include fully connected neural networks (FCNs) [30], recurrent neural networks (RNNs), e.g., long short-term memory (LSTM) layers [31, 32, 33], and convolutional neural networks (CNNs) [8, 34, 35]. It is noticeable that both CNNs and RNNs have shown excellent capability in improving the perceptual evaluation of speech quality (PESQ) score [36], the extended short-time objective intelligibility (ESTOI) score [37], and the signal-to-distortion ratio (SDR) [38], and they respectively have properties of requiring much fewer trainable parameters and standout generalization ability [39], and, thus, these two kinds of models have become the most widely used models in the last decade. More recently, the integration of CNN and RNN has arisen, such as convolutional recurrent networks (CRNs) [11] and gated convolutional recurrent networks (GCRN) [12]. These models can benefit from the advantages of the CNN and RNN models simultaneously.

To be emphasized, most of these state-of-the-art methods focus on extracting only one desired speaker. In many practical speech communication applications, multiple speakers may speak at the same time and all of them need to be enhanced and then transferred to far end, while this problem has not been well studied. Besides, many aforementioned NN-based methods have been proved that the phase recovery is also important for speech enhancement in improving perceptual speech quality in low SNR scenarios and high reverberation cases [22, 27, 28]. Thus, a two-stage complex network (TSCN) has been proposed in this literature [40]. The TSCN is made up entirely of CNN layers and decouples the optimization of magnitude and phase to achieve better performance. Note that it mainly focuses on noise reduction applications in this previous work. In this chapter, we compare one-stage and multi-stage CNN-based models to tackle the simultaneous denoising and dereverberation problem. The advantage and disadvantage are clearly revealed when applying these two CNN-based models on both one desired speaker and multiple desired speakers scenarios.

This chapter is organized as follows. Section 3.2 describes the signal model, and we also formulate the problem in this section. We give the details of the one-stage and multi-stage CNN models in Section 3.3. Section 3.4 presents the experimental setup, followed by the experimental results and some analysis in Section 3.5. Finally, we draw some discussions and give some conclusions in Section 3.6.

3.2 Signal Model and Problem Formulation

3.2.1 Signal Model

For indoor applications, the time-domain noisy-reverberant speech received at a microphone can be given by

$$y(t) = \sum_{c=1}^{C} s^{(c)}(t) * h^{(c)}(t) + n(t), \tag{3.1}$$

where $s^{(c)}(t)$, $h^{(c)}(t)$, and $n(t)$ denote the cth clean speech source, its corresponding room impulse response (RIR), and the environmental noise at the time index t, respectively. $C \geq 1$ is the number of clean speech sources. $*$ stands for the convolution operator. For many practical applications, it is common to divide $h^{(c)}(t)$ into two parts, namely, the direct impulse response $h_d^{(c)}(t)$ and the reverberation impulse response $h_r^{(c)}(t)$. Therefore, the noisy and reverberant speech signal can be expressed as

$$\begin{aligned} y(t) &= \sum_{c=1}^{C} \left(s^{(c)}(t) * h_d^{(c)}(t) + s^{(c)}(t) * h_r^{(c)}(t) \right) + n(t) \\ &= \sum_{c=1}^{C} x^{(c)}(t) + \sum_{c=1}^{C} r^{(c)}(t) + n(t) \\ &= x(t) + r(t) + n(t), \end{aligned} \tag{3.2}$$

where $x(t)$ and $r(t)$ represent the desired direct speech components and the reverberant speech components of C desired speakers, respectively.

3.2.2 Feature Extraction

Instead of recovering speech in the time domain directly, various studies have shown that it is more advantageous to implement speech enhancement in the T-F domain [5, 12, 30]. With short-time Fourier transform (STFT), the spectral patterns of the noise components and the speech components can be effectively decomposed and, thus, can be distinguished more easily, which can be beneficial for modeling training. Therefore, in this chapter, we study to handle the reverberation and the noise in the T-F domain. Specifically, taken the STFT on both sides of the eqn (3.2), the time-frequency (T-F) signal model can be given by

$$
\begin{aligned}
Y(k,l) &= \sum_{c=1}^{C}\left(S^{(c)}(k,l)H_d^{(c)}(k,l) + S^{(c)}(k,l)H_r^{(c)}(k,l)\right) + N(k,l) \\
&= \sum_{c=1}^{C} X^{(c)}(k,l) + \sum_{c=1}^{C} R^{(c)}(k,l) + N(k,l) \\
&= X(k,l) + R(k,l) + N(k,l),
\end{aligned}
\tag{3.3}
$$

where $Y(k, l)$ is the complex spectrum of $y(t)$, and the remaining variables in eqn (3.3) have similar definitions. k and l denote the frequency index and the time index, respectively. The index (k, l) pair will be omitted in the following when no confusion arises.

In [22], compressed complex spectra are introduced instead of the original complex spectra as the input as well as the output of the CNN model. Note that only the magnitudes of the complex spectra are compressed while the phase information is unchanged. This is because phase spectra almost have no regular structure, and it is often difficult to recover them accurately. With the compressed scheme, the complex spectrum with magnitude compression can be expressed as $Y^{\circ} = |Y|^{\gamma} e^{i\theta_Y}$, which can also be written as

$$
\begin{aligned}
Y^{\circ} &= Y_r^{\circ} + Y_i^{\circ} \\
&= |Y|^{\gamma}\cos\theta_Y + j|Y|^{\gamma}\sin\theta_Y,
\end{aligned}
\tag{3.4}
$$

where $j = \sqrt{-1}$, $(\cdot)_r$, and $(\cdot)_i$ indicate extraction of the real part and the imaginary part of a complex spectrum, respectively. γ denotes the power compression parameter and θ_Y is the phase of Y. Pairs of $\{Y_r^{\circ}, Y_i^{\circ}\}$ are chosen as the input features throughout this chapter.

3.2.3 Problem Formulation

For denoising only, we only suppress the noise without handling reverberant speech components, and the target of this task is to extract $x(t) + r(t)$ or $X + R$ from $y(t)$ or Y. On the contrary, for dereverberation only, we only suppress reverberant speech components without suppressing the noise, and the target of dereverberation is to extract $x(t) + n(t)$ or $X + N$ from $y(t)$ or Y. When both denoising and dereverberation are necessary, the target is to extract $x(t)$ or X from $y(t)$ or Y. For many applications, such as automatic speech recognition systems, hands-free speech communication devices, and hearing-assistive devices, it is highly desirable to suppress both the noise and late reverberant speech components.

In this study, we take the approach of compressed complex spectrum mapping for simultaneously denoising and dereverberation, and a neural network with fully CNN layers is utilized to recover the complex spectrum of direct speech signal. The input features of the network are the compressed complex spectrum of the noisy-reverberant speech, i.e., $\{Y_r^\circ, Y_i^\circ\}$, and the target for network training is the compressed complex spectrum of the direct speech, i.e., $\{X_r^\circ, X_i^\circ\}$. The mapping process can be written as

$$\{\widetilde{X}_i^\circ, \widetilde{X}_r^\circ\} = \mathcal{G}(Y_r^\circ, Y_i^\circ; \Phi), \tag{3.5}$$

where $\mathcal{G}(\cdot, \cdot; \Phi)$ denotes the mapping function of the network with the parameter set Φ, and the symbol $\widetilde{(\cdot)}$ denotes the estimated value.

Finally, we can recover the uncompressed complex spectrum from the estimated compressed complex spectrum, which can be given by

$$\begin{aligned} \widetilde{X} &= \widetilde{X}_r + \widetilde{X}_i \\ &= (|\widetilde{X}|^\circ)^{1/\gamma} \cos\theta_{\widetilde{X}} + j(|\widetilde{X}|^\circ)^{1/\gamma} \sin\theta_{\widetilde{X}}. \end{aligned} \tag{3.6}$$

Because complex spectrum compression can be regarded as the pre-processing and post-processing of NN-based denoising and dereverberation algorithms, we omit the superscript $\circ$ and the compression parameter γ in the following to facilitate notations in this study. The same as [41], we set $\gamma = 0.5$.

3.3 One-stage CNN-based Speech Enhancement

As mentioned above, CRNs become popular for better capturing the short-term and long-term dependencies of speech and parameter efficiency. To further improve parameter efficiency, squeezed temporal convolutional modules (S-TCM) were introduced to replace the LSTM modules in GCRN [40], where we name this network as gated convolutional TCM Net (GCT-Net). GCT-Net consists entirely of CNNs and is adopted as the basic network for both one-stage and multi-stage speech enhancement in this study.

3.3.1 The Architecture of GCT-Net

GCT-Net, as shown in Figure 3.1(a), was trained to map the complex spectrum of the direct speech from that of its corresponding noisy-reverberant

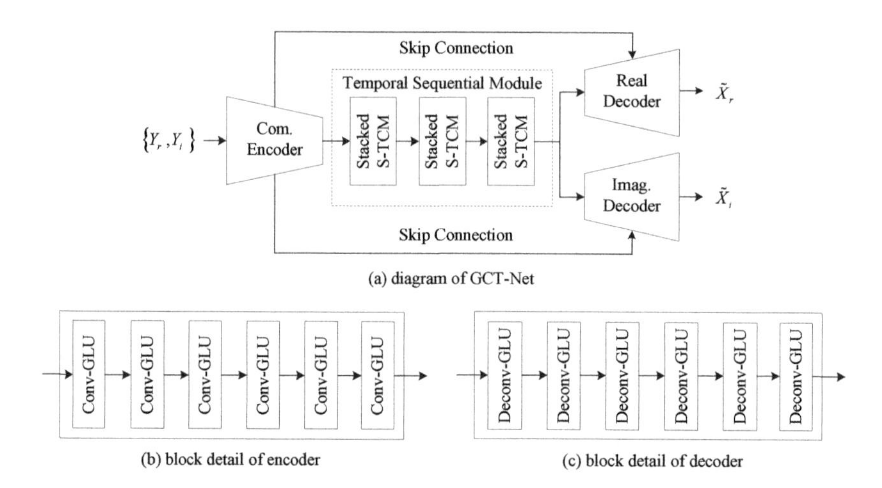

Figure 3.1 (a) Diagram of the GCT-Net. (b) The encoder of GCT-Net having six Conv-GLU blocks. (c) The decoder of GCT-Net having six Deconv-GLU blocks.

version, which consists of three modules, namely encoder, stacked S-TCMs, and decoder. Specifically, the encoder comprises six convolutional gated linear unit (Conv-GLU) layers for complex spectral feature extraction as shown in Figure 3.1(b), and the two decoders have the same structure as the encoder, where both of the two decoders contain six deconvolutional gated linear unit (Deconv-GLU) layers as shown in Figure 3.1(c). These two decoders output the real and imaginary parts of the complex spectrum of the direct speech, respectively. Between the encoder and decoder modules, three groups of stacked S-TCMs are adopted to capture the sequential information, each of which contains six S-TCM units with exponentially increasing dilation rate to enable the model to have a large temporal receptive field. Moreover, we introduce skip connections to concatenate the output of each encoder layer with the input of its corresponding decoder layer. In this GCT-Net, all convolutional layers and deconvolutional layers are causally implemented. In other words, there is not any future information being applied to infer the current output.

With the one-stage speech enhancement, the mapping process can be formulated as

$$\{\widetilde{X}_i, \widetilde{X}_r\} = \mathcal{G}_{\mathrm{GCT}}(Y_r, Y_i; \Phi_{\mathrm{GCT}}), \tag{3.7}$$

where $\mathcal{G}_{\mathrm{GCT}}(\cdot, \cdot; \Phi_{\mathrm{GCT}})$ is the mapping function of GCT-Net, and Φ_{GCT} is the corresponding parameter set.

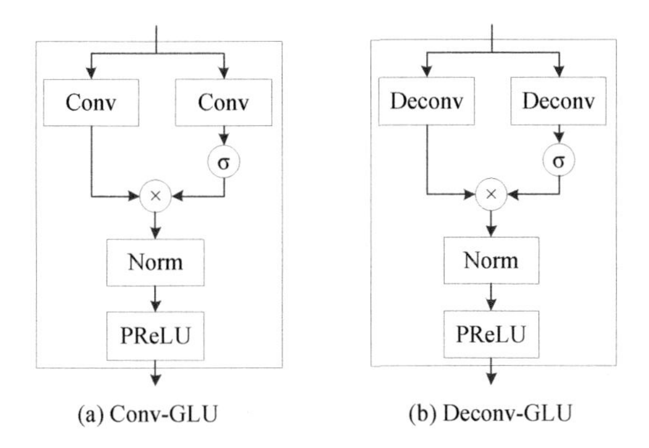

Figure 3.2 (a) The structure of Conv-GLU. (b) The structure of Deconv-GLU. $\sigma(\cdot)$ denotes a *sigmoid* function.

3.3.2 Gated Linear Units

Gate structure can control the data flow in the network, so that it allows for modeling more complex networks. For example, by endowing different gates with specific functions, RNNs can build long-term memory in modeling sequential data. Accordingly, the gated linear units (GLU) were proposed in [42], which can be given by

$$\mathbf{v} = (\mathbf{W}_1\mathbf{u} + \mathbf{b}_1) \otimes \sigma(\mathbf{W}_2\mathbf{u} + \mathbf{b}_2), \tag{3.8}$$

where $\mathbf{u}$ and $\mathbf{v}$ are the input and output of the gated linear units. $\mathbf{W_1}$, and $\mathbf{W_2}$ are the convolution weights and $\mathbf{b_1}$, and $\mathbf{b_2}$ are their corresponding biases. $\sigma(\cdot)$ denotes a *sigmoid* function, which compresses the value from 0 to 1. In this study, GLUs are integrated into both the convolutional layers and deconvolutional layers in the encoder and the two decoders. The Conv-GLU block is illustrated in Figure 3.2(a), where parametric ReLU (PReLU) and normalization (Norm) layers are applied after the multiplication of eqn (3.8). The Deconv-GLU block has a structure similar to the Conv-GLU, as plotted in Figure 3.2(b).

3.3.3 S-TCMs

TCMs have already been widely used in speech separation [13, 43] and speech enhancement [40, 41]. It can even perform better than LSTM in the aspect of temporal sequence modeling, and due to the parallel convolution

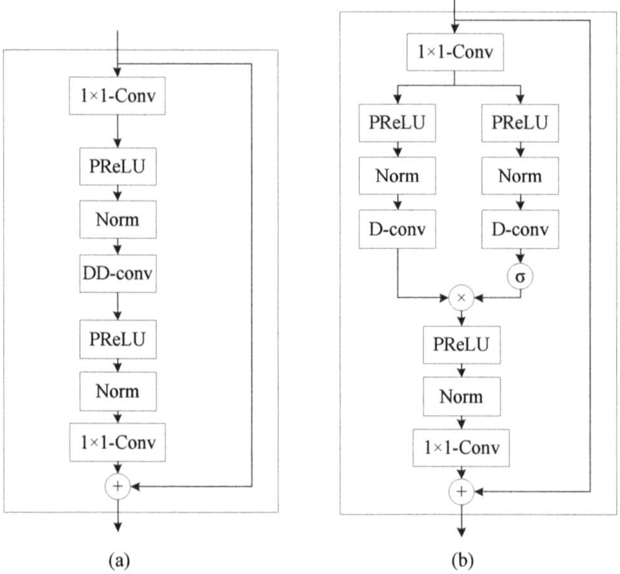

Figure 3.3 Diagram of an S-TCM. $\sigma(\cdot)$ denotes a *sigmoid* function.

mechanism, it can reduce inference time remarkably. Figure 3.3(a) presents the architecture of TCM, which is the same as [41].

Although stacked TCMs have obtained satisfactory performance, the parameter burden problem is still noticeable, which is not suitable for some portable devices. Thus, we adopt a squeezed TCM (S-TCM) [41], substituting depth-wise convolution with regular convolution. The diagram of this S-TCM unit is shown in Figure 3.3(b). Different from TCM, the proposed S-TCM unit squeezes the input channel into a lower dimension by a regular dilated 1×1-Conv to decrease the number of parameters. Besides, inspired by GLUs, we also introduce another gated branch. Its structure is similar to the main branch except that the *sigmoid* function is adopted, which is beneficial for the back propagation of gradients in the network.

3.3.4 Framework Details

The detailed parameters of GCT-Net are presented in Table 3.1. The size of input features is given as (*TimeStep* $\times$ *FeatureSize*) and (*ChannelNum* $\times$ *TimeStep* $\times$ *FeatureSize*). The hyper-parameters are given as (*KernelSize*,

Table 3.1 Detailed parameter setup for GCT-Net and ME-Net, where δ denotes the number of input channels, and T denotes the *TimeStep*

	Layer name	Input size	Hyper-parameters	Output size
Encoder	Conv-GLU_1	$\delta \times T \times 257$	$2 \times 5, (1,2), 64$	$64 \times T \times 127$
	Conv-GLU_2	$64 \times T \times 127$	$2 \times 3, (1,2), 64$	$64 \times T \times 63$
	Conv-GLU_3	$64 \times T \times 63$	$2 \times 3, (1,2), 64$	$64 \times T \times 31$
	Conv-GLU_4	$64 \times T \times 31$	$2 \times 3, (1,2), 64$	$64 \times T \times 15$
	Conv-GLU_5	$64 \times T \times 15$	$2 \times 3, (1,2), 64$	$64 \times T \times 7$
	Conv-GLU_6	$64 \times T \times 7$	$2 \times 3, (1,2), 64$	$64 \times T \times 3$
	Reshape_Size_1	$64 \times T \times 3$	-	$T \times 192$
Temporal sequential module	S-TCMs	$T \times 192$	$3 \times \left\{ \begin{pmatrix} 1,1,64 \\ 5,1,64 \\ 1,1,192 \end{pmatrix} \begin{pmatrix} 1,1,64 \\ 5,2,64 \\ 1,1,192 \end{pmatrix} \begin{pmatrix} 1,1,64 \\ 5,4,64 \\ 1,1,192 \end{pmatrix} \begin{pmatrix} 1,1,64 \\ 5,8,64 \\ 1,1,192 \end{pmatrix} \begin{pmatrix} 1,1,64 \\ 5,16,64 \\ 1,1,192 \end{pmatrix} \begin{pmatrix} 1,1,64 \\ 5,32,64 \\ 1,1,192 \end{pmatrix} \right.$	$T \times 192$
	Reshape_Size_2	$T \times 192$	-	$64 \times T \times 3$
Decoder	skip_connect_1	$64 \times T \times 3$	-	$128 \times T \times 3$
	Deconv-GLU_1	$128 \times T \times 3$	$2 \times 3, (1,2), 64$	$64 \times T \times 7$
	Skip_Connect_2	$64 \times T \times 7$	-	$128 \times T \times 7$
	Deconv-GLU_2	$128 \times T \times 7$	$2 \times 3, (1,2), 64$	$64 \times T \times 15$
	Skip_Connect_3	$64 \times T \times 15$	-	$128 \times T \times 15$
	Deconv-GLU_3	$128 \times T \times 15$	$2 \times 3, (1,2), 64$	$64 \times T \times 31$
	Skip_Connect_4	$64 \times T \times 31$	-	$128 \times T \times 31$
	Deconv-GLU_4	$128 \times T \times 31$	$2 \times 3, (1,2), 64$	$64 \times T \times 63$
	Skip_Connect_5	$64 \times T \times 63$	-	$128 \times T \times 63$
	Deconv-GLU_5	$128 \times T \times 63$	$2 \times 3, (1,2), 64$	$64 \times T \times 127$
	Skip_Connect_6	$64 \times T \times 127$	-	$128 \times T \times 127$
	Deconv-GLU_6	$128 \times T \times 127$	$2 \times 5, (1,2), 64$	$64 \times T \times 257$
	Reshape_Size_3	$1 \times T \times 257$	-	$T \times 257$
	Linear_1(Softplus)	$T \times 257$	257	$T \times 257$

Stride, *ChannelNum*) for the encoder and the decoder, and (*KernelSize*, *DilationRate*, *ChannelNum*) for S-TCMs. δ refers to the number of input channels.

More specifically, within each Conv-GLU block, the kernel size in the time axis is set to 2, and that in the frequency axis is set to 5 in the first block and 3 in the following blocks. The stride is set to $(1, 2)$, so that the frequency size can be halved gradually, while the time size keeps unchanged. The number of channels in each convolution layer is 64. Besides, instance normalization and PReLU layers are adopted after each convolutional layer. The decoder has an architecture similar to the encoder, which replaces all Conv-GLUs by Deconv-GLUs. Moreover, skip connection is introduced between each Conv-GLU and Deconv-GLU pair.

In the bottleneck of GCT-Net, three groups of stacked S-TCMs are adopted to gradually capture the long-range temporal information, and each of them has six S-TCM units with their dilation rates exponentially increased, i.e., (1,2,4,8,16,32). Note that the module consisting of stacked S-TCMs is causal because no future frames are necessary to infer current frame output.

3.3.5 Loss Function

The mean square error (MSE) is one of the most widely used loss functions for training neural network models, even in the application of complex spectrum mapping [12]. Recently, an integration of the magnitude constraint and the complex spectrum constraint has been proven to be essential for spectral recovery [41, 44]. In this study, the following loss function is considered:

$$\mathcal{L}_{\text{gct}} = \alpha\mathcal{L}_{\text{gct}}^{\text{RI}} + (1-\alpha)\mathcal{L}_{\text{gct}}^{\text{Mag}}, \tag{3.9}$$

where $\mathcal{L}_{gct}^{RI}$ and $\mathcal{L}_{gct}^{Mag}$ denote the constraint of the complex spectrum and that of the magnitude, respectively, which are given by

$$\mathcal{L}_{\text{gct}}^{\text{RI}} = \left\|X_r - \widetilde{X}_r\right\|_2^2 + \left\|X_i - \widetilde{X}_i\right\|_2^2, \tag{3.10}$$

$$\mathcal{L}_{\text{gct}}^{\text{Mag}} = \left\|\sqrt{X_r^2 + X_i^2} - \sqrt{\widetilde{X}_r^2 + \widetilde{X}_i^2}\right\|_2^2, \tag{3.11}$$

where $\alpha \in [0, 1]$ is a tuning parameter to make a trade-off between the RI loss and the spectral magnitude loss. We empirically set $\alpha = 0.5$ in this study.

3.4 Multi-Stage CNN-based Speech Enhancement

Figure 3.4(a) plots the two-stage network based on the (compressed) complex spectrum mapping (CTS-Net). It shows superior performance in the second deep noise suppression (DNS) challenge [40]. The details of CTS-Net can be found in [41], where ME-Net aims to estimate the raw magnitude, denoted by $|\widetilde{X}^{me}|$, and CS-Net aims to reduce the residual noise components and recover the real and imaginary parts of the residual complex spectrum, denoted by $\widetilde{X}^{cs}$, which is finally added together to obtain the estimation of the complex spectrum. The whole procedure of the two-stage network can be formulated as

$$|\widetilde{X}^{me}| = \mathcal{G}_1(|Y|; \Phi_1), \tag{3.12}$$

$$\widetilde{X}_r^{me} = \Re(|\widetilde{X}^{me}|e^{j\theta_Y}), \widetilde{X}_i^{me} = \Im(|\widetilde{X}^{me}|e^{j\theta_Y}), \tag{3.13}$$

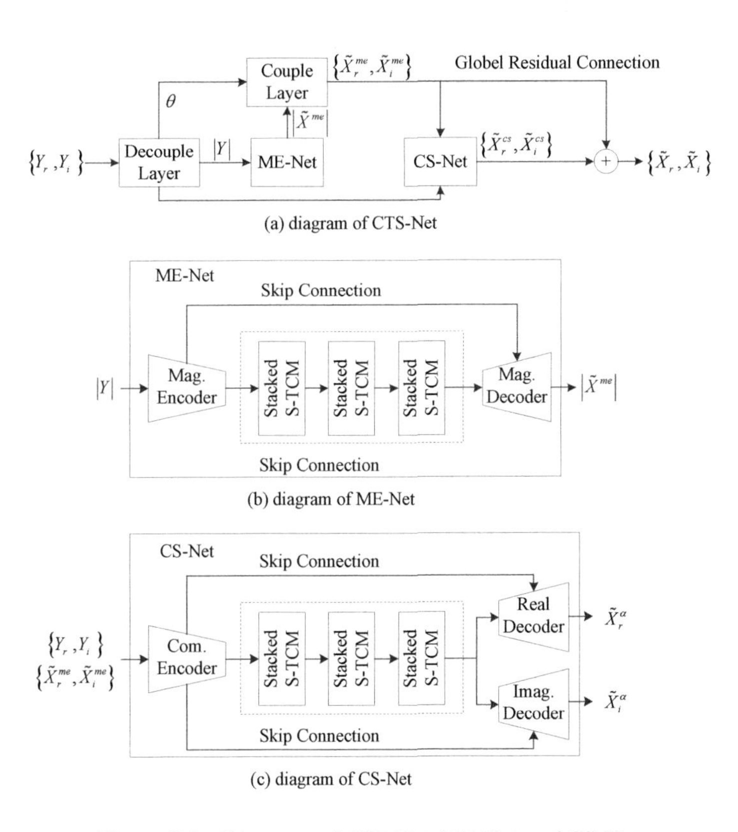

Figure 3.4 Diagrams of CTS-Net, ME-Net, and CS-Net.

$$\{\widetilde{X}_i^{cs}, \widetilde{X}_r^{cs}\} = \mathcal{G}_2(Y_r, Y_i, \widetilde{X}_r^{me}, \widetilde{X}_i^{me}; \Phi_2), \tag{3.14}$$

$$\widetilde{X}_r = \widetilde{X}_r^{me} + \widetilde{X}_r^{cs}, \widetilde{X}_i = \widetilde{X}_i^{me} + \widetilde{X}_i^{cs}, \tag{3.15}$$

where θ_Y denotes the original noisy phase, and $\Re$ and $\Im$ denote to extract the real and imaginary parts, respectively. $\mathcal{G}_1(\cdot; \Phi_1)$ and $\mathcal{G}_2(\cdot, \cdot, \cdot, \cdot; \Phi_2)$ represent the nonlinear mapping function of ME-Net and that of CS-Net, respectively.

3.4.1 Framework Structure

The detailed structure of ME-Net and that of CS-Net are presented in Figures 3.4(b) and 3.4(c), respectively. One can see that ME-Net has similar topology to GCT-Net, except that ME-Net has only one decoder for magnitude estimation. Besides, the topology of CS-Net is the same as that of GCT-Net, but it aims to estimate the residual RI components rather than the desired RI components. Therefore, the hyper-parameters setups for ME-Net and CS-Net are nearly the same as GCT-Net, which is presented in Table 3.1. Note that $\delta = 1$ for ME-Net because there is only one input channel. Moreover, because the range of the spectral magnitude $|\widetilde{X}^{me}| \in [0, \infty)$, *Softplus* is chosen as the activation function of the last layer. Different from ME-Net, the linear outputs are adopted in CS-Net because both the real and imaginary parts of the complex spectrum have unbounded ranges. In addition, as the inputs of CS-Net contain both original and coarsely estimated complex spectra, the number of input channels is set to $\delta = 4$.

3.4.2 Loss Function

We apply a two-stage training strategy to train CTS-Net. First, ME-Net is trained with the MSE loss, which is

$$\mathcal{L}_{me} = \left\| |X| - |\widetilde{X}^{me}| \right\|_2^2. \tag{3.16}$$

Second, ME-Net and CS-Net are trained jointly, and the loss function can be finally expressed as

$$\mathcal{L} = \mathcal{L}_{cs} + \lambda \mathcal{L}_{me}, \tag{3.17}$$

where λ is a positive real value, and the loss $\mathcal{L}_{cs}$ is defined as

$$\mathcal{L}_{cs} = \alpha \mathcal{L}_{cs}^{RI} + (1 - \alpha) \mathcal{L}_{cs}^{Mag}, \tag{3.18}$$

where

$$\mathcal{L}_{cs}^{RI} = \left\| X_r - \widetilde{X}_r \right\|_2^2 + \left\| X_i - \widetilde{X}_i \right\|_2^2, \tag{3.19}$$

$$\mathcal{L}_{cs}^{Mag} = \left\| \sqrt{X_r^2 + X_i^2} - \sqrt{\widetilde{X}_r^2 + \widetilde{X}_i^2} \right\|_2^2. \tag{3.20}$$

In this study, set $\alpha = 0.5$ and $\lambda = 0.1$ which is the same as [41] for joint denoising and dereverberation.

3.5 Experimental Setup

3.5.1 Datasets

In this study, we conduct all the experiments on the DNS-Challenge clean speech dataset [45], which is taken from the public audio books dataset called Librivox. 45,000, and 3,000 utterances are randomly selected for training and validation, respectively. 150 utterances of other untrained speakers are selected for models evaluation. To simulate noisy environment, we randomly selected around 550 noise processes from the DNS-Challenge datasets [45] as the noise set for training and validation.

As for simulating indoor reverberated environment, three sets of reverberant rooms are generated, namely small room set, medium room set, and large room set. Each room set contains 200 rooms of different sizes. Similar to [46], the width and length of the rooms in small room set, medium room set, and large room set are sampled uniformly from 1 to 10 m, 10 to 30 m, and 30 to 50 m, respectively. The height of each room ranges from 3 m to 5 m, and room absorption coefficient ranges from 0.2 to 0.8. In each room, a receiver position is randomly sampled and then 100 RIRs are randomly generated according to different speaker positions. The image method is used to simulate RIRs [47]. Besides, as small size rooms have more complicated acoustical environment and have been widely concerned in the field of speech dereverberation, extra RIRs of 28 small rooms are simulated according to uniformly sampled T_{60} from 0.3 to 1.0 s. Therefore, we obtain totally 62,800 RIRs for simulating training utterances.

To generate training set, we consider one desired speaker and multiple desired speakers speech communication scenarios; so two types of datasets are prepared for both training and validation sets, namely *one-speaker* dataset and *multi-speaker* dataset.

In the *one-speaker* dataset, each clean utterance for training and validation is convolved with a random RIR to generate a reverberated speech. The reverberated speech is then added with a randomly selected noise under a randomly chosen SNR ranging from 10 to 30 dB. As a result, we totally generate 45,000 and 3000 utterances in *one-speaker* dataset for training and validation, whose duration is about 100 and 7 hours, respectively.

As for the *multi-speaker* dataset, it contains half *one-speaker* dataset and half *two-speaker* dataset. The *one-speaker* dataset is generated in the same way as above, which has totally 22,500 and 1500 utterances for training and validation, respectively. To generate the *two-speaker* dataset, two reverberated speech signals of the same room are randomly selected and mixed first, which is then added with a randomly selected noise under the SNR ranging from 10 to 30 dB. As the same with *one-speaker* dataset, the *two-speaker* dataset has totally 22,500 and 1500 utterances for training and validation, respectively. Thus, there are around 100 and 7 hours of speech data are generated for training and validation in both *one-speaker* dataset and *multi-speaker* dataset.

3.5.2 Parameter Configuration

We resample all the utterances to 16 kHz and chunk them to 8 s. 32 ms Hanning window is applied before STFT, with the frame shift of 8 ms. 512-point fast Fourier transform (FFT) is applied to extract the spectral features. Each model is trained for 50 epochs using the Adam optimizer [48]. For the one-stage GCT-Net and the first stage of the CTS-Net, the learning rate is set to 0.001, which is halved when the validation loss increases for consecutive three epochs. In the second stage of the CTS-Net, the learning rate of ME-Net is fine-tuned as 0.0001, and that of CS-Net is set to 0.001. We set the batch size to 16.

3.6 Results and Analysis

To evaluate and compare the performances of the one-stage and the multi-stage CNN models, we test GCT-Net and CTS-Net on a prepared test set. This evaluation also aims to investigate the advantage and disadvantage of these models in extracting one and/or multiple desired speech signals. There are seven values of T_{60} in this dataset, i.e., $\{0, 0.2, 0.4, 0.6, 0.8, 1.0, 1.2\}$ s, and for each T_{60}, 150 mixed speech signals and 150 pairs of RIRs of untrained small room configurations are generated. Both the seen and unseen noise scenarios are considered when generating the test set, where the unseen noise are taken from NOISEX-92 [49], namely babble and cafe.

Similar to the training set and the validation set, two types of datasets are generated in the test set, namely *one-speaker* test dataset and *two-speaker* test dataset. In *one-speaker* test dataset, an RIR is randomly chosen from the 150 generated RIRs and is convolved with clean utterances under each T_{60}. Then it is mixed with a randomly chosen seen or unseen noise with SNR ranging from 10 to 30 dB. As for *two-speaker* test dataset, a pair of reverberated utterances is generated by randomly selecting two clean utterances and calculating the convolution of the two utterances and a pair of RIRs from the same room. Later, the two reverberated utterances are mixed first, and then the mixed utterance is mixed with a randomly chosen noise again, with the SNR ranging from 10 to 30 dB. Therefore, there are totally 1050 utterances in both *one-speaker* test dataset and *two-speaker* test dataset. During the experiments, all of these evaluation models are trained using Pytorch [50] with only one Nvidia Tesla V100.

In the following, we demonstrate speech spectrograms using different CNN models and different training strategies for different values of T_{60}. And then, we choose the perceptual evaluation of speech quality (PESQ) score [36], the signal-to-distortion ratio (SDR) [38], and the extended short-time objective intelligibility (ESTOI) score [37] as objective measurements to measure the perceptual speech quality, speech intelligibility, and speech distortion, respectively.

3.6.1 Spectrograms

To intuitively study the denoising and dereverberation performance of different models and different training strategies, we plot speech spectrograms of the enhanced speech signals with different models and training strategies in Figure 3.5 with the T_{60} value being equal to 0.6 s, where the mixed speech contains only one desired speaker, and the noise is unseen. In Figure 3.5, (a) is the clean speech spectrogram, (b) is the noisy-reverberant speech spectrogram, (c) is the spectrogram of the enhanced speech with GCT-Net trained using *one-speaker* training set, (d) is the enhanced speech spectrogram with GCT-Net trained using *multi-speaker* training set, (e) is the enhanced speech spectrogram with CTS-Net trained using *one-speaker* training set, and (f) is the enhanced speech spectrogram with CTS-Net trained using *multi-speaker* training set.

One can see from Figure 3.5 that both GCT-Net and CTS-Net have impressive dereverberation performance. By comparing Figures 3.5(c) and 3.5(d), we see that the enhanced speech with CTS-Net has less reverberation than that with GCT-Net. For example, the temporal smearing phenomenon in Figure 3.5(e) is less obvious than that in Figure 3.5(c). Moreover, the time-frequency structure of the enhanced speech with CTS-Net is more clear than that with GCT-Net, e.g., in the time interval [0.9 1.2] s. From Figures 3.5(e)

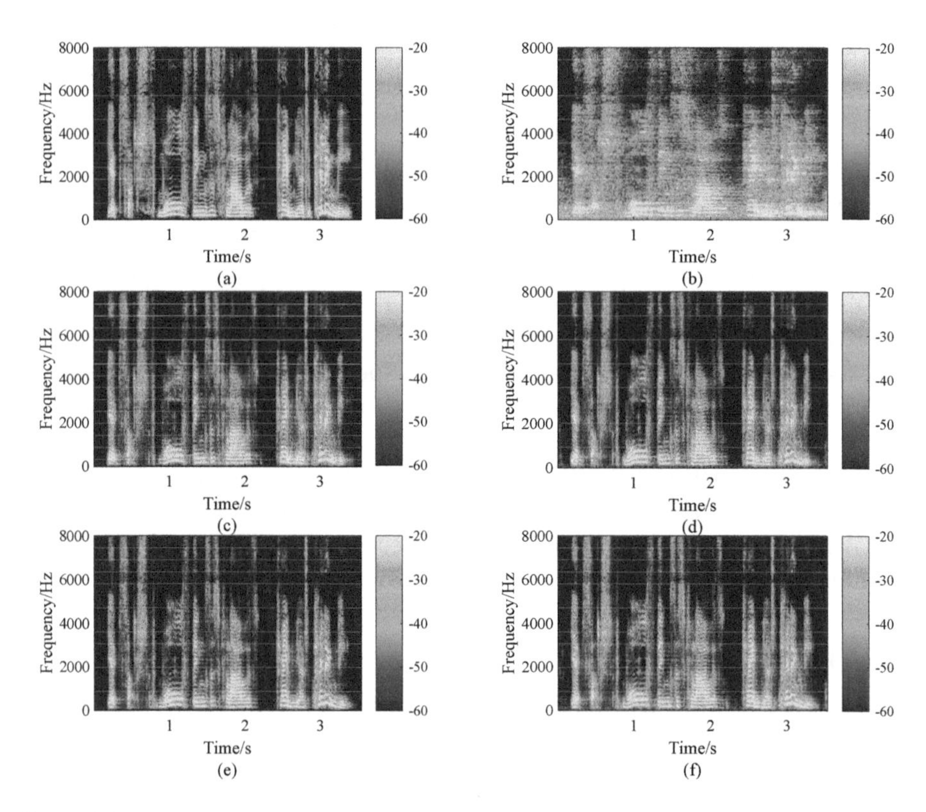

Figure 3.5 Speech spectrograms with the reverberation time $T_{60} = 0.6$ s containing only one desired speaker of (a) clean speech, (b) noisy-reverberant speech (PESQ $= 1.98$), (c) the enhanced speech with GCT-Net trained by *one-speaker* training set (PESQ $= 2.50$), (d) the enhanced speech with GCT-Net trained by *multi-speaker* training set (PESQ $= 2.48$), (e) the enhanced speech with CTS-Net trained by *one-speaker* training set (PESQ $= 2.56$), and (f) the enhanced speech with CTS-Net trained by *multi-speaker* training set (PESQ $= 2.54$).

and 3.5(f), one can see that CTS-Net trained by *one-speaker* training set can suppress slightly more reverberant components than that trained by *multi-speaker* training set.

In Figure 3.5, there is only one desired speaker. While in Figure 3.6, we test different CNN models and different training strategies for two desired speakers cases, and the T_{60} is set to 0.6 s. The same as the above three figures, the speech spectrograms of Figures 3.6(c) and 3.6(d) demonstrate that CTS-Net outperforms GCT-Net in terms of denoising and dereverberation because both desired speech signals can be better recovered. For example,

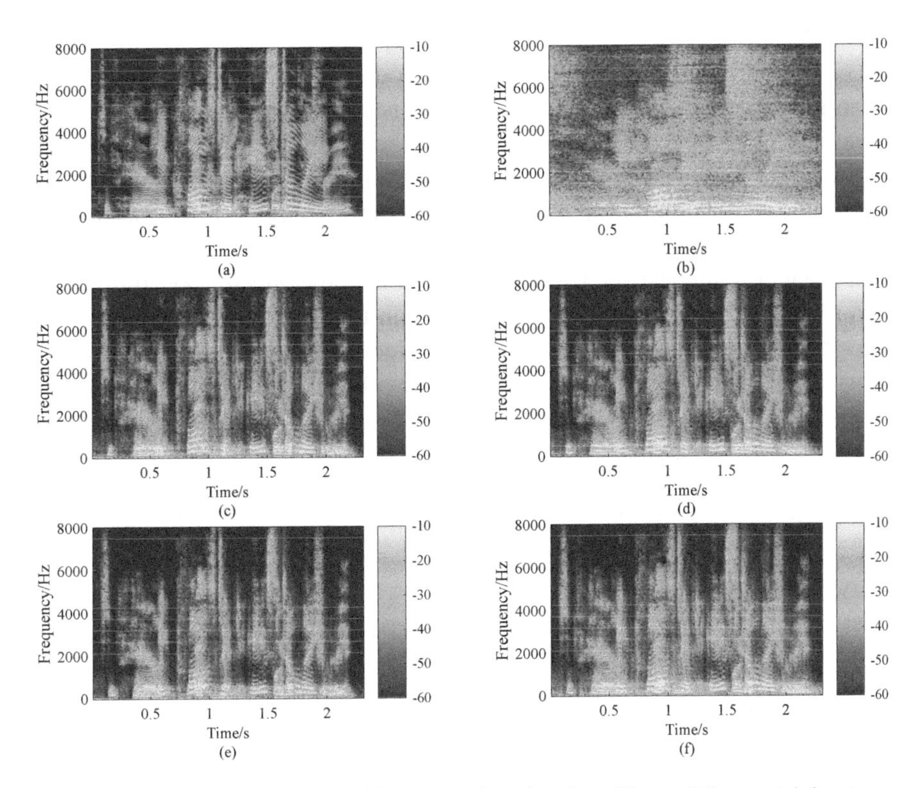

Figure 3.6 Speech spectrograms with the reverberation time $T_{60} = 0.6$ s containing two desired speakers of (a) clean speech, (b) noisy-reverberant speech (PESQ $= 1.42$), (c) the enhanced speech with GCT-Net trained by *one-speaker* training set (PESQ $= 1.64$), (d) the enhanced speech with GCT-Net trained by *multi-speaker* training set (PESQ $= 1.96$), (e) the enhanced speech with CTS-Net trained by *one-speaker* training set (PESQ $= 1.69$), and (f) the enhanced speech with CTS-Net trained by *multi-speaker* training set (PESQ $= 2.08$).

around 1 s in Figure 3.6(c), the spectral component of the weaker speech signal is almost removed, but it can be better reconstructed in Figure 3.6(e). Moreover, by comparing Figures 3.6(e) and 3.6(f), one can observe that, in the time interval [1.4, 1.5] s, the spectral harmonic structure in Figure 3.6(f) is more clear, indicating that CNN models trained by *multi-speaker* training set perform better than that trained by *one-speaker* training set, especially when two desired speakers exist. To investigate the performance of different CNN models as well as different training strategies more comprehensively, objective measures are conducted in the following.

3.6.2 PESQ Scores

Table 3.2 gives the PESQ scores of different models in the seen noise test set. One can get that the performance of CTS-Net is better than that of GCT-Net in all situations. Thus, we can conclude that the two-stage training strategy shows much more efficiency than the one-stage training strategy in improving speech quality. By comparing the results of the *one-speaker* test set and the *two speaker* test set in Table 3.2, one can see that, if there is only one desired speaker in the test set, CTS-Net trained with the *one-speaker* training set can obtain slightly higher PESQ scores than that trained with the *multi-speaker* training set. On the contrary, GCT-Net trained with *multi-speaker* training set performs better than that trained with *one-speaker* training set.

Table 3.2 The PESQ scores of different models in the **seen** noise test set

Metrics	T_{60} (ms)	One speaker test set					Two speaker test set				
			One speaker training set		Mix-speaker training set			One speaker training set		Mix-speaker training set	
		Mix	GCT	CTS	GCT	CTS	Mix	GCT	CTS	GCT	CTS
PESQ	1200	1.62	2.06	**2.16**	2.09	**2.16**	1.41	1.58	1.61	1.75	**1.79**
	1000	1.71	2.17	**2.28**	2.20	**2.28**	1.50	1.69	1.71	1.87	**1.91**
	800	1.80	2.30	**2.41**	2.33	2.40	1.60	1.80	1.83	2.00	**2.04**
	600	1.94	2.49	**2.60**	2.52	2.58	1.74	1.98	2.04	2.21	**2.25**
	400	2.17	2.77	**2.89**	2.80	2.87	1.99	2.24	2.31	2.47	**2.52**
	200	2.70	3.25	**3.34**	3.28	3.33	2.58	2.80	2.84	3.05	**3.09**
	0	3.12	3.69	**3.72**	**3.72**	3.70	3.38	3.72	3.72	**3.84**	3.81

Table 3.3 The PESQ scores of different models in the **unseen** noise test set.

Metrics	T_{60} (ms)	One speaker test set					Two speaker test set				
			One speaker training set		Mix-speaker training set			One speaker training set		Mix-speaker training set	
		Mix	GCT	CTS	GCT	CTS	Mix	GCT	CTS	GCT	CTS
PESQ	1200	1.63	2.00	**2.10**	2.02	2.08	1.41	1.54	1.56	1.70	**1.74**
	1000	1.70	2.11	**2.22**	2.13	2.19	1.48	1.61	1.64	1.80	**1.84**
	800	1.76	2.18	**2.29**	2.19	2.27	1.57	1.74	1.76	1.93	**1.97**
	600	1.91	2.36	**2.47**	2.37	2.43	1.72	1.90	1.97	2.12	**2.17**
	400	2.10	2.61	**2.73**	2.61	2.69	1.95	2.14	2.21	2.37	**2.42**
	200	2.57	3.04	**3.13**	3.04	3.11	2.48	2.67	2.69	2.88	**2.93**
	0	2.90	3.41	**3.43**	3.39	3.40	3.03	3.39	3.35	**3.46**	**3.46**

This is because GCT-Net and CTS-Net trained with *one-speaker* set can suppress more reverberation than that trained with *multi-speaker*, but more speech distortion may be introduced. Furthermore, when there are two desired speakers at the same time, the PESQ scores obtained by training the *multi-speaker* set are much higher than those obtained by training the *one-speaker* set, no matter which CNN model is used. An explanation for this result is that, when two desired speakers co-exist, the speech distortion of CTS-Net trained with *one-speaker* may heavily reduce PESQ scores. Therefore, the CTS-Net that is trained by *multi-speaker* can obtain higher speech quality than that trained by *one-speaker*.

We also test the PESQ metrics in the unseen noise test set to verify the generalization ability of different models in Table 3.3. The results of Table 3.3 are similar to Table 3.2, when the target signal has only one desired speaker, CTS-Net trained by *one-speaker* is the best one. This is because it can not only effectively suppress the noise and the reverberation but also retain the detailed spectral structure of the desired speech. Moreover, in the case that two desired speakers co-exist, CTS-Net trained by *multi-speaker* can obtain the highest PESQ scores. This is because the former one can retain the weaker speech components. Note that CTS-Net performs better than GCT-Net in most cases.

3.6.3 ESTOI scores

Table 3.4 summarizes ESTOI scores of different models in the seen noise test set. By comparing the ESTOI scores of GCT-Net and CTS-Net in all the situations, it can be seen that CTS-Net can obtain higher speech intelligibility

Table 3.4 The ESTOI scores of different models in the **seen** noise test set

Metrics	T_{60} (ms)	One speaker test set					Two speaker test set				
			One speaker training set		Mix-speaker training set			One speaker training set		Mix-speaker training set	
		Mix	GCT	CTS	GCT	CTS	Mix	GCT	CTS	GCT	CTS
ESTOI (%)	1200	30.11	55.95	**58.93**	55.84	58.20	28.47	45.63	46.83	47.70	**49.27**
	1000	33.24	59.61	**62.58**	59.62	62.16	31.98	49.28	50.73	51.76	**53.61**
	800	39.07	63.97	**66.90**	64.19	66.60	37.87	54.12	55.29	56.91	**58.63**
	600	46.92	69.12	**71.69**	69.03	71.44	45.33	59.60	61.07	62.91	**64.74**
	400	58.56	76.57	**78.93**	76.55	78.64	56.65	66.66	67.89	70.40	**71.91**
	200	77.93	86.47	**88.14**	86.45	87.92	76.12	78.56	79.33	82.35	**83.58**
	0	93.49	96.28	96.45	96.20	**96.47**	95.96	95.54	95.66	97.30	**97.49**

Table 3.5 The ESTOI scores of different models in the **unseen** noise test set

Metrics	T_{60} (ms)	One speaker test set					Two speaker test set				
			One speaker training set		Mix-speaker training set			One speaker training set		Mix-speaker training set	
		Mix	GCT	CTS	GCT	CTS	Mix	GCT	CTS	GCT	CTS
ESTOI (%)	1200	30.11	54.19	**56.91**	53.69	55.92	27.87	44.01	45.13	46.12	**47.62**
	1000	34.24	58.87	**61.81**	58.55	61.01	31.07	47.19	48.62	49.59	**51.35**
	800	37.56	61.38	**64.20**	61.29	63.60	36.79	52.37	53.41	55.03	**56.65**
	600	45.34	66.76	**69.38**	66.70	68.98	44.37	57.62	59.02	61.04	**62.78**
	400	55.86	73.68	**76.15**	73.58	75.64	55.09	64.52	65.67	68.36	**69.94**
	200	74.27	83.59	**85.39**	83.44	85.04	73.58	76.35	76.92	80.24	**81.37**
	0	89.42	93.57	**93.66**	93.42	93.53	91.99	92.21	91.94	94.45	**94.60**

improvement than GCT-Net, verifying the advantage of two-stage CNN models. Besides, when there is only one desired speaker and the CNN model to be tested is GCT-Net, the ESTOI scores are comparable. As for CTS-Net, the ESTOI scores of *one-speaker* training set are higher than those of *multi-speaker* training set, especially when the T_{60} is relatively large. In the case that the number of desired speakers is two, both GCT-Net and CTS-Net can perform better if trained by using *multi-speaker* training set. By focusing on the results of CTS-Net, one can see that the ESTOI score is higher when the number of desired speakers in the test set matches with that in the training set, and this may due to the overfitting problem of deep learning.

Table 3.5 presents the ESTOI scores of different CNN models and different training strategies in the unseen noise test set. It can be observed that CTS-Net is better than GCT-Net in improving speech intelligibility. Besides, when the number of the desired speakers in the test set matches with the training set, we can get the highest PESQ scores for both one-speaker and two-speaker scenarios. The results as well as the reasons are similar to Table 3.4, and the main reason is the same as that of PESQ.

3.6.4 SDR

Table 3.6 gives the SDR results in the seen noise test set. It can be seen from Table 3.6 that CNN models trained with *multi-speaker* set can obtain higher SDR values in most cases. Besides, when training with *multi-speaker* set and the T_{60} is relatively short, e.g., less than 0.6 s, the SDR values of CTS-Net are higher than those of GCT-Net, indicating that GCT-Net may cause more speech distortion. As the T_{60} increases, the SDR values of GCT-Net becomes

Table 3.6 The SDR values of different models in the **seen** noise test set

Metrics	T_{60} (ms)	One speaker test set					Two speaker test set				
			One speaker training set		Mix-speaker training set			One speaker training set		Mix-speaker training set	
		Mix	GCT	CTS	GCT	CTS	Mix	GCT	CTS	GCT	CTS
SDR (dB)	1200	-2.96	-0.25	-0.35	**0.13**	-0.03	-5.39	-3.94	-3.99	**-3.23**	-3.28
	1000	-2.05	0.52	0.55	**0.99**	0.86	-4.73	-3.41	-3.51	**-2.69**	-2.85
	800	1.80	2.30	**2.41**	2.33	2.40	-3.46	-2.66	-2.79	**-1.88**	-1.92
	600	1.30	2.17	2.27	**2.61**	2.53	-1.69	-1.42	-1.48	**-0.53**	-0.54
	400	4.78	4.69	5.22	4.95	**5.44**	0.86	0.20	0.20	1.26	**1.30**
	200	**13.14**	9.93	10.96	10.06	11.07	**5.57**	3.89	3.69	5.23	5.30
	0	18.98	23.42	24.13	23.31	**24.34**	20.47	19.34	19.14	23.31	**24.25**

Table 3.7 The SDR values of different models in the **unseen** noise test set

Metrics	T_{60} (ms)	One speaker test set					Two speaker test set				
			One speaker training set		Mix-speaker training set			One speaker training set		Mix-speaker training set	
		Mix	GCT	CTS	GCT	CTS	Mix	GCT	CTS	GCT	CTS
SDR (dB)	1200	-2.66	-0.16	-0.07	**0.18**	0.17	-5.39	-4.01	-4.05	**-3.34**	-3.37
	1000	-1.63	0.84	0.92	**1.16**	1.07	-4.72	-3.49	-3.48	**-2.80**	-2.90
	800	-0.65	1.30	1.51	**1.73**	**1.73**	-3.45	-2.75	-2.85	**-1.98**	-2.01
	600	1.41	2.44	2.51	**2.72**	**2.72**	-1.68	-1.46	-1.50	**-0.60**	-0.61
	400	4.54	4.69	5.12	4.92	**5.25**	0.88	0.10	0.09	1.17	**1.21**
	200	12.72	9.47	10.15	9.51	**10.26**	5.56	3.71	3.43	5.06	**5.13**
	0	19.59	21.39	21.80	21.18	**21.81**	19.65	16.82	16.38	20.38	**20.87**

higher than those of CTS-Net. This is because the residual reverberation component of GCT-Net is less than that of CTS-Net.

In Table 3.7, the SDR values of different CNN models that are trained by different sets in the unseen noise set are given. It can be seen from Table 3.7 that CNN models trained by *multi-speaker* training set can obtain higher SDR values than those trained by *one-speaker* training set, indicating that the former one can cause less speech distortion. Moreover, CTS-Net gets higher SDR values than GCT-Net when T_{60} is smaller than 0.6 s, but the results will be reversed when the T_{60} is not smaller than 0.6 s. In can be found that this is similar to the results in Table 3.6.

To sum up, we demonstrate the testing results of PESQ scores, ESTOI scores, and SDR values for different CNN models as well as different training strategies in both seen and unseen noise sets. From these quantitative results, we can see that CTS-Net has better denoising and dereverberation performance than GCT-Net in most cases, indicating the validation of the two-stage CNN models. Besides, the CNN models trained by *one-speaker* training set is more suitable than those trained by *multi-speaker* training set when there is only one desired speaker, as it can lead to higher speech quality as well as higher speech intelligibility improvement. On the contrary, when two desired speakers co-exist, the CNN model trained by *multi-speaker* training set can perform better. Because the differences of the PESQ and ESTOI scores are more significant in the latter case, we can conclude that the CNN models trained by *multi-speaker* training set is more robust. Moreover, by testing the seen and unseen noise scenarios, the results of all the objective metrics seem not to be affected remarkably; this validates the generalization ability of GCT-Net and CTS-Net when they are trained with numerous noise sets.

3.6.5 Subjective Listening Test

In this section, we utilize AB listening tests for the subjective evaluation, whose procedure is similar to [41]. The experiment is conducted at the Institute of Acoustics, Chinese Academy of Sciences, and there are 12 audiologically normal-hearing subjects participating in the listening test. They are graduate students or teachers, whose ages range from 25 to 35. When conducting the listening test, two groups of speech signals are compared, namely the enhanced speech of GCT-Net versus that of CTS-Net, and CTS-Net trained by *one-speaker* set versus that trained by *multi-speaker* set. Twenty pairs of utterances are tested in each group, which are randomly chosen from the test set. All the speech signals are played by a PC soundcard and reproduced with a circumaural earphone (Sennheiser HD 380 Pro). Note that the order of the methods to be compared is shuffled. Before the listening test, all the participants are told to distinguish the speech quality and speech intelligibility of each pair. When it comes to the second group, they are told to focus on the continuity and naturality of both the two speech signals. During the listening test, the original noisy speech and the two processed speech signals are sequentially provided to each participant. Later, they are asked to select their preference utterance. The "Equal" option is provided in case they could not distinguish the difference between the two speech signals.

There groups of subjective evaluation results are shown in Figure 3.7. Group (a) shows the preference percentages of CTS-Net and GCT-Net in one desired speaker case. It can be seen from the comparison of group (a) that the preference percentage of CTS-Net is far larger than that of GCT-Net.

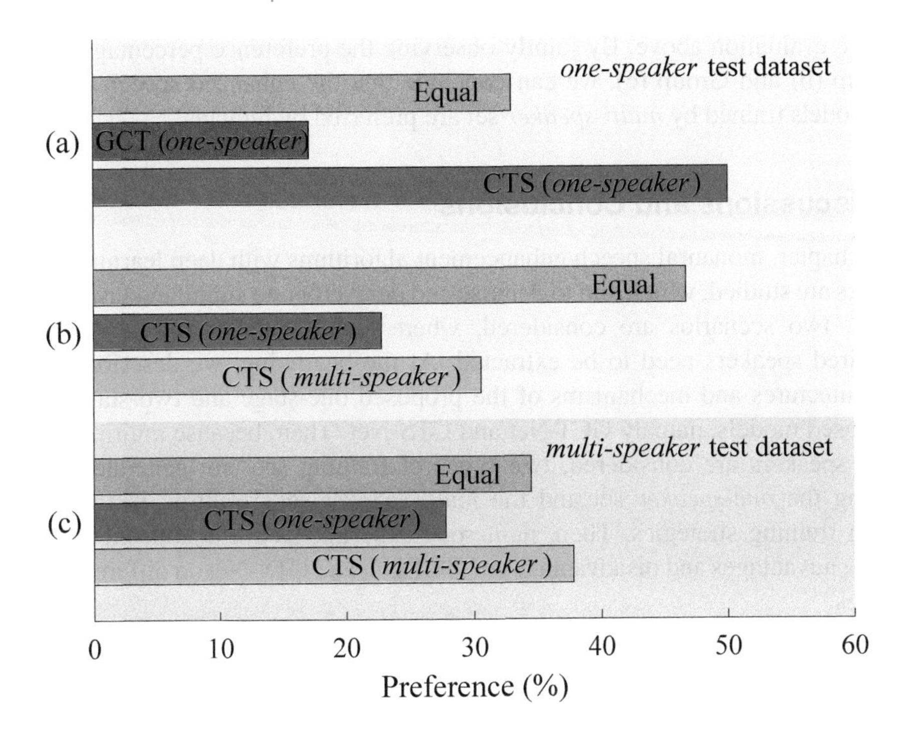

Figure 3.7 Subjective comparison with AB listening test between (a) CTS-Net and GCT-Net trained by *one-speaker* set in only one desired speaker scenario, (b) CTS-Net trained by *one-speaker* set and that trained by *multi-speaker* set in only one desired speaker scenario, and (c) CTS-Net trained by *one-speaker* set and that trained by *multi-speaker* set in two desired speakers scenario.

This result indicates that CTS-Net has better denoising and dereverberation performance than GCT-Net, namely the two-stage CNN model outperforms the one-stage CNN model for speech enhancement. Group (b) compares the preference percentages of the CTS-Net trained by *one-speaker* set and by *multi-speaker* set. Interestingly, one can see that although the CTS-Net trained by *one-speaker* set can get higher PESQ and ESTOI scores than that trained by *multi-speaker* set in the one desired speaker case, as shown in Tables 3.3 and 3.5, the preference percentage of the latter training strategy is larger than that of the former one. Group (c) shows the result when two desired speakers co-exist, where the preference percentage of CTS-Net trained by *multi-speaker* set is larger than CTS-Net trained by *one-speaker*.

This result is in accordance with the PESQ and ESTOI scoring in the objective evaluation above. By jointly observing the preference percentages of Group (b) and Group (c), we can conclude that the enhanced speech of CNN models trained by *multi-speaker* set are preferred by humans.

3.7 Discussions and Conclusions

In this chapter, monaural speech enhancement algorithms with deep learning networks are studied, which aim to denoise and dereverberate simultaneously. Besides, two scenarios are considered, where single speaker and multiple desired speakers need to be extracted. At the beginning, we described the architectures and mechanisms of the proposed one-stage and two-stage CNN-based models, namely GCT-Net and CTS-Net. Then, because multiple desired speakers are considered, two types of training sets are generated, including the *one-speaker* set and the *multi-speaker* set, resulting in two types of training strategies. Then, numerous experiments are conducted to show the advantages and disadvantages of GCT-Net and CTS-Net on different scenarios.

The experiments reveal that CTS-Net shows better denoising and dereverberation performance than GCT-Net. On the one hand, CTS-Net can not only suppress more noise and reverberation but also recover more speech components than GCT-Net, which can be observed in spectrograms of the enhanced speech signals. Meanwhile, the values of different metrics in objective evaluation, including PESQ scores, ESTOI scores, and SDR values, as well as the preference percentage in subjective evaluation, indicate that CTS-Net outperforms GCT-Net in most cases.

Moreover, CNN models trained by using different training set have advantages on different aspects. It can be seen that GCT-Net or CTS-Net trained by *one-speaker* set can achieve more denoising and dereverberation amount than that trained by *multi-speaker* set; on the contrary, the two models trained by *multi-speaker* set can reduce the speech distortion of desired speakers as much as possible. Consequently, when there is only one desired speaker, CNN models trained by *one-speaker* set show slightly better performance than those trained by *multi-speaker* set, but when the number of the desired speaker increases, the former training strategy may cause significant speech distortion. By contrast, the training strategy that uses *multi-speaker* set shows moderate performance in the one speaker scenario and shows much better performance in two-speaker scenario, illustrating that this training strategy is more robust to multiple speakers scenarios.

References

[1] Michael Berouti, Richard Schwartz, and John Makhoul. Enhancement of speech corrupted by acoustic noise. In *IEEE International Conference on Acoustics, Speech, and Signal Processing*, volume 4, pages 208–211. IEEE, 1979.

[2] Pascal Scalart et al. Speech enhancement based on a priori signal to noise estimation. In *IEEE International Conference on Acoustics, Speech, and Signal Processing Conference Proceedings*, volume 2, pages 629–632. IEEE, 1996.

[3] Yariv Ephraim and David Malah. Speech enhancement using a minimum mean-square error log-spectral amplitude estimator. *IEEE Transactions on Acoustics, Speech, and Signal Processing*, 33(2):443–445, 1985.

[4] Israel Cohen and Baruch Berdugo. Speech enhancement for non-stationary noise environments. *Signal Processing*, 81(11):2403–2418, 2001.

[5] Yi Hu and Philipos C. Loizou. A generalized subspace approach for enhancing speech corrupted by colored noise. *IEEE Transactions on Speech and Audio Processing*, 11(4):334–341, 2003.

[6] Yuxuan Ke, Andong Li, Chengshi Zheng, Renhua Peng, and Xiaodong Li. Low-complexity artificial noise suppression methods for deep learning-based speech enhancement algorithms. *EURASIP Journal on Audio, Speech, and Music Processing*, 2021(1):1–15, 2021.

[7] DeLiang Wang and Guy J. Brown. *Computational Auditory Scene Analysis: Principles, Algorithms, and Applications*. Wiley-IEEE Press, 2006.

[8] DeLiang Wang and Jitong Chen. Supervised speech separation based on deep learning: An overview. *IEEE/ACM Transactions on Audio, Speech, and Language Processing*, 26(10):1702–1726, 2018.

[9] Soumitro Chakrabarty and Emanuël A. P. Habets. Time–frequency masking based online multi-channel speech enhancement with convolutional recurrent neural networks. *IEEE Journal of Selected Topics in Signal Processing*, 13(4):787–799, 2019.

[10] Yuxuan Ke, Yi Hu, Jian Li, Chengshi Zheng, and Xiaodong Li. A generalized subspace approach for multichannel speech enhancement using machine learning-based speech presence probability estimation. In *Audio Engineering Society Convention 146*. Audio Engineering Society, 2019.

[11] Ke Tan and DeLiang Wang. A convolutional recurrent neural network for real-time speech enhancement. In *Interspeech 2018*, pages 3229–3233, 2018.

[12] Ke Tan and DeLiang Wang. Learning complex spectral mapping with gated convolutional recurrent networks for monaural speech enhancement. *IEEE/ACM Transactions on Audio, Speech, and Language Processing*, 28:380–390, 2019.

[13] Yi Luo and Nima Mesgarani. Conv-tasnet: Surpassing ideal time–frequency magnitude masking for speech separation. *IEEE/ACM Transactions on Audio, Speech, and Language Processing*, 27(8):1256–1266, 2019.

[14] Chengshi Zheng, Renhua Peng, Jian Li, and Xiaodong Li. A constrained MMSE LP residual estimator for speech dereverberation in noisy environments. *IEEE Signal Processing Letters*, 21(12):1462–1466, 2014.

[15] Takuya Yoshioka, Tomohiro Nakatani, Keisuke Kinoshita, and Masato Miyoshi. Speech dereverberation and denoising based on time varying speech model and autoregressive reverberation model. In *Speech Processing in Modern Communication*, pages 151–182, 2010.

[16] Takuya Yoshioka and Tomohiro Nakatani. Generalization of multi-channel linear prediction methods for blind mimo impulse response shortening. *IEEE Transactions on Audio, Speech, and Language Processing*, 20(10):2707–2720, 2012.

[17] Hirokazu Kameoka, Tomohiro Nakatani, and Takuya Yoshioka. Robust speech dereverberation based on non-negativity and sparse nature of speech spectrograms. In *2009 IEEE International Conference on Acoustics, Speech and Signal Processing*, pages 45–48. IEEE, 2009.

[18] Rita Singh, Bhiksha Raj, and Paris Smaragdis. Latent-variable decomposition based dereverberation of monaural and multi-channel signals. In *2010 IEEE International Conference on Acoustics, Speech and Signal Processing*, pages 1914–1917. IEEE, 2010.

[19] Kun Han, Yuxuan Wang, DeLiang Wang, William S. Woods, Ivo Merks, and Tao Zhang. Learning spectral mapping for speech dereverberation and denoising. *IEEE/ACM Transactions on Audio, Speech, and Language Processing*, 23(6):982–992, 2015.

[20] Eric W. Healy, Masood Delfarah, Eric M. Johnson, and DeLiang Wang. A deep learning algorithm to increase intelligibility for hearing-impaired listeners in the presence of a competing talker and reverberation. *The Journal of the Acoustical Society of America*, 145(3):1378–1388, 2019.

[21] Yan Zhao, DeLiang Wang, Buye Xu, and Tao Zhang. Monaural speech dereverberation using temporal convolutional networks with self attention. *IEEE/ACM Transactions on Audio, Speech, and Language Processing*, 28:1598–1607, 2020.

[22] Andong Li, Chengshi Zheng, Renhua Peng, and Xiaodong Li. On the importance of power compression and phase estimation in monaural speech dereverberation. *JASA Express Letters*, 1(1):014802, 2021.

[23] Yong Xu, Jun Du, Li-Rong Dai, and Chin-Hui Lee. An experimental study on speech enhancement based on deep neural networks. *IEEE Signal Processing Letters*, 21(1):65–68, 2013.

[24] Tomohiro Nakatani and Keisuke Kinoshita. A unified convolutional beamformer for simultaneous denoising and dereverberation. *IEEE Signal Processing Letters*, 26(6):903–907, 2019.

[25] Tomohiro Nakatani and Keisuke Kinoshita. Simultaneous denoising and dereverberation for low-latency applications using frame-by-frame online unified convolutional beamformer. In *Interspeech*, pages 111–115, 2019.

[26] Andong Li, Wenzhe Liu, Xiaoxue Luo, Guochen Yu, Chengshi Zheng, and Xiaodong Li. A simultaneous denoising and dereverberation framework with target decoupling. *arXiv preprint arXiv:2106.12743*, 2021.

[27] Hyeong-Seok Choi, Hoon Heo, Jie Hwan Lee, and Kyogu Lee. Phase-aware single-stage speech denoising and dereverberation with u-net. *arXiv preprint arXiv:2006.00687*, 2020.

[28] Donald S. Williamson and DeLiang Wang. Time-frequency masking in the complex domain for speech dereverberation and denoising. *IEEE/ACM Transactions on Audio, Speech, and Language Processing*, 25(7):1492–1501, 2017.

[29] Jiaqi Su, Zeyu Jin, and Adam Finkelstein. HiFi-GAN: High-fidelity denoising and dereverberation based on speech deep features in adversarial networks. *arXiv preprint arXiv:2006.05694*, 2020.

[30] Yuxuan Wang, Arun Narayanan, and DeLiang Wang. On training targets for supervised speech separation. *IEEE/ACM Transactions on Audio, Speech, and Language Processing*, 22(12):1849–1858, 2014.

[31] Jitong Chen, Yuxuan Wang, Sarah E. Yoho, DeLiang Wang, and Eric W. Healy. Large-scale training to increase speech intelligibility for hearing-impaired listeners in novel noises. *The Journal of the Acoustical Society of America*, 139(5):2604–2612, 2016.

[32] Xiaofei Li and Radu Horaud. Online monaural speech enhancement using delayed subband LSTM. In *Interspeech 2020*, pages 2462–2466, 2020.

[33] Nils L. Westhausen and Bernd T. Meyer. Dual-signal transformation LSTM network for real-time noise suppression. In *Interspeech 2020*, pages 2477–2481, 2020.

[34] Yanxin Hu, Yun Liu, Shubo Lv, Mengtao Xing, Shimin Zhang, Yihui Fu, Jian Wu, Bihong Zhang, and Lei Xie. DCCRN: Deep complex convolution recurrent network for phase-aware speech enhancement. In *Interspeech 2020*, pages 2472–2476, 2020.

[35] Maximilian Strake, Bruno Defraene, Kristoff Fluyt, Wouter Tirry, and Tim Fingscheidt. A fully convolutional recurrent network (FCRN) for joint dereverberation and denoising. In *Interspeech 2020*, pages 2467–2471, 2020.

[36] Antony W. Rix, John G. Beerends, Michael P. Hollier, and Andries P. Hekstra. Perceptual evaluation of speech quality (PESQ)-a new method for speech quality assessment of telephone networks and codecs. In *2001 IEEE IEEE International Conference on Acoustics, Speech, and Signal Processing. Proceedings (Cat. No. 01CH37221)*, volume 2, pages 749–752. IEEE, 2001.

[37] Jesper Jensen and Cees H. Taal. An algorithm for predicting the intelligibility of speech masked by modulated noise maskers. *IEEE/ACM Transactions on Audio, Speech, and Language Processing*, 24(11):2009–2022, 2016.

[38] Emmanuel Vincent, Hiroshi Sawada, Pau Bofill, Shoji Makino, and Justinian P. Rosca. First stereo audio source separation evaluation campaign: data, algorithms and results. In *International Conference on Independent Component Analysis and Signal Separation*, pages 552–559. Springer, 2007.

[39] Jitong Chen and DeLiang Wang. Long short-term memory for speaker generalization in supervised speech separation. *The Journal of the Acoustical Society of America*, 141(6):4705–4714, 2017.

[40] Andong Li, Wenzhe Liu, Xiaoxue Luo, Chengshi Zheng, and Xiaodong Li. Icassp 2021 deep noise suppression challenge: Decoupling magnitude and phase optimization with a two-stage deep network. In *ICASSP 2021-2021 IEEE International Conference on Acoustics, Speech and Signal Processing (ICASSP)*, pages 6628–6632. IEEE, 2021.

[41] Andong Li, Wenzhe Liu, Chengshi Zheng, Cunhang Fan, and Xiaodong Li. Two heads are better than one: A two-stage complex spectral mapping approach for monaural speech enhancement. *IEEE/ACM Transactions on Audio, Speech, and Language Processing*, 29:1829–1843, 2021.

[42] Yann N. Dauphin, Angela Fan, Michael Auli, and David Grangier. Language modeling with gated convolutional networks. In *International Conference on Machine Learning*, pages 933–941. PMLR, 2017.

[43] Cunhang Fan, Jianhua Tao, Bin Liu, Jiangyan Yi, Zhengqi Wen, and Xuefei Liu. End-to-end post-filter for speech separation with deep attention fusion features. *IEEE/ACM Transactions on Audio, Speech, and Language Processing*, 28:1303–1314, 2020.

[44] Zhong-Qiu Wang, Peidong Wang, and DeLiang Wang. Complex spectral mapping for single-and multi-channel speech enhancement and robust ASR. *IEEE/ACM Transactions on Audio, Speech, and Language Processing*, 28:1778–1787, 2020.

[45] Chandan K. A. Reddy, Harishchandra Dubey, Vishak Gopal, Ross Cutler, Sebastian Braun, Hannes Gamper, Robert Aichner, and Sriram Srinivasan. ICASSP 2021 deep noise suppression challenge. In *ICASSP 2021-2021 IEEE International Conference on Acoustics, Speech and Signal Processing (ICASSP)*, pages 6623–6627. IEEE, 2021.

[46] Tom Ko, Vijayaditya Peddinti, Daniel Povey, Michael L. Seltzer, and Sanjeev Khudanpur. A study on data augmentation of reverberant speech for robust speech recognition. In *2017 IEEE International Conference on Acoustics, Speech and Signal Processing (ICASSP)*, pages 5220–5224. IEEE, 2017.

[47] Emanuel A. P. Habets. Room impulse response generator. *Technische Universiteit Eindhoven, Tech. Rep*, 2(2.4):1, 2006.

[48] Diederik P. Kingma and Jimmy B. A. Adam: A method for stochastic optimization. *arXiv preprint arXiv:1412.6980*, 2014.

[49] Andrew Varga and Herman J. M. Steeneken. Assessment for automatic speech recognition: Ii. noisex-92: A database and an experiment to study the effect of additive noise on speech recognition systems. *Speech Communication*, 12(3):247–251, 1993.

[50] Adam Paszke, Sam Gross, Soumith Chintala, Gregory Chanan, Edward Yang, Zachary DeVito, Zeming Lin, Alban Desmaison, Luca Antiga, and Adam Lerer. Automatic differentiation in pytorch. 2017.

4

Edge Computing and Controller Area Network (CAN) for IoT Data Classification using Convolutional Neural Network

Sivanantham kalimuthu[1,*], Arumugam Ganesan[2], S. Sathish[3], S. Ramesh Aravindh[4], Sivaneshwaran Shanmugam[5], and Leena Bojaraj[6]

[1]Tech Lead, HCL Technologies, Coimbatore, Tamil Nadu, India
[1,2,3,4]Research & Development Team, Crapersoft, India
[5]Iquants Engineering, India
[6]KGiSL Institute of Technology, India
E-mail: sivanantham.k@hcl.com
*Corresponding Author

Abstract

Emotions are incredibly vital in the mental existence of humans. There are a few universal emotions that any intelligent system with finite processing resources can be trained to recognize or synthesize as needed, including neutral, anger, happiness, and sadness. In the modern era, newline workers have lost their interest or concentration in work activity, and these factors lead to an effect on the productivity of concerned industries; so it is necessary to monitor the employee's emotional activities with the help of IoT technology for smart newline industrial environment. Furthermore, despite the rapid rise of IoT, in the newline field of the modern industrial era, current IoT-based systems notably lack cognitive newline industrial, implying that they are unable to meet the needs of industrial services. Deep learning has become one of the most widely used approaches in various diagnosis and predictions of applications and studies. While it is mostly used for content-based newline image retrieval, it can still be improved by using it in various computer vision newline applications. From the various conventional approaches and their research analysis, new machine learning algorithms such as deep learning or CNN-based approaches need to address the existing issue by developing an employee's emotional predictions. The proposed research work mentors and counsels workers by monitoring their newline behavior in the workplace. The goal of this research was to develop a CNN model and controller area protocol-based emotional intelligence system (EIS) to automatically categorize expressions in the Facial Expression Recognition newline (FER2013) and kaggle databases. The proposed CNN-based EIS prediction system achieves 96.75% better efficiency in 36.65-second time duration. The proposed system produces better performance results compared to the existing support vector machine, decision tree, and artificial neural network algorithm.

Keywords: Facial expression recognition, emotional intelligence, deep learning, emotional intelligence system, convolution neural networks.

4.1 Introduction

Data mining refers to computer systems that are modeled after the human brain and are capable of natural language processing, learning from the past, organically cooperating with humans and assisting in decision-making. Researchers developed computers that ran at a faster rate than the human brain at the turn of the twenty-first century, resurrecting the data mining approach (Kalimuthu Sivanantham *et al.*, 2021). Furthermore, academics have begun to utilize the phrase data mining prediction, which combines technology and biology in an attempt to reverse engineering. Brain activities are the most efficient and effective computers on the planet (Toneva and Leila, 2019). The following sections explain the relationship between edge computing and CNN classification techniques.

Figure 4.1 explains the general architecture diagram for IoT data transferred using CAN protocol and edge-computing-based classification. CAN protocol architecture layer and edge computing take part as analytical, storage, addressing the quality of services and sensing connectivity; finally, four stages of CNN deep learning techniques are explained. The Internet of Things (IoT) data prediction refers to the integration of cognition into IoT and all the characteristics of IoT apply to the cognitive Internet of Things as well (Qihui *et al.*, 2014). As a result, an assumption is that the terms "IoT data" and "security" are used in the same way in the architecture diagram. Although

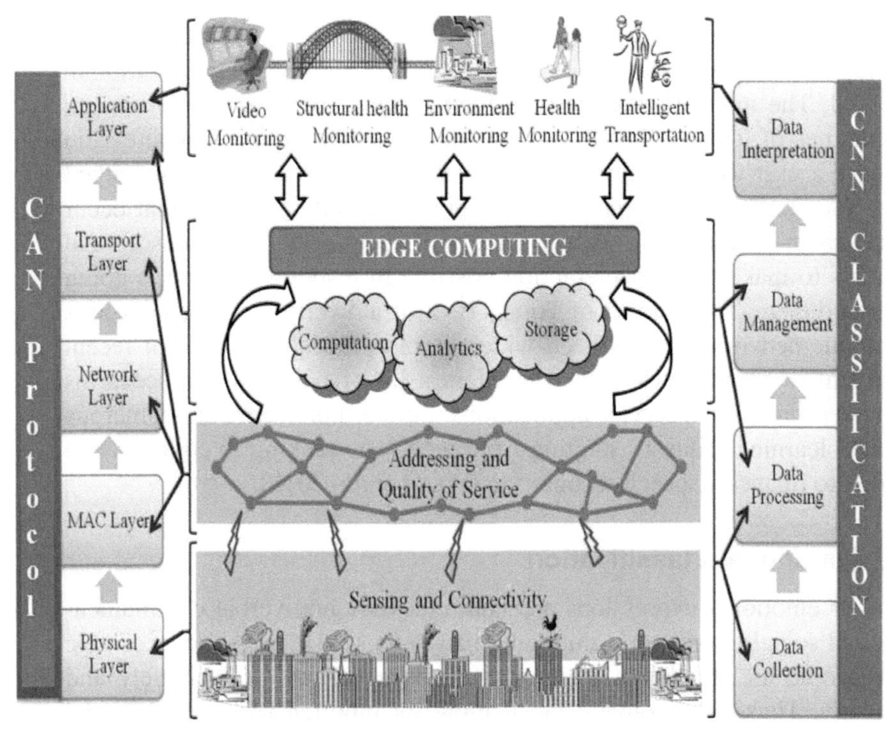

Figure 4.1 General architecture diagram for IoT data transferred using CAN protocol and edge-computing-based classification.

Kevin Ashton, a digital innovation expert, is credited with the initial use of the phrase IoT, other groups later defined the term "IoT" based on the widespread belief that the Internet's original version was about data provided by people, while the next version is about data created by things. Definition of the "Internet of Things" was generated previously, which was non-trivial, taking into account the extensive experience and required technologies, ranging from sensing objects, data aggregation and pre-processing, and communication systems to object instantiation and finally service provision. However, being a worldwide notion, it necessitates a standard definition (Hiller *et al.*, 2018).

4.1.1 Internet of Things (IoT)

The Internet of Things (IoT) is the next wave of innovation, stemming from the Internet's and smart gadgets' convergence. Smart things are objects equipped with appropriate sensors and actuators, as well as communication technologies such as RFID or NFC. IoT will produce and use the data to help people in their daily lives, such as costs, time, and improving optimization in any field. The innovative framework for fostering collaboration to operate the Internet of Things is still lacking. Smart things unlike computers do not have a user to set them to various conditions; instead, they must adapt themselves to their environment and respond correctly to events that occur around them (Nalbandian, 2015). They will require the right sensors and actuators to make the best prediction possible in response to what is going on around them. The Internet of Things (IoT) is a diverse and heterogeneous ubiquitous network that has been widely explored in the domain of recent intelligent or smart services, and it offers a lot of potential and scenarios for modern intelligent service applications. Furthermore, there are numerous machine learning requests, ranging from web page ranking to collaborative filtering to image or speech recognition (Liang *et al.*, 2018).

4.1.2 Emotional Classification

Facial or emotional expressions are vital clues for non-verbal communications and social interactions among human beings. It is merely conceivable since people are intelligent to identify their moods pretty accurately and efficiently. Thus, an automatic facial mood or emotion recognition model is a vital module in the interaction of humans with machines. Rather than the commercial uses of such a system, it would be advantageous to integrate some clues as of the biotic neural system. In this proposed model, it is used to improve supplementary perceptions into the cognitive or intellectual processing capability of the human brain (Javier *et al.*, 2013).

One of the hallmarks of emotional intelligence, a component of human intelligence that has been considered to be even more essential than mathematical and verbal intelligence, is the ability to perceive emotion. Emotion identification according to researchers is the step toward machine emotion intelligence (Scherer and Ursula, 2011). Emotion can be recognized through a variety of that, including facial expressions, speech, and writing. When it comes to emotional intelligence, there are several factors to consider (E.I.). In this research, quickly associate it with feelings and declarations of feelings. Joyful knowledge is a type of understanding that encompasses the ability to perceive and influence emotions (Claudia, 2007). The limit of recognizing and conveying the sensation, assimilating it to the idea, understanding and prevailing upon it, and having the capacity to regulate it in yourself and others according to the joyful insight. According to some authors, emotional intelligence is linked to the observation and processing of emotions because humans think and behave based on their life experiences, which are prompted by current or past situations (Sivanantham, 2021).

Machine learning is making significant strides in every industry, including medicine, oil and gas, education, energy, weather forecasting, and the stock market, to name a few. Machine learning is affecting not only technology but also the lives of ordinary people, as evidenced by these actions (Shen and Tongda, 2012). Because of machine learning, a regular person has evolved into a tech-aware individual, a gadget man. However, existing IoT lacks intelligence, implying that it cannot be used for industrial service application requirements. Furthermore, IoT is now based on out-of-date stationary models and architectures that lack sufficient recent intelligence services such as mood recognition and are unable to meet the increasing performance demands of companies. It is possible to develop a unique notion of IoT by incorporating emotional intelligence concepts into IoT (Fatima *et al.*, 2020). Cooperative IoT strategies for enhancing enactment and reaching modern intelligence IoT can detect current network circumstances, assess perceived knowledge, make intelligent judgments, and take adaptive actions to improve network performance. The Internet of Things (IoT) is a global network architecture made up of countless connected objects that rely on sensing, communication, networking, and data processing technologies (Francesco *et al.*, 2019). Basic universal expressions as proposed by Ekman *et al.* (1976) are listed as follows:

- Happiness
- Anger
- Sadness
- Surprise
- Disgust
- Fear

Some of the non-basic expressions are as follows:

- Irritation
- Despair
- Boredom
- Panic
- Shame
- Excitement

Novel paradigms dependent on IoT applications are emerging in a variety of industrial applications. However, due to heterogeneous and uncertain omnipresent networks, which have a wide range of applications in the field of modern intelligent services and discrepancies between offering and application demand, today's industrial systems face many issues. Accordingly, the researchers feel that current technologies and approaches, particularly the Internet of Things (IoT), are lacking in cognitive intelligence and hence cannot deliver the predicted upgrades and smart industry advancements (Zhaozong *et al.*, 2016). As a result, the primary goal of this article is to study notions of IoT-based settings as well as to review technologies, platforms, and developments in cognitive and smart industrial systems. Following that, we examined the research problems and open topics to improve knowledge accumulation in the field of the Internet of Things (IoT) (Da and Shancang, 2014). As a result, the driving force behind the efforts is to develop an emotional intelligence system capable of recognizing and regulating people's facial expressions and emotions during social interactions. As a result, in the era of factual time judgment making, this model will help industry professionals track and govern their customer's and workers' real-time feelings and behaviors (Catherine *et al.*, 2020). Technologies at the core of this impact are machine learning and deep learning. These change-initiating agents alter the environment in which human beings live and communicate, like the industrial revolution. Though still budding, they are getting into our daily needs of healthcare advancements, power grid creation, agricultural yield improvements, smartphone technology, and climate change monitoring. ML algorithms are used to build a mathematical model based on training samples, to make decisions in the future about testing samples. ML is a subset of artificial intelligence. They are used in various computer vision tasks (El Naq, Issam, and Murphy, 2015).

4.1.3 Applications

Following some emotional prediction, application applied to improve manpower, to reduce the cost of producing goods. Because emotion is our biologically innate ability and a part of our evolutionary history, we all can read them. It is an ability that gets better from the experience in our everyday lives. Because of the prominence of human–computer interface (HCI) nowadays, understanding the facial visual curves of an individual by machine is in need. It can be made well that by understanding the cues of a human, a robot can enhance its value to perform its various tasks.

This proposed work can be used to evaluate and respond appropriately to human feelings by providing empathic responses in the fields of emotion handling which is an important aspect of people's well-being.

- Preventive medicine
- Remote monitoring of patient's physiology
- E-learning
- Social monitoring
- Online games

In the domains of computer analysis, lie detection, airport security, non-verbal communication, and even the importance of expressions in art, the study and comprehension of human facial expression have various applications.

Recognizing a man's expression can help in a variety of fields, including medical science, where a doctor can be notified if a patient is in extreme agony while asleep (Rhawn, 1988). It aids in taking quick action at the appropriate time. Teachers can tell a lot about a student's learning status by looking at their facial expressions. As a result, vision-based expression analysis is also useful in e-learning (Maryam and Montazer, 2019). A computer vision system can be used to automatically assess learners in remote education to discern non-verbal face expressions and determine their learning state (Chunfeng, 2016). It is beneficial to both teachers and students because it can aid in the improvement of the teaching and learning process. Facial reactions reveal a lot about a person's reaction to a stimulus. Facial coding has been a common method in market research and media measurement in recent years because it may unobtrusively record information about an individual's response at the moment. We can acquire self-reports with quantifiable assessments of more unconscious emotional responses to a product or service by tracking people's facial expressions when they give comments about it. Market segmentation may be examined and target audiences can be determined using facial expression analysis to optimize items on the market.

The remainder of this chapter is organized as follows. Section 4.2 dicusses emotional prediction using facial expression recognition and related work. The planned controller area network (CAN) is discussed in Section 4.3. Section 4.4 compares proposed and existing systems' experimental outcomes for IoT emotional data in conventional neural networks. Finally, Section 4.5 concludes with some thoughts on the proposed edge computing and CAN for IoT data classification utilizing convolutional neural networks as well as its future scope.

4.2 Literature Review

Smart manufacturing has emerged as a vital engine of research, innovation, productivity, and export growth in recent years, because of rapid technological advancements, particularly in cognitive sensor technology. The goal is to achieve a new level of productivity, security, safety, and optimizations, as well as the transformation of data into insightful and timely information, allowing decision-makers across the enterprise to gain new visibility into operations, improve their ability to respond to market and business challenges, and eliminate operational inefficiencies (Alejandra *et al.*, 2019).

Shen *et al.* (2017) have proposed bidirectional LSTM and convolutional neural networks (BLSTM-CNN) technique, which could learn the vibrant appearances and shapes of areas of the facial units and provisions for vital look for the form of face action units. Using the fractious-modality reliable deterioration method which reasonably adjusts the CNN model, the models for combined graphic written sentimentality investigation of social multimedia were designed. The authors also studied a method to construct huge scale datasets for image feeling recognition by the deep convolutional neural network models. A similar model for effective computing of neuromorphic using the deep CNN method was designed. A facial mood recognizer using the technique referred to as softmax deeper regression code namely MNIST and a 2D convolutional neural network also developed by Chang *et al.* (2018).

Using the deep CNN technique, a model for emotion recognitions in insignificant data had been established. Consequently, the outcomes showed that the sufficient adjusting cascading technique realized improved outcomes than a sole phase fine-tuning. Furthermore, a model that achieved higher accuracy for substance-free feeling recognition from expressions facials via pooled CNN and DBN was developed. An end-to-end speech emotion identification system was proposed by integrating CNN and LSTM networks. The authors proposed an RNN platform for recognizing feelings in datasets or video using the CNN hybrid RNN (CNN-RNN) model. An emotion identification model was also proposed in the wild using CNN and mapped binary patterns (Meiyin and Chen, 2015).

Robotized liquid industrial estimation is one of the models that can employ IoT organizations to redesign execution and ordinariness over liquid industrial regions, increasing the notion of strategy estimates and limiting the negative impact using imperative control techniques. An IoT design for industrial applications is demonstrated in this chapter. The proposed RSS-based security system, which is based on a cloud system and background analysis, collects and sends the essential information. The RSS-based control program makes cloud imperative to disclose the information security algorithm with current communication problems. Confirmation, information security, respectability and fill in this information as a complete guide to get the highest level of assurance in the clouds. Performance analysis of flowrate measurement is depicted in Table 4.1. As compared with conventional methods like fuzzy and PID, the proposed method delivers perfect results. Imperative, fuzzy, and PID have SSE and settling time values of 0.048 (seconds), 0.039 (seconds), and 0.049 (seconds) and 21 (seconds), 26 (seconds), and 32 (seconds), respectively, explained in Kanagaraju and Nallusamy (2019).

As a result, a deep CNN model might be utilized to promote deep learning grounded facial action unit incidence and power approximation. The CNN model was also used to construct a strategy for identifying semantics of facial features and the extreme sharing native dissimilarity stabilization approach. A multimodal emotional state recognition system was built, with a 91.3% accuracy rate (Latha and Mohana, 2016). For imbalanced multimedia data categorization, a CNN model was combined with a bootstrapping strategy. The authors created unique hardware that reduced the amount of time needed to train neural networks and CNN, as well as applied it to video analytics in smart cities. Chronological deep learning for human action recognition was expressed using the 3D CNN and improved recognition accuracy by 92.17%–94.39%. It was created as a source for EEG-based emotion or feeling categorization that would be properly acquired to classify the data using the CNN model (Zhang and Dongrui, 2019). From a pool of literature evaluations on the convolutional neural network technology, the deep learning technique employed contributions and classification, accuracy attained, and limitations.

The 2D images are constrained to work due to various factors such as imaging conditions, pose, etc. Most of the research works in emotion recognition heavily depend on the landmarks on the face such as eyebrows, mouth, nose tip, etc. Locating the most crucial landmarks on the face that contribute to expression is a tedious task. This is because of the face complexity, and it needs human intervention for better accuracy that results in more processing time. Neutral facial scan of the face as a reference for the model to fitting or find the displacements of the landmarks to identify the expressions (Xiaoguang and Jain, 2008).

Many manufacturing systems rely on technology, which is merging to support and enable IoT applications in the direction of cognitive MS as a rising sector. IoT architecture, networks technology, software and algorithms, security, trust, reliability, privacy, interoperability, and standardization are only a few of the technologies covered (Asma *et al.*, 2016). However, the significance of it cannot be emphasized due to many impediments, and the issues that manufacturing faces now are not the same as those that it faced previously. Many manufacturing processes rely on technology, which is coming together to support and enable IoT applications in the direction of cognitive MS as a growing industry. Only a few of the technologies covered are IoT architecture, network technology, software and algorithms, security, trust, reliability, privacy, interoperability, and standardization (Asma *et al.*, 2016). However, due to several constraints, the significance of it cannot be overstated, and the difficulties that manufacturing faces now are not the same as those that it faced earlier. For establishing cognitive or smart manufacturing systems, bridging the gap between approaches from many disciplines is a critical challenge. Manufacturers, too, seek practical assistance; yet, the majority of academic research is unrelated to their needs. Academics investigate technological frontiers such as artificial intelligence and deep learning without contemplating how they will be used in the future (Fotis, 2020). Manufacturers want to know what kind of data to collect, which sensors to use, and where to place them on the production line. As a result, the study is required to find the optimal sensor setups. Open concerns in smart manufacturing innovation, for example, include adopting strategies, improving data collecting, utilization, sharing, designing predictive models, studying generic predictive models, connecting factories, and controlling processes, among others. Because of these characteristics, industrial production applications differ from lightweight and centralized monitoring CIoT-based applications (Long *et al.*, 2019).

Motivated by the above-mentioned fact, the proposed work aims to embed emotional intelligence with machine intelligence in industrial environments to access the essential working condition of employees and the behavior of employees of the industry in making better decisions about correlations of emotional states and the performance. In addition to this, in this proposal, a novel approach has been introduced to create a variety of policies to track the emotion and monitor the behavior of employees in industrial environments. However, as described in the first section of the proposal's introduction section, the proposed project also describes difficult issues that are unique for obtaining trustworthy emotional data and collecting a large set of image data from employees trying to elicit and experience each of six emotional states. To expose the prospects for manufacturing, healthcare, and automotive industries to create a set of use cases based on a rough integration of cognitive technologies, cognitive architectures and models, based employees emotions prediction with this help to improve their products and performance.

4.3 System Design

Facial or emotional expressions are vital clues for non-verbal communications and social interactions among human beings. It is merely conceivable since people are intelligent to identify moods pretty accurately and efficiently (Giardini and Michael, 2008). Thus, an automatic facial mood or emotion recognition model is a vital module in the interaction of humans with machines. Rather than commercializing such a system, it would be more beneficial to incorporate certain clues from the biotic neural system into our model and use it to improve supplemental judgments of the human brain's cognitive or intellectual processing capability. A DL-based model for improving the performance of IoT-based systems that detect human emotional states is proposed in this chapter. The concept allows data collected by IoT devices to be utilized to detect and analyze employee emotional states. Cécile *et al.* (2016) found that the machine learns the rules of the industry and links success with employee sentiment.

4.3.1 Featured Image Formation

Instead of using the raw facial images to learn the facial features, we used two featured images, as they represent some prominent features at the initial level. The process of two featured image formations is explained in the subsequent sections.

The distance between all the landmark points in the referred image and the corresponding landmark points in each expressed image is calculated. The difference between the corresponding landmark points of both the images is the vector of displacement having various feature points. Figure 4.2 explains the flowchart for the proposed system design feature point extraction (displacement) for facial values.

The largest edge direction magnitude, E, is computed from the provided picture I, which has a size of $m \times n$.

$$E_{i,j} = max_{x=1,2,3} \sum_{l=m-1}^{1} \sum_{k=-1}^{1} k_{i,j}^{z} * i_{1+i,k+j} \qquad (4.1)$$

where i and j are integers ranging from 2 to $m - 1$ and 2 to $n - 1$, respectively. It is derived from a set of eight Kirsch kernels, k in each of the eight

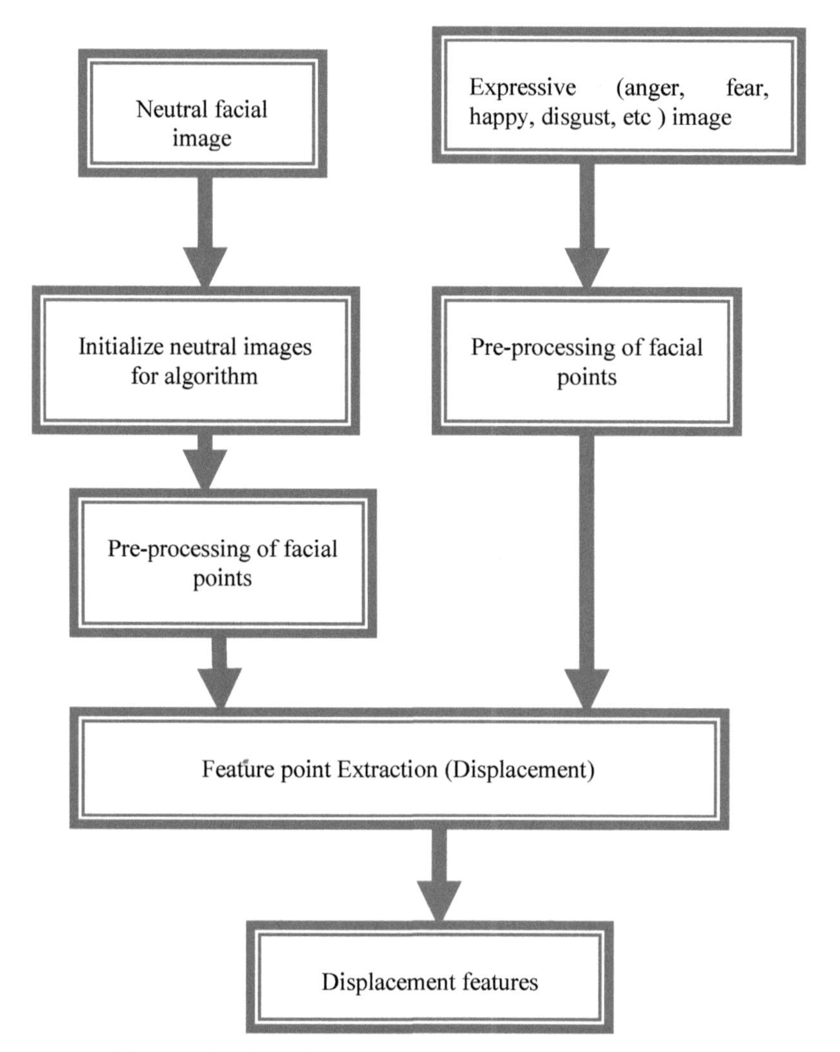

Figure 4.2 Flow diagrams for displacement feature extraction.

compass directions. It is determined by the mask that produces the largest edge magnitude. Let denote the replies obtained after applying each of the eight Kirsch Kernels, K_i, i = 1, 2, ..., 8, respectively. Assume that maximum is the highest obtained edge magnitude.

$$LDP = \sum_{I=0}^{7} b(kirsch - kirsch_{max}) * 2^i \tag{4.2}$$

where

$$b(x) = \left\{ \begin{array}{ll} 1; & x > 0 \\ 0; & \text{otherwise} \end{array} \right\}. \tag{4.3}$$

The image is robust since it is formed by taking into account all eight directions. LBP is an image operator that converts an image into an array or image of integer labels, as previously mentioned. These labels or their statistics (most commonly, the histogram) are used for further picture processing. In this case, a modified LBP approach is applied.

The input image is separated into many fixed-size blocks. In a 3 × 3 window, each pixel's eight neighbors (top, bottom, left, right, top left, bottom left, top right, and bottom right) are determined. Instead of comparing the value of the central pixel with the intensity values of the eight neighbors, the average intensity values of the neighbors are computed and compared with the intensity values of the eight neighbors. If the condition is met, the value is 1, or else it is 0. This yields an eight-bit binary number, which is used as a feature. The neighbors can be processed in either clockwise or anti-clockwise directions.

The generated binary numbers (also known as LBP codes) represent the image's local primitive properties. They show the many forms of curving edges, spots, flat sections, and so on. Figure 4.3 explained the facial image

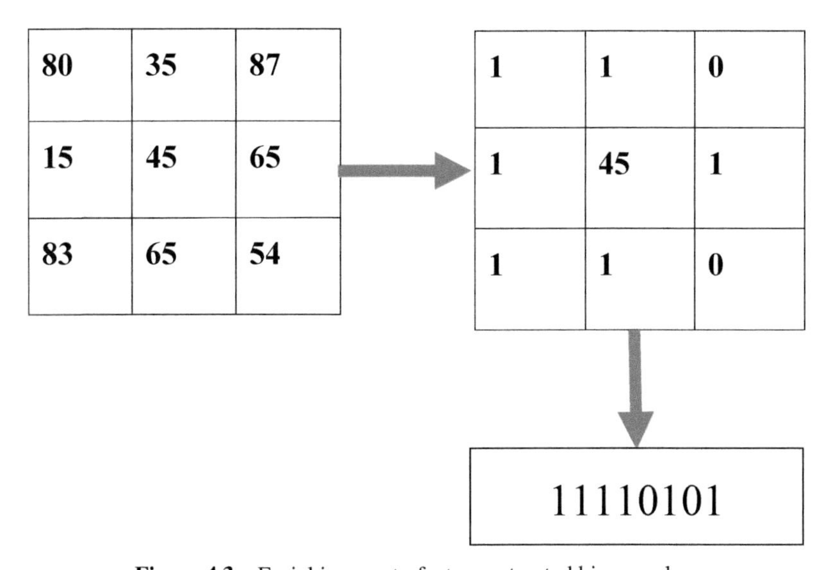

Figure 4.3 Facial images to feature extracted binary value.

converted into binary futures details. The proposed study uses a neural network to classify Indian facial traits, using a multimodal feature fusion dataset. Using CNN-based training and recognition, these characteristics were used to classify the signals. According to the psychological review, there is a need to address the fusion or integration of multimodal emotional data for optimal man–machine communication. It is because the human sensory system can analyze and extract accurate information about an individual's emotional state from a collection of face traits.

4.3.2 CNN Classification

The suggested deep learning framework for employee emotional prediction architecture is depicted in Figure 4.4. Using a typical neural network, the face regions such as the left eye, right eye, and mouth are extracted and trained. On the face images, histogram equalization, rotation correction, and spatial

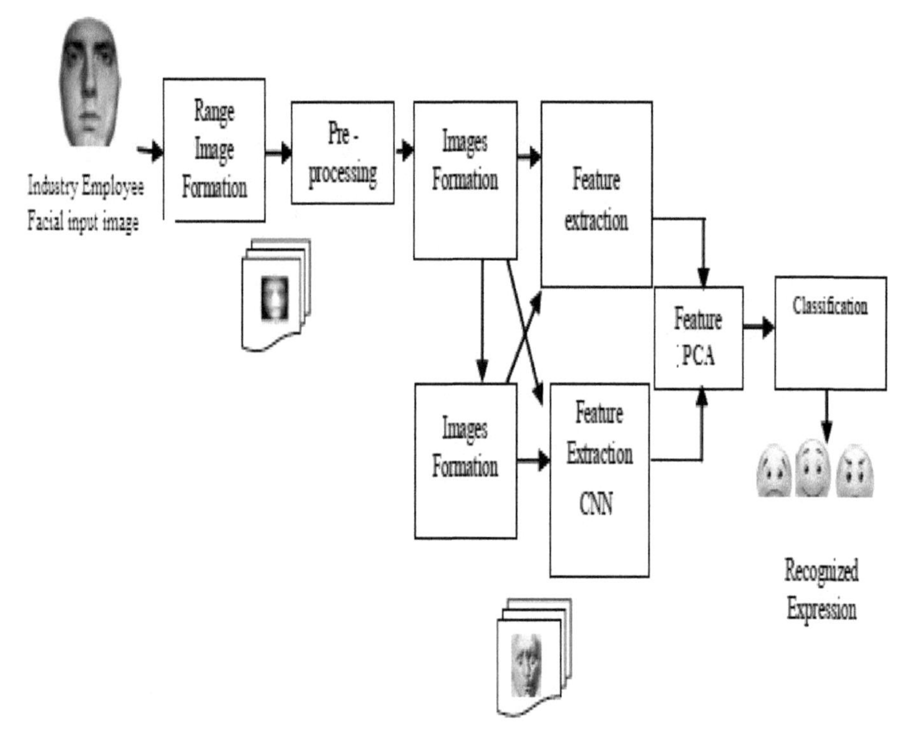

Figure 4.4 Proposed deep learning framework for employee's emotional prediction architecture.

normalizing are conducted to obtain representable features. The ultimate result of expression recognition is determined via majority voting of the three CNN's results, using a decision-level fusion procedure.

Image creation, pre-processing, feature extraction, and feature classification for emotion recognition are among the five processes in this architecture. Furthermore, the rigorous theoretical and practical analysis to be carried out to analyze six physiological signals of employees in industrial environments which exhibit a problematic day-to-day variation of employee faces or facial expressions helps to detect and interpret the emotions (Gouizi *et al.*, 2011). On the same day, the traits of distinct emotions tend to cluster more densely than the features of the same feeling on different days. A fresh emotional intelligent method must be created to deal with the daily variances.

Furthermore, the proposed regulations are not limited to monitoring the conduct of industry employees; they are easily applicable to stressed individuals working in a multinational firm who can be mentored and counseled by the company's manager. Emotional intelligence policies can be applied to a variety of settings, including homes, offices, conference rooms, hospitals, control centers, and retail, among others. There is plenty of evidence that our visual processing architecture is divided into layers. Each stage alters the data in a way that makes the visual task easier to complete. The ability to share features or sub-features is another interesting feature of deep learning models. It has also been demonstrated that insufficiently deep structures might be exponentially inefficient in terms of computation. When they developed a very effective method for training multilayer neural networks (Zhong *et al.*, 2017), deep learning was revolutionized. Figure 4.5 explains that proposed systems consist of three units, such as input, processing unit, and output unit. The employee details psychological input battery input, ground, and camera used to receive the facial input of employees designed in the input unit. These details are transferred using the controller area network (CAN) in the microcontroller and converted details are processed in the processing unit; these values are stored in EEPROM memory. The final values like features are stored in the output section.

Figure 4.6 explains the input values, configuration, and facial message signal are transferred CAN communication protocol to the controller to calculate the facial signal information is stored as a feature to the output memory storage devices. Figure 4.7 explains the CAN communications for signal stored and message signal attributes, and the number of bytes consumed in each signal for employ emotions prediction. Totally, six signals are processed

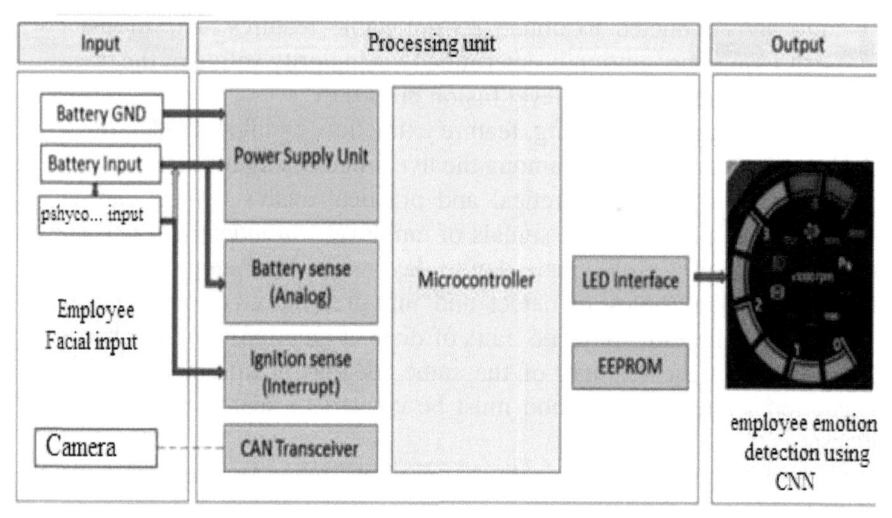

Figure 4.5 Proposed system architecture diagram.

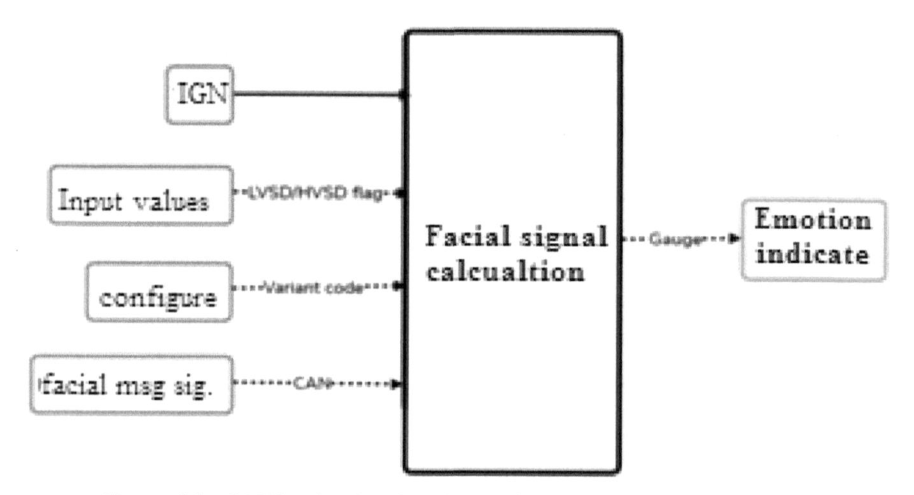

Figure 4.6 CAN bus bar functionality configuration emotional indications.

and sent to the CAN communications protocol with the number of bytes consumed, and transfer bit rates are also defined.

Figure 4.8 explains the analyzer and DBCAN+ software tool used to manually check the signal before hardware design. In this tool displayed, each of the employee's signals in Figure 4.5 motioned was transferred and displayed in color representation. Each color defines the different signals of employees' signal monitoring in concerns.

Signal name	Human_Pressure	
Byte	2	3
Bit	7..0	7..0

Resolution and unit	0.01 km/h
Linearization	Y = 100 * X
Example	0x2710 = 100km/h

Signal name	Facial_signal_1
Byte	4
Bit	7..6

0b00	Diagnosis in progress
0b01	Alarm not active
0b10	Alarm active
0b11	Coding alarm

Signal name	Human_EEG_Signal1
Byte	6
Bit	5..4

0b00	Reserved
0b01	Alarm not active
0b10	Alarm active
0b11	Reserved

Signal name	Human_EEG_Signal2
Byte	6
Bit	3..2

0b00	Diagnosis in progress
0b01	Alarm not active
0b10	Alarm active
0b11	Coding alarm

Signal name	Facial_signal_2
Byte	6
Bit	1..0

0b00	Reserved
0b01	Alarm not active
0b10	Alarm active
0b11	Reserved

Signal name	Facial_signal_3
Byte	6
Bit	4

0b0	ABS idle
0b1	Sign_working

Figure 4.7 CAN bus bar message and signal design configuration for emotional indications.

The CAN transferred signals are stored and passed through the following CNN based approaches. The following CAN data received from humans such as pressure, facial emotion using controller area network (CAN), data was stored and applied to the seven categorizations emotional AlexNet is a first successful CNN for a large training dataset, ImageNet (Tom *et al.*, 2014). Its layout is depicted as relatively simple as compared to other recent architectures. It has five convolutional layers, three fully connected layers along with max-pooling, and dropout layers. The traditional activation function Tanh is given by the following equation:

$$f(x) = T\text{anh}(x) = 2 * \frac{1}{1 + e^{-2x}} - 1 \tag{4.4}$$

This Tanh function is a saturating nonlinearity, which is slow to train as the exponential function is computationally expensive. So, a no saturating nonlinearity, ReLU, as given by eqn (4.5), is used by AlexNet, which is faster to train and computationally efficient (Krizhevsky *et al.*, 2012).

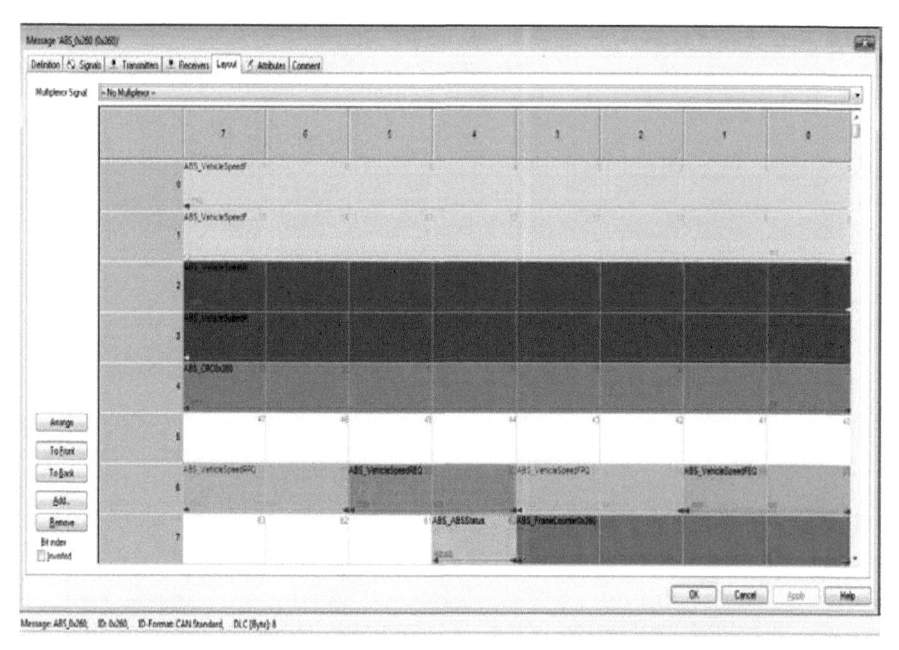

Figure 4.8 CAN signal transmission for emotional indications.

$$f(x) = ReLU(x) = \max(0, x). \tag{4.5}$$

ReLUs are said to be six times faster, and they do not need input normalization to prevent saturation. But a local response normalization called brightness normalization is done to achieve generalization. This normalization results in a form of lateral inhibition, which is found in real neurons. AlexNet uses ReLU instead of Tanh for adding the nonlinearities, which boosts up the speed for the same accuracy level. To deal with the overfitting issue, it uses dropout. To reduce overfitting to the training data and to learn robust features, with a probability of 0.5, the output of each hidden neuron is set to zero. Because the dropout operation doubles the number of repetitions necessary to converge, it is only used on the first two fully linked layers of the AlexNet design. The neurons that are dropped out will not participate in forwarding pass and, hence, in backpropagation. Thus, it reduces the co-adaptations of neurons, as each neuron cannot rely on other neurons. The pooling layer is used to summarize the neighboring neuron's output in the same kernel map. To reduce the size of the network, it overlaps the pooling layers. AlexNet models were trained using stochastic gradient descent that minimizes the cross-entropy loss function with particular values for momentum as well as weight decay. All the layer's learning rates are first set to 0.01 and then manually modified when the validation error rate stops increasing versus the current learning rate (Shanthi and Sabeenian, 2019). Here, we trained a single CNN for the two different featured images, instead of multiple CNN models. This single network training can save both time and memory, thus preserving accuracy. There is no need to learn different weights for different networks that correspond to score-level fusion strategy. Instead, we adapted a feature-level fusion strategy. Compared to other structures of deep learning, CNN is commonly used as it demonstrates better performance in recognition tasks due to its ability to extract and learn robust features. Moreover, CNN uses a small amount of bias and weight values as compared to other structures of deep learning.

The algorithm combines the original image with the virtual image, which is the outcome of the original image's nonlinear modification. Pixels with a moderate intensity are boosted, whereas pixels with high or low intensity are altered. For both the test and training sets of photos, the distance between the original image and the virtual image is computed using statistical metrics. Furthermore, in expressing face images, the value of distinct pixels fluctuates. As a result, different weights might be assigned to different pixels. The weight is then calculated using the two least distance values, yielding an adaptive weight that is different for each image. Then, to recognize the faces, score level fusion is used, which overcomes the illumination caused by the image's lighting. According to the findings of the experiments, the proposed method outperforms earlier work in recognizing accuracy.

4.4 Result and Discussion

The construction and implementation of convolutional neural networks to categorize seven fundamental emotion types from face picture datasets in employees were presented in this study. The proposed system was created in Matlab2019b with the IDE, which includes deep learning packages and libraries such as AlexNet. We used CNN to build our model, which was assisted by an embedded camera backend and a PCA feature extraction module. The face image can be classified into one of the following emotions using a neural network system that includes convolution layers, pooling, and fully connected networks: neutral, angry, disgust, fear, sad, surprised, or joyful. Preparation of datasets The dataset, the facial expression image from the Facial Expression Recognition newline (FER2013), and the kaggle

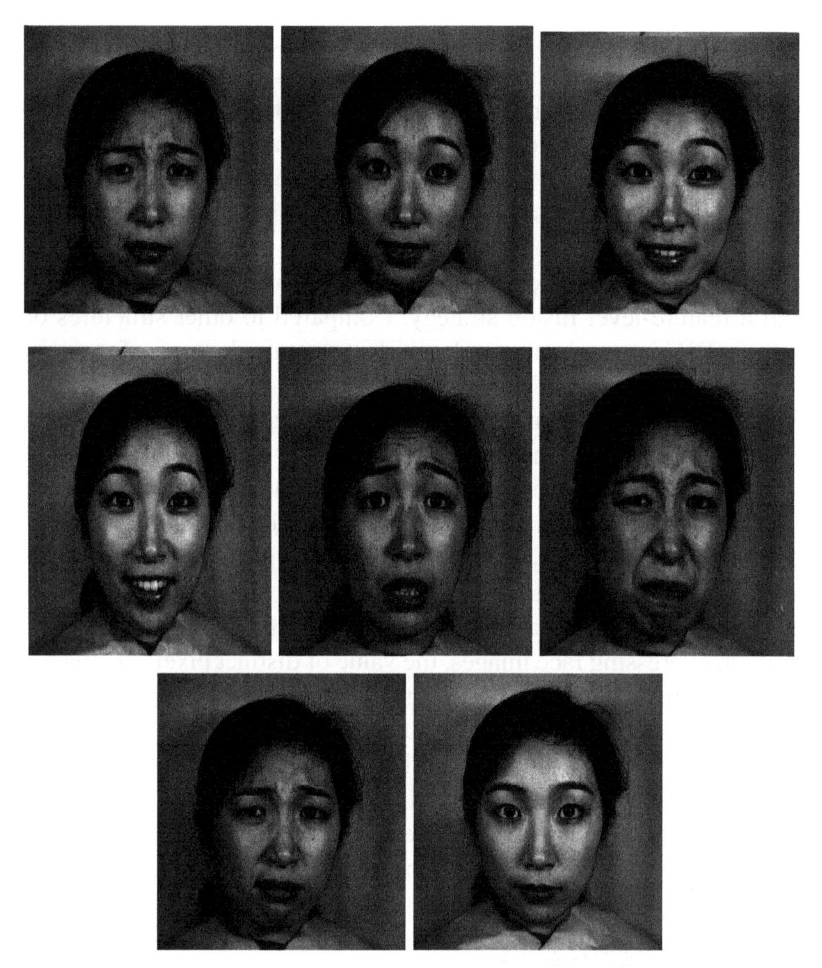

Figure 4.9 Input images used in the employee emotional predictions.

image database was organized and extracted (Choi and Byung, 2020). The test image from the simulation is obtained in the form of JPG format with the range of 512 × 512 pixels captured from a real-time HD camera and applied in the experiments. Furthermore, our model had also been constructed to recognize the emotions of human beings from a live video. The several multiplicative and additive noises can be reduced using the denoising work, and also the filtered image conception is enhanced to a very great extent. Figure 4.9 explains the proposed system described image in various types such as happy, fear, guilty, disgust, normal, pain, and surprise dataset images.

Table 4.1 Employee emotional prediction performance results

Methods	Efficiency	Precision	Error rate
Support vector machine	79.35	85.60	14.4
Decision tree	81.25	89.70	9.3
Artificial neural network algorithm	86.50	90.85	9.15
Proposed convolutional neural network system	96.75	95.65	4.35

The type of parameters taken from the facial image determines the performance of a neural network or CNN. The performance is also influenced by how the parameter data is processed before being presented to the networks. The system created a model with 25 features and 19 facial points based on frontal photos of the face and 10 points based on profile images of the face. The geometric face model based on 30 feature characteristic points has now been completed. The identification rate of most other methods, such as feature point tracking, Gabor wavelet analysis, and optical flow tracking, was comparable to or slightly better than the identification rate of the seven real-valued and eight binary parameters utilized in the study.

Negative, positive, or no significant divergence from the neutral value was seen in real-valued parameters. For diverse expressions, the trend of variation of different parameters concerning neutral values aids in the effective training of neural networks to recognize specific expressions. Each expression is defined by the real-valued and binary parameters (Table 4.1). However, for some expressions, some parameters do not show a significant variation from neutral value, and, hence, they do not contribute to detecting that expression. Employee images in Figure 4.7 were utilized, and it was determined that the mean average % for all the parameters, as well as the average percentage existence of all binary parameters for all seven expressions, deviates from their respective neutral values.

Table 4.1 explains that the model based on proposed CNN yields the maximum efficiency, precision, and minimum error rate for Facial Expression Recognition newline (FER2013) dataset of (96.75, 95.65, and 4.35), at nearer neural network algorithm (Khatun and Turzo, 2020) provides only 86.50 of efficiency, 90.85 of precision, and 9.15 of error rate values.

Comparatively, the proposed approach provides a better result in the different terms among 5.25, 3.80, and 4.49 values. The SVM offers the least efficiency values of 79.35%, precision of 86.5, and error rate values of 14.4 (Healy *et al.*, 2018). The decision tree offers the least efficiency values of

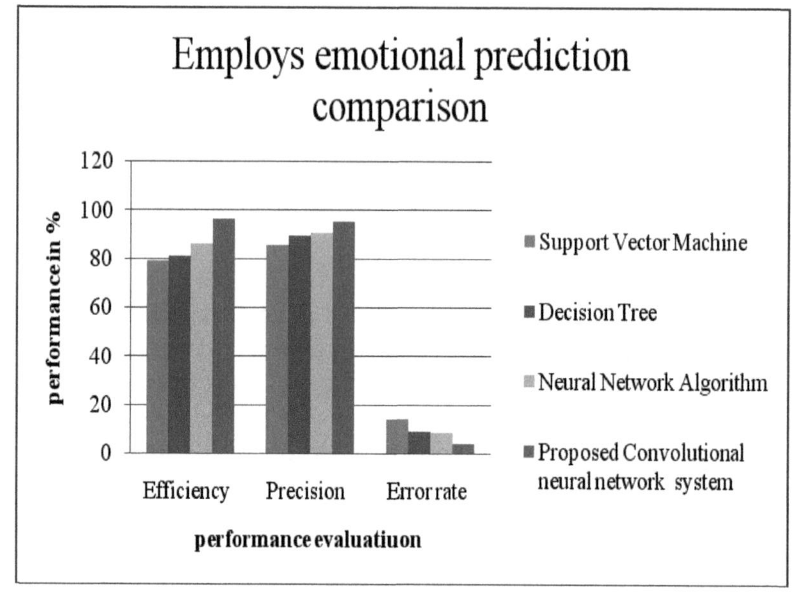

Figure 4.10 Employee emotional prediction performance results.

Table 4.2 Employee emotional prediction time performance results.

Methods	Support vector machine	Decision tree	Artificial neural network algorithm	Proposed convolutional neural network system
Time in seconds	38.25	48.52	56.55	36.65

81.25% precision for 89.70 and error rate values of 9.3 (Jamil and Hamzah, 2018). Whereas, the proposed CNN yields better quality matrix values for the Facial Expression Recognition newline (FER2013) dataset as explained in Figure 4.8.

Table 4.2 explains the model based on the proposed CNN algorithm time taken efficiency, precision, and minimum error rate for Facial Expression prediction Recognition newline (FER2013) dataset of (36.65 seconds), decision tree algorithm achieves efficiency of 81.25% for 48.52 seconds taken, support vector machine algorithm achieves 79.35% efficiency in 38.25 seconds and artificial neural network algorithm achieves 86.50% efficiency in 56.55 seconds. Comparatively, the proposed approach provides a better result in terms of time: 1.6 seconds in SVM, 11.87 seconds in decision tree, and 19.9

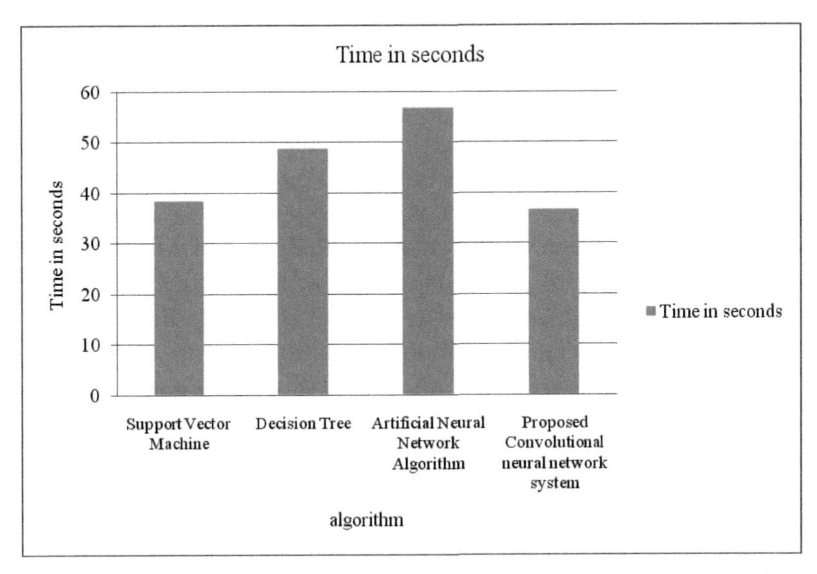

Figure 4.11 Employee emotional prediction time duration performance results.

seconds in artificial neural network algorithm. Whereas, the proposed CNN yields better quality matrix values for the Facial Expression Recognition newline (FER2013) dataset explained in Figure 4.9. The quality efficiency values subjected to the suggested CNN are superior to those exposed to SVM, DT, and NN. The proposed CNN system removes unrelated and superfluous structures from the data, and the structure chosen will aid in the enhancement of the presentation of learning models if the data is reduced for classification.

The experimental results of the proposed systems are analyzed with different databases and it has been observed that the proposed methods show significant improvement in genetic face recognition quality and intelligibility.

4.5 Conclusion

In image processing applications, face recognition is a promising area of research. For security purposes and in a variety of applications such as biometric authentication, human–computer interaction, video surveillance, credit card verification, automatic indexing of images, and criminal identification, a facial recognition system can verify or identify an individual from a digital image. The proposed method allows for a model-based safety framework for industrial automation systems, allowing management to better understand employees' current feelings about their work. The current method, however, is limited to discrete-time models. This issue will be addressed in this paper by using a CNN approach to simulate the continuous dynamics of monitoring in conjunction with the discrete character of the control logic. It discusses a reconfigurable smart WSNs unit based on the Internet of Things for monitoring technical safety factors in the workplace. The system can intelligently collect sensor data. It was created with the use of wireless communication in mind. It is well suited to the real-time and practical needs of a high-speed emotional data gathering system in an IoT setting. Emotions are recognized automatically by facial emotion recognition systems. Healthcare, tailored learning, robotics, event detection, and surveillance are just a few of the applications. On the other hand, major facial traits that are discriminative have been identified, and this continues to be a challenge because each emotion has its variability and nuance that must be represented. The primary goal of this project is to address these issues. When characteristics involving motion sequence are evaluated, an appropriate alternative is an optical flow, and the attributes of PCA are the histogram of motion orientation, which are weighted through the amplitude of motion.

This research's future focus will be on this issue, and it will offer a system that will considerably improve emotional identification for a larger number of images and with a longer linage of face images. Finally, this study could be expanded by employing a larger number of face photographs from various perspectives to demonstrate an improvement in the performance of identifying emotions in faces. Also, the future of industrial management will be centered on meeting climate change expectations, safeguarding human and machine safety, and measuring and improving performance. Though no one predicts that safety engineers or industrial hygienists will go out of business in the next decade, our respondents are almost unanimous in their belief that businesses will continue to need professionals who can handle a wide range of industrial functions and even allied responsibilities.

References

M. F. Ali, M. Khatun, & N. A. Turzo. "Facial emotion detection using neural network." *The International Journal of Scientific and Engineering Research* 2020.

Carmona, Alejandra M., Ana I. Chaparro, Ricardo Velásquez, Juan Botero-Valencia, Luis Castano-Londono, David Marquez-Viloria, and Ana M. Mesa. "Instrumentation and data collection methodology to enhance productivity in construction sites using embedded systems and IoT technologies." In *Advances in Informatics and Computing in Civil and Construction Engineering*, pp. 637-644. Springer, Cham, 2019.

Chang, Peter D., Edward Kuoy, Jack Grinband, Brent D. Weinberg, Matthew Thompson, Richelle Homo, Jefferson Chen *et al.* "Hybrid 3D/2D convolutional neural network for hemorrhage evaluation on head CT." *American Journal of Neuroradiology* 39, no. 9 (2018): 1609-1616.

Choi, Dong Yoon, and Byung Cheol Song. "Semi-supervised learning for continuous emotion recognition based on metric learning." *IEEE Access* 8 (2020): 113443-113455.

Da Xu, Li, Wu He, and Shancang Li. "Internet of Things in industries: A survey." *IEEE Transactions on Industrial Informatics* 10, no. 4 (2014): 2233-2243.

Delcourt, Cécile, Dwayne D. Gremler, Allard C. R. van Riel, and Marcel J. H. Van Birgelen. "Employee emotional competence: Construct conceptualization and validation of a customer-based measure." *Journal of Service Research* 19, no. 1 (2016): 72-87.

Ekman, P., and W. V. Friesen. "Measuring facial movement." *Environmental Psychology and Nonverbal Behavior* 1, no. 1 (1976): 56-75.

El Naqa, I., and M. J. Murphy. "What is machine learning?." In *Machine Learning in Radiation Oncology*, pp. 3-11. Springer, Cham, 2015.

Fernandez, Claudia S. P. "Emotional intelligence in the workplace." *Journal of Public Health Management and Practice* 13, no. 1 (2007): 80-82.

Foukalas, Fotis. "Cognitive IoT platform for fog computing industrial applications." *Computers & Electrical Engineering* 87 (2020): 106770.

Giardini, Angelo, and Michael Frese. "Linking service employees' emotional competence to customer satisfaction: A multilevel approach." *Journal of Organizational Behavior: The International Journal of Industrial, Occupational and Organizational Psychology and Behavior* 29, no. 2 (2008): 155-170.

Gouizi, Khadidja, F. Bereksi Reguig, and Choubeila Maaoui. "Emotion recognition from physiological signals." *Journal of Medical Engineering & Technology* 35, no. 6-7 (2011): 300-307.

Haroon, Asma, Munam Ali Shah, Yousra Asim, Wajeeha Naeem, Muhammad Kamran, and Qaisar Javaid. "Constraints in the IoT: The world in 2020 and beyond." *Constraints* 7, no. 11 (2016): 252-271.

M. Healy, R. Donovan, P. Walsh, and H. Zheng. "A machine learning emotion detection platform to support affective well being." *In 2018 IEEE*

International Conference on Bioinformatics and Biomedicine (BIBM), pp. 2694-2700. IEEE, December 2018.

Hiller, Jens, Martin Henze, Martin Serror, Eric Wagner, Jan Niklas Richter, and Klaus Wehrle. "Secure low latency communication for constrained industrial IoT scenarios." In *2018 IEEE 43rd Conference on Local Computer Networks (LCN)*, pp. 614-622. IEEE, 2018.

Hu, Long, Yiming Miao, Gaoxiang Wu, Mohammad Mehedi Hassan, and Iztok Humar. "iRobot-Factory: An intelligent robot factory based on cognitive manufacturing and edge computing." *Future Generation Computer Systems* 90 (2019): 569-577.

Hussain, Fatima, Rasheed Hussain, Syed Ali Hassan, and Ekram Hossain. "Machine learning in IoT security: Current solutions and future challenges." *IEEE Communications Surveys & Tutorials* 22, no. 3 (2020): 1686-1721.

Imani, M., & G. A. Montazer. "A survey of emotion recognition methods with emphasis on E-Learning environments." *Journal of Network and Computer Applications* 147 (2019): 102423.

Joseph, R. "The right cerebral hemisphere: Emotion, music, visual–spatial skills, body–image, dreams, and awareness." *Journal of Clinical Psychology* 44, no. 5(1988): 630-673.

Kalimuthu, Sivanantham, Farid Naït-Abdesselam, and B. Jaishankar. "Multimedia data protection using hybridized crystal payload algorithm with chicken swarm optimization." In *Multidisciplinary Approach to Modern Digital Steganography*, pp. 235-257. IGI Global, 2021.

Kalimuthu, Sivanantham. "Sentiment analysis on social media for emotional prediction during COVID-19 pandemic using efficient machine learning approach." *Computational Intelligence and Healthcare Informatics* 215 (2021).

Kanagaraju, P., and R. Nallusamy. "Registry service selection based secured Internet of Things with imperative control for industrial applications." *Cluster Computing* 22, no. 5 (2019): 12507-12519.

Krizhevsky, Alex, Ilya Sutskever, and Geoffrey E. Hinton. "Imagenet classification with deep convolutional neural networks." *Advances in Neural Information Processing Systems* 25 (2012): 1097-1105.

Latha, Charlyn Pushpa, and Mohana Priya. "A review on deep learning algorithms for speech and facial emotion recognition." *APTIKOM Journal on Computer Science and Information Technologies* 1, no. 3 (2016): 92-108.

Liu, C., R. A. Calvo, and R. Lim. "Improving medical students' awareness of their non-verbal communication through automated non-verbal behavior feedback." *Frontiers in ICT* 3, no. 11 (2016).

Lu, X., and A. Jain. "Deformation modeling for robust 3D face matching." *IEEE Transactions on Pattern Analysis and Machine Intelligence* 30, no. 8, (2008): 1346-1357.

Meng, Zhaozong, Zhipeng Wu, Cahyo Muvianto, and John Gray. "A data-oriented M2M messaging mechanism for industrial IoT applications." *IEEE Internet of Things Journal* 4, no. 1 (2016): 236-246.

Nalbandian, S. "A survey on Internet of Things: Applications and challenges. In *2015 International Congress on Technology, Communication and Knowledge (ICTCK)*, pp. 165-169. IEEE, November 2015.

Paine, Tom Le, Pooya Khorrami, Wei Han, and Thomas S. Huang. "An analysis of unsupervised pre-training in light of recent advances." *arXiv preprint arXiv:1412.6597* (2014).

Piccialli, Francesco, Salvatore Cuomo, Vincenzo Schiano di Cola, and Giampaolo Casolla. "A machine learning approach for IoT cultural data." *Journal of Ambient Intelligence and Humanized Computing* (2019): 1-12.

Prentice, Catherine, Sergio Dominique Lopes, and Xuequn Wang. "Emotional intelligence or artificial intelligence–an employee perspective." *Journal of Hospitality Marketing & Management* 29, no. 4 (2020): 377-403.

Rázuri, J. G., D. Sundgren, R. Rahmani, & A. M. Cardenas. "Automatic emotion recognition through facial expression analysis in merged images based on an artificial neural network." In *2013 12th Mexican International Conference on Artificial Intelligence*, pp. 85-96. IEEE, November 2013.

Saad, M. M., N. Jamil, and R. Hamzah. "Evaluation of support vector machine and decision tree for emotion recognition of malay folklores." *Bulletin of Electrical Engineering and Informatics* 7, no. 3 (2018): 479-486.

Scherer, Klaus R., and Ursula Scherer. "Assessing the ability to recognize facial and vocal expressions of emotion: Construction and validation of the emotion recognition index." *Journal of Nonverbal Behavior* 35, no. 4 (2011): 305-326.

Shanthi, T., and R. S. Sabeenian. "Modified Alexnet architecture for classification of diabetic retinopathy images." *Computers & Electrical Engineering* 76 (2019): 56-64.

Shen, Qianzi, Zijian Wang, and Yaoru Sun. "Sentiment analysis of movie reviews based on CNN-BLSTM." In *International Conference on Intelligence Science*, pp. 164-171. Springer, Cham, 2017.

Shen, Shunrong, Haomiao Jiang, and Tongda Zhang. "Stock market forecasting using machine learning algorithms." *Department of Electrical Engineering, Stanford University, Stanford, CA* (2012): 1-5.

Toneva, Mariya, and Leila Wehbe. "Interpreting and improving natural-language processing (in machines) with natural language-processing (in the brain)." *arXiv preprint arXiv:1905.11833* (2019).

Wu, Meiyin, and Li Chen. "Image recognition based on deep learning." In *2015 Chinese Automation Congress (CAC)*, pp. 542-546. IEEE, 2015.

Wu, Qihui, Guoru Ding, Yuhua Xu, Shuo Feng, Zhiyong Du, Jinlong Wang, and Keping Long. "Cognitive Internet of Things: A new paradigm beyond connection." *IEEE Internet of Things Journal* 1, no. 2 (2014): 129-143.

Xiao, Liang, Xiaoyue Wan, Xiaozhen Lu, Yanyong Zhang, and Di Wu. "IoT security techniques based on machine learning: How do IoT devices use AI to enhance security?" *IEEE Signal Processing Magazine* 35, no. 5 (2018): 41-49.

Yin, Zhong, Mengyuan Zhao, Yongxiong Wang, Jingdong Yang, and Jianhua Zhang. "Recognition of emotions using multimodal physiological signals and an ensemble deep learning model." *Computer Methods and Programs in Biomedicine* 140 (2017): 93-110.

Zhang, Xiao, and Dongrui Wu. "On the vulnerability of CNN classifiers in EEG-based BCIs." *IEEE Transactions on Neural Systems and Rehabilitation Engineering* 27, no. 5 (2019): 814-825.

Zhang, Zhi-Kai, Michael Cheng Yi Cho, Chia-Wei Wang, Chia-Wei Hsu, Chong-Kuan Chen, and Shiuhpyng Shieh. "IoT security: Ongoing challenges and research opportunities." In *2014 IEEE 7th International Conference on Service-Oriented Computing and Applications*, pp. 230-234. IEEE, 2014.

5

Assistive Smart Cane for Visually Impaired People Based on Convolutional Neural Network (CNN)

Farhatullah[1], Ajantha Devi[2], Muhammad Abul Hassan[3], Iziz Ahmad[4], Muhammad Sohail[5], Muhammad Awais Mahoob[6], and Hazrat Junaid[7]

[1]School of Automation, China University of Geosciences, Wuhan 430074, China
[2]Research Head, AP3 Solution Chennai, TN, India
[3,4]Department of Computing and Technology, Abasyn University Peshawar, Pakistan
[5]Department of Computer Science, Bahria University Islamabad Campus, Pakistan
[6]Department of Computer Science, University of Engineering & Technology Texila, Pakistan
[7]Department of Computer science and Information Technology, University of Malakand, Pakistan
E-mail: farhatkhan8398@gmail.com; ap3solutionsresearch@gmail.com; abulhassan900@gmail.com; izazahmad445@gmail.com; sohailbahria02@gmail.com; awaisntu@gmail.com; abidj3692@gmail.com

Abstract

According to the World Health Organization, there are millions of visually impaired persons throughout the world who struggle to move freely. They always need help from people with normal sight. For visually impaired persons (VIP), finding their way to their desired destination in a new area is a huge difficulty. This research aimed to assist these individuals in resolving their challenges in moving to any place on their own. To this end, we proposed a method for VIP using a convolutional neural network (CNN) to recognize the condition and scene objects automatically. The proposed system consists of Arduino UNO, ultrasonic sensors, a camera, breadboards, jumper wires, a buzzer, and an earphone. Breadboards are used to connect the sensors with the help of Arduino UNO and jumper wires. The sensors are used for the detection of obstacle and potholes while the camera performs as a virtual eye for the visually impaired people by recognizing these obstacles in any direction (i.e., front, left, and right). An important feature is provided by this system, in which the blind receives the scene object, the system automatically calculates how far he is away from the obstacles, and a voice message alerts him and directs him via earphone. The obtained experimental results show that the CNN yielded impressive results of 99.56% accuracy and has a loss validation of 0.0201%.

Keywords: Convolutional neural network, Arduino UNO, ultrasonic sensors, wayfinding system, situation awareness, activity instruction.

5.1 Introduction

According to the World Health Organization (WHO), at least more than two billion people have sight problems, with 1 billion avoiding or leaving untreated [1]. The world faces numerous challenges in the field of eye care, including treatment, a shortage of trained eye care providers, and a lack of integration of eye care into the healthcare system. The study was released on World Sight Day to alert the public about the increasing number of vision deficiency and blindness problems caused by numerous eye disorders such as presbyopia, which affects 1.8 billion people and occurs at a young age, myopia, which affects 2.6 billion people, cataract, which affects 65.2 million people, and corneal opacity, which affects 6.9 million people [2]. A visually impaired person faces numerous challenges that necessitate the assistance of a sighted person for him/her to find his/her way. Because of the unfamiliar environment, blind people are unable to find their way. VIPs typically use a walking cane. Raspberry Pi, Arduino UNO, and ultrasonic sensors can be used to assist visually impaired people. Arduino UNO technology was employed in this system to design a smart blind device. The design not only provides detection of barriers but also guides them along from a specific range to locate themselves easily. The ultrasonic system emits energy waves that reflect from obstacles on any side (i.e., left, right, and front) to assist the visually impaired person in detecting the obstacle within the defined range. The distance between visually impaired people and objects is calculated using the ultrasonic sensor's starting and ending pulses. A buzzer sensor is used to alert the blind user to potential hazards. If the obstacles are too close to the blind user, the proposed system will generate a voice message and activate a buzzer to alert the visually impaired person to the obstacles. With a voice message, this proposed design can provide full support against obstacle avoidance. The proposed module will be beneficial to people who are blind. It is easier for them to find ways in daily activities that do not require the use of the standard mobility aids and are available to individuals with this disability. Mobility aids are less expensive, smaller in size, and easier to transport. The remaining of the chapter is divided into five sections: Section 5.2 contains a literature review, Section 5.3 contains the details of the proposed methodology [convolutional neural network (CNN)], Section 5.4 contains study and discussion of the experimental results, and Section 5.5 concludes.

5.2 Literature Review

Many attempts are being made all over the world to propose designs that will assist visually impaired people in detecting obstacles using various electronic technologies. Some of these works are focused on microcontrollers, and some of them use the technology of the global positioning system (GPS) and global system for mobile communications (GSM), but the majority of them use ultrasonic sensors to detect obstacles, according to a literature review. All of these efforts are focused on assisting blind people in detecting obstacles, rather than assisting them in navigating on their own. Batavia *et al.* [3] suggest a distance-dependent method based on a camera, with background motion measured using homo-graphic renovates. The barriers are classified as critical or normal based on their proximity to a particular distance, after which they are detected and identified. Hesch *et al.* [4] proposed that a design model has been consisting of a two-dimensional laser scanner, foot mounted pedometer, and three axis gyroscopes for the aid of VIP in an inside environment. They offered two-layered estimators in the first estimated layer, the location of the blind cane was monitored, and the second layer was estimated to determine the person's location. Pradeep *et al.* [5] suggested an environment perception device that uses an RGB-D camera. In the proposed system, they impose three things: self-localization, obstacle detection, and object recognition. In self-localization, the depth has been perceived based on the tracking technique on color information. Obstacle detection and recognition provide meaningful information to the visually impaired people and recognized the obstacle such as stairs, walls, vehicles, doors, etc. Rodríguez *et al.* [6] proposed a method for obstacle prevention devices to help visually impaired persons. They have implemented an incremental map of the environment with the help of optical SLAM techniques, to provide spatial direction and location of the visually impaired people at the same. The proposed design also provided audio feedback to alert a blind person to the existence of potential impediments. Tapu *et al.* [1] designed smartphone-based obstacle to detect and classify to VIP to walk freely and carefully in an inside and outside environment. The Lucas Kanade algorithm has been used to extract the feature from images. Shahdib *et al.* [7] designed a model that involves the head of mounted stereo camera to search ground planes and break up six-degrees-of-freedom into ground smooth and planar motion which assist VIP. Due to the investigation of the variation range, they evaluated the ground smooth using optical/visible data or with the IMU reading. Maidenbaum *et al.* [8] designed a system, based on ultrasonic sensors and a camera for obstacle detection and recognition, where the camera has been used to recognize and measure the size of obstacles. For VIP, Leung *et al.* [9] suggested a helmet, audio-based guidance system. They have used visible odometry and feature-based SLAM in the proposed design system to make a three-dimensional map for obstacle detection. Xiao [10] set out a system for vision-impaired people to use for context-aware navigation services. To incorporate sophisticated intelligence into navigation, the user must be aware of the semantic qualities of the things in their surroundings. This interaction is critical for improving communication about things and places so that better travel decisions may be made. Suba Raja *et al.* [11] created a system for visually impaired people that use mobile computing to detect and recognize their faces. This mobile system is supported by a server-based support system. Vlaminck *et al.* [12] used three ultrasonic sensors and a microcontroller to detect the object range. The audio and vibration systems also have been used to warn visually impaired people to avoid obstacles. Eunjeong *et al.* [13] proposed a novel mobile navigation device that can identify the condition and scene objects during walking time. The proposed framework classifies a user's current condition in terms of their position by analyzing streaming images. Then, using computer vision techniques, only the appropriate background objects are identified and interpreted based on the current situation. Ramadhan *et al.* [14] presented a wearable intelligent system to help VIP to go along the roads themselves, navigating in public places and seeking assistance. The system tracks the path and alerts the user to any obstacles using a set of sensors. The user is warned about the sound of a bolt and vibrations on the wrist, which can be useful in a noisy setting or when the user has a hearing loss.

Table 5.1 Evaluation of previous reviewed systems with some limitations

System name	Type of the sensors	Disadvantage	Techniques
Smart cane	Water ultrasonic sensors	If the water is less than 0.5 inches deep, the water sensor will not detect it, and the buzzer will not stop until the water is completely dry.	Ultrasonic
Eye substitution	Vibrator motors with two ultrasonic sensors	For haptic feedback, the team employed three motors. To offer more detailed feedback, they may use a 2D array of these actuators with restricted use by Android smartphones.	Ultrasonic with GPRS, GPS, and GSM
Ultrasonic navigation cane	Ultrasonic sensor with Arduino	Cannot detect items that come out of nowhere	Ultrasonic
Assistive ultrasonic headset	Four ultrasonic sensors	Only a few guidelines are given, and the headgear muffles outside sounds.	Ultrasonic
Navigation system for outdoor assistive	Three axial accelerometer magnetometer sensors (AAMS)	Because of its restricted range, the GPS receiver must be connected through Bluetooth to work.	GPS technology
Obstacle avoidance using auto-adaptive thresholding	Kinect's depth camera	When the distance between source and target increases, the accuracy of the Kinect depth image diminishes. After 2500 mm, the auto-adaptive threshold was unable to distinguish between the floor and the item.	Auto-adaptive thresholding

Table 5.1 Continued.

System name	**Type of the sensors**	**Disadvantage**	**Techniques**
A mobility device for visually impaired people using dynamic vision sensors (DVS)	DVS	Further intense testing is needed to demonstrate the performance of object avoidance and navigation strategies, as the test focused mostly on item identification in the scene's center region.	Event-based
Obstacle avoidance using laser rangefinder (LR)	Novint Falcon, photo sensors, and supplementary sensors	It was difficult to pinpoint the exact position of barriers and angles.	Laser rangefinder
A path force	Kinect sensor	This design's detecting range is too short, and the user must be educated to distinguish between vibration patterns for each cell.	Infrared and GPS
Artificial vision	Optical sensors	Assure its performance, the technology has not been tested or connected with navigation systems, and it is uncertain whether it will improve navigation systems as claimed by the creators.	GPS and GIS vision-based positioning
Guidance system	Kinect sensor	The system's detection of spatial markers must be improved, as well as the stability of the rebuilt walking plane.	Stereo vision, canny filter, vanishing point, and fuzzy rules .
Navigation-based system	Monocular camera	Their predetermined image sizes based on category may make it difficult to recognize	RANSAC algorithm, HOG descriptor, BoVW

Table 5.1 Continued.

System name	Type of the sensors	Disadvantage	Techniques
		the same item at different sizes.	vocabulary development, and SVM classifier
Vision assistance using RGB-camera	RGB-camera sensor	The infrared's sensitivity to sunlight can have a detrimental impact on the system's functionality when used outside or during the day.	RANSAC detection algorithm with infrared technology and density images
Ultrasonic sensors for obstacles detection	Four ultrasonic sensors	Above the waist level, the system is unable to identify impediments. There is no information about how to get about. The detecting range is limited.	Ultrasonic sensors and vision-based obstacles detection with SVM classifiers

Nivedita *et al.* [15] designed an electronic aid device that includes a Raspberry Pi device, an ultrasonic sensor, a webcam microphone, and a light dependent resistor (LDR) sensor. For obstacle detection, an ultrasonic sensor with a camera is used. The LDR sensor has been used to detect the brightness of the environment to determine whether it is dark or bright. The proposed system detects objects in the environment for visually impaired people and provides feedback by voice using an earphone. Souza *et al.* [16] presented a Raspberry Pi based system for blind people with unique features such as tracking objects and sending feedback via voice, as well as giving environmental information. The most crucial aspect of this work is the tracking of blind person's location and notifying the caretaker to ensure safety. Wafa Elmannai *et al.* [17] presented model comparison of wearable and portable assistive equipment for people with visual problems to demonstrate the progress of this group of people in assistive technology. Charis *et al.* [18] developed a new smart assistive systems framework for VIP that employs a user-cantered design method to evaluate a variety of operational and optional system characteristics. The outcomes of a series of interviews and surveys with visually impaired and non-visually impaired persons were analyzed to create the system's criteria for both on-site and distant usage. Yosra *et al.* [18] helped these communities to overcome their difficulties in moving independently to any location. This research specifies three paths for visually impaired students in the University of Gezira to easily reach different places on campuses.

The data acquired from the surrounding environment (through laser scanners, detected camera, or sonar) and sent to the user via tactile, auditory, and most electronic devices that provide services for VIP employ either or both methods. Different perspectives on whether approach gives greater feedback are currently being debated. Furthermore, despite the fact that numerous systems have been presented over the last decade, none of them is regarded a comprehensive solution capable of assisting VI individuals in all parts of their lives. As a result, this chapter highlights some of the work that has been completed.

5.3 Proposed Methodology

This research aims to detect and recognize the obstacles for VIP based on the deep convolutional network. The operation algorithm for detecting the obstacles mechanism in the paths is shown in Figure 5.1. Also, Figure 5.1 depicts the process for choosing the desired path as well as the steps involved in getting to the correct area. If the blind face up with obstacles, the device will keep track of how far he has strayed and warn him to change to a path.

5.3.1 Assistive Algorithm

An assistive algorithm describes the destination of a visually impaired user. Every obstacle (i.e., front, left, and right) is recognized in every direction and each obstacle is formatted as a statement of instructions. Thus, it is necessary to define the action first and then estimate the corresponding parameters, e.g., counts step, direction, current position, etc. The ultrasonic sensor information is used to calculate those parameters for this module.

5.3.2 Data Acquisition

The dataset was created using recorded videos with an Arduino UNO camera, and each frame was extracted with a width and height of (640,480). The key rule of deep learning is to split the data into two phases: training and testing, with 70% of the dataset being used for training and 30% for testing.

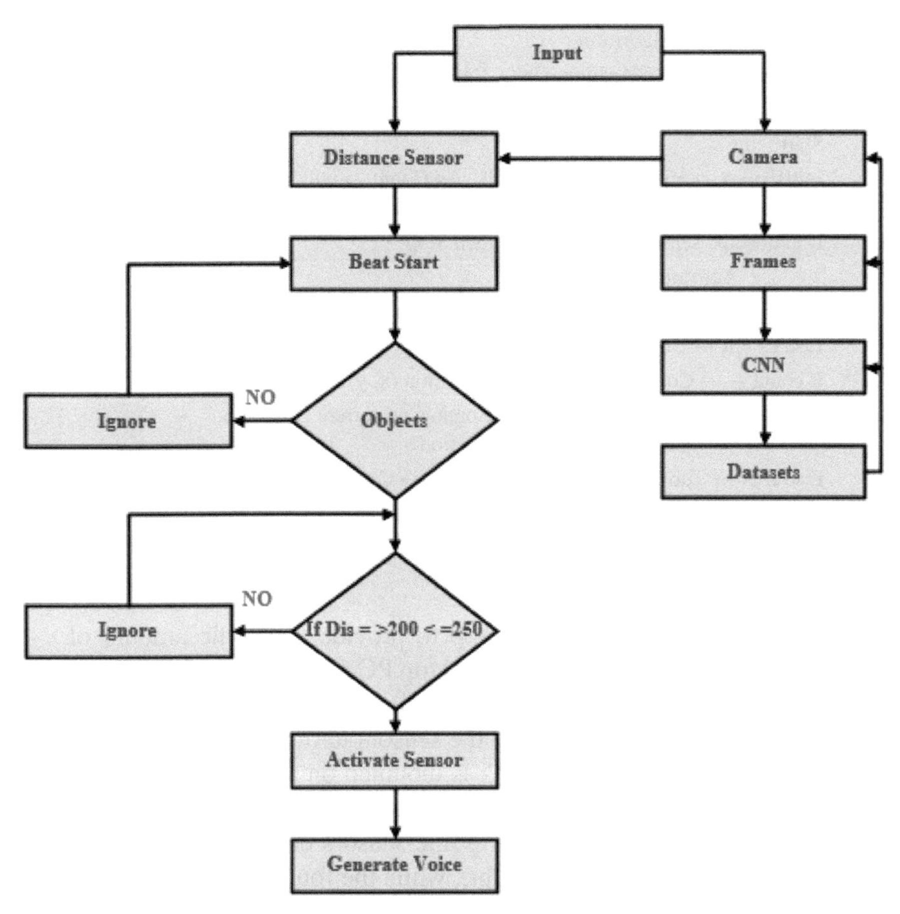

Figure 5.1 Framework of the proposed system.

5.3.3 Device Architecture

Figure 5.1 shows the design block diagram, which includes an Arduino UNO earphone, ultrasonic sensors, camera, breadboard, jumper wires, buzzer, and an earphone that comprise the device architecture. A breadboard is a critical component in the construction of a circuit. The breadboard acts as a link between the sensors and the Arduino UNO. The jumper wires are used to connect the sensor to the Arduino UNO indirectly. The Arduino UNO has

Algorithm 1: Assistive Algorithm	
Input: Ultrasonic sensor U, Arduino UNO, Camera, Buzzer	
Output: Contains the set of instructions, I (Turn, Count, Direction, Current position)	
1.	Start
2.	Initialize A → null, Direction → null, step Count → null, current position (x, y) → null;
3.	If Ultrasonic sensor U <= 200 cm, then Stop
4.	Activate Buzzer
5.	Else if Current Direction – Previous Direction > 250 → Turn
6.	Else Count → Continue A → Stop Buzzer
7.	If count A → Continue then Current position (x, y)
8.	Count → Count + 1, x → Count. Cos current Direction, y → Count. Sin Current Direction
9.	Else If A → Turn, then previous Direction → Current Direction
10.	If position → Destination → Terminate
11.	Else Go step 3

a power bank device mounted on the top to provide a specific amount of power when compared to a full-fledged desktop PC. The Arduino UNO has 4 ports, 40 GPIO pins, a memory card, a camera interface (CSI), and an HDMI port. Jumper wires are used to connect the sensors to the Arduino UNO. To power the Arduino UNO, a power bank is mounted on a wooden cane. The four ultrasonic sensors are linked to an Arduino UNO, which requires a 5-V power supply. Three of the four ultrasonic sensors detect obstacles from three directions (i.e., front, left, and right), while the fourth detects potholes. The visual sensor has a resolution of 5 megapixels and is directly connected to the Arduino UNO via the camera port to detect obstacles from the front. The buzzer sensor is linked to the Arduino UNO to alert the blind person to obstacles that require three voltages of power. The earphone is used for audio feedback and transmits a voice message to warn the VIP of the existence of obstacles. It also calculates the direction and distance from the obstacles.

5.3.4 Arduino and Its Interfacing

On the Arduino UNO, there are 14 digital input/output pins (six of which may be used as PWM outputs), six analog inputs, a 16-MHz crystal oscillator, a USB connection, a power connector, an ICSP header, and a reset button. It comes with everything you need to get started with the microcontroller; simply plug it into a computer via USB or use an AC-to-DC converter or

Figure 5.2 Arduino UNO.

battery to power it. The UNO varies from prior boards in that it does not use an FTDI USB-to-serial driver chip, instead using an Atmega8U2 intended to operate as a USB-to-serial converter. The term "UNO" is derived from the Italian word "UNI," meaning "one." It was chosen to commemorate the upcoming release of Arduino 1.0. The Uno and version 1.0 will be the reference versions of Arduino in the future, while the Arduino UNO is the newest in a series of USB Arduino boards that serve as the platform's standard model for comparing past iterations.

5.3.5 Power

The Arduino UNO may be powered through USB or an external power supply, and the power source is chosen automatically. External (non-USB) power can be supplied via an AC-to-DC adaptor (wall-wart) or a battery. To connect the adapter, a 2.1-mm center-positive plug may be inserted into the board's power connector. The power connection's Gnd and VIN pin headers can be inserted with battery leads. The board may be powered by an external source ranging from 6 to 20 V. If less than 7 V is supplied, the 5-V pin may deliver less than 5 V, causing the board to become unstable. If more than 12 V is utilized, the voltage regulator may overheat and kill the board.

VIN is the Arduino board's input voltage when utilizing an external power source (rather than 5 V via a USB connection or other regulated power source). If power is supplied via the power connector, this pin can be used to supply or access voltage. The CPU and other components on the board

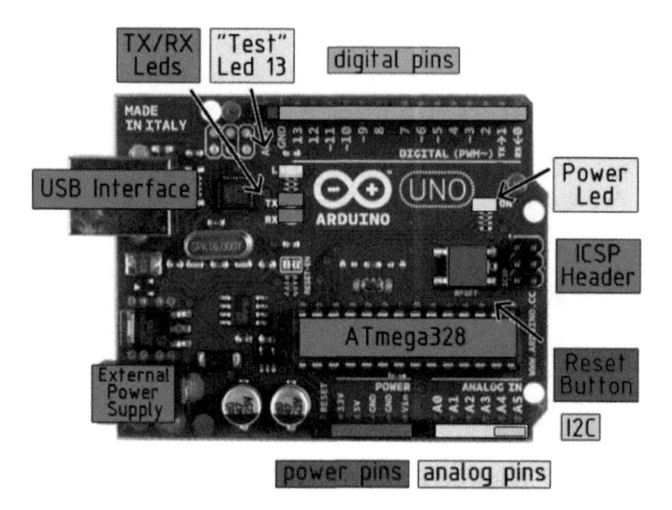

Figure 5.3 Arduino pin description.

are powered by a 5-V regulated power source. This can come from VIN or another controlled 5-V source, or it can come from USB or another regulated 5-V supply. The on-board regulator generates a 3.3-V supply. The maximum current usage is 50 milliamperes.

5.3.6 Memory

The Atmega328 is equipped with 32 KB of flash memory (0.5 KB for the boot loader), 2 KB of SRAM, and 1 KB of EEPROM for code storage (which can be read and written with the EEPROM library).

5.3.7 Deep Convolutional Neural Network (CNN)

Deep learning has proven to be a particularly effective technology in recent decades due to its ability to handle large amounts of data. Hidden layers have eclipsed traditional approaches in popularity when it comes to pattern recognition. Convolutional neural networks (CNNs) are a common type of deep neural network. Researchers have been working on a system that can recognize visual input since the 1950s, when artificial intelligence was in its infancy. This subject was dubbed computer vision in the years that followed. In 2012, a group of researchers from the University of Toronto developed an AI model that outperformed the best picture recognition algorithms by a significant margin, ushering in a new era in computer vision.

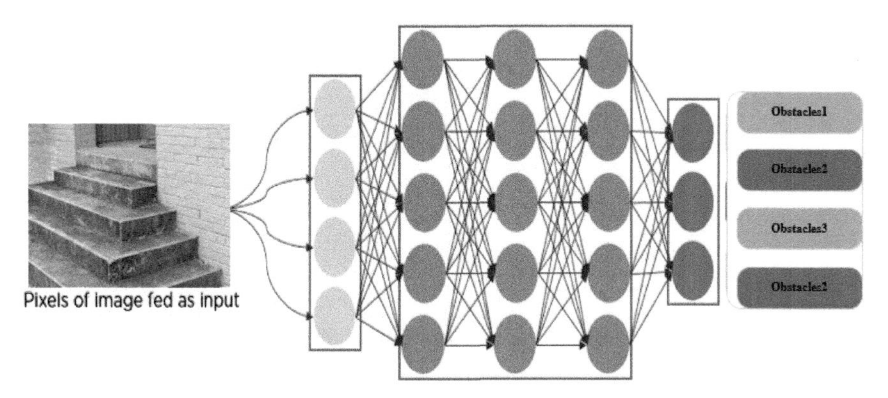

Figure 5.4 CNN architecture.

The Alex-Net AI system (named after its creator, Alex Krizhevsky) won the 2012 ImageNet computer vision challenge with an incredible 85% accuracy. The runner-up scored a respectable 74% on the test. Alex-Net used convolutional neural networks, a kind of neural network that approximates human vision. CNNs have become an essential component of many computer vision applications over time and are thus included in every online computer vision course. So, let us take a closer look at how CNNs work. In the following session, we will go through numerous pre-trained models for image creation (IC) in computer vision applications such as image captioning, neural style transfer (NST), anomaly detection, and image categorization.

5.3.8 Alex-Net Architecture

Deep CNN-based Alex-Net architecture is used in this study for real-time detection and recognition of obstacles such as vehicles, doors, pillars, and stairs for visually impaired people. Alex-Net, a pre-trained model, is used to extract deep features for the classification of complex images that cannot be classified using simple handcrafted features [6].

5.3.9 Xception Model

Francois Chollet proposes the Xception Model, which is an extension of the inception architecture in which depth-wise separable convolutions replace

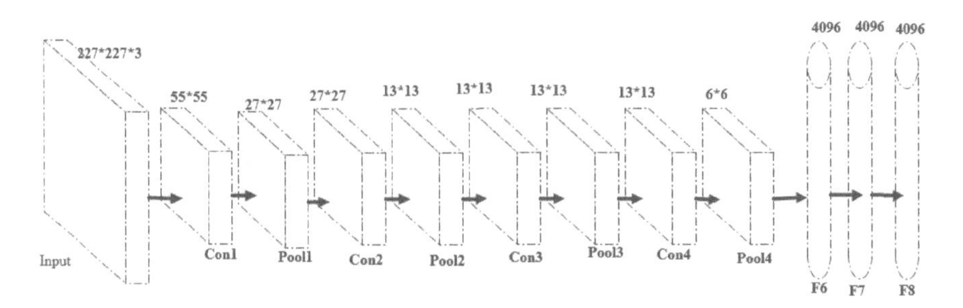

Figure 5.5 Alex-Net architecture.

the normal inception modules. It is a 71-layer deep pre-trained version of a convolutional neural network trained on more than a million photos from the ImageNet database. The network can categorize images into 1000 object categories, like pillar, car, stair, and a variety of other obstacles. As a consequence, the network has learnt rich feature representations for a wide range of images, which may be used to categorize barriers for persons who are visually impaired.

5.3.10 Visual Geometry Group (VGG16,19)

ImageNet, a massive visual database project utilized in visual object recognition software, uses VGG16, a basic and widely used convolutional neural network (CNN) architecture. Karen Simonyan and Andrew Zisserman of the University of Oxford created and launched the VGG16 architecture in their essay "Very Deep Convolutional Networks for Large-Scale Image Recognition" in 2014. The term "VGG" stands for Visual Geometry Group, a group of scholars at the University of Oxford who designed this architecture, and the number "16" indicates that there are 16 levels in this architecture. Many deep learning image classification approaches employ VGG16, which is popular owing to its ease of use. Because of its advantages, VGG16 is often utilized in learning applications. VGG16 is a CNN architecture that won the ImageNet Large Scale Visual Recognition Challenge (ILSVRC) in 2014 and is still considered one of the best vision architectures available today. VGG19 has

19 layers, including three blocks of extra convolution layers. Because of their deep layers, both VGG16 and VGG19 succeed in the image competition [10].

5.3.11 Residual Neural Network (ResNet)

A residual neural network (ResNet) is a form of artificial neural network (ANN) that is based on the cerebral cortex's pyramidal cell structures. Skip connections, sometimes known as shortcuts, are utilized by residual neural networks to bypass certain layers. The bulk of ResNet models include double- or triple-layer skips interspersed with nonlinearities (ReLU) and batch normalization. An additional weight matrix can be used to learn the skip weights, which is recognized as HighwayNets. Skipping connections is done for two reasons: first, to prevent the issue of vanishing gradients, and, second, to avoid the deterioration that occurs when adding extra layers to a sufficiently deep model leads to increased training error. ResNet architecture comes in a variety of forms, each with the same principle but a different number of layers. ResNet-18, ResNet-164, ResNet-34, ResNet-101, ResNet-1202, ResNet-110, ResNet-152, ResNet-164, ResNet-50, and others are examples of variations [9, 35].

5.3.12 Inception (V2, V3, InceptionResNet)

GoogLeNet [34], the winner of the ImageNet Large Scale Visual Recognition Challenge (ILSVRC) 2014, has 22 layers and an inception network. It also emphasizes the significance of depth. GoogLeNet, on the other hand, uses 12 times fewer parameters than Alex-Net and achieves human-like performance. Inception V2 researchers compute the mean and standard deviation of all feature maps at the output of a layer and use these values to normalize the responses. Later, Inception V3 is created by carefully building networks and using 3*3 and 1*1 filters rather than other.

Google provides InceptionResNet, which was inspired by ResNet's performance. Residual is added to the output of the convolution operation of the inception module. After convolution, the depth is increased, and this model achieves a top-5 error on ImageNet classification. With 467 layers, InceptionResNet combines the notions of an inception network with a deep residual network to speed up training and improve accuracy.

5.3.13 MobileNet

MobileNet employs depth-wise separable convolutions. Depth-wise convolution and point-wise convolution are the two techniques that make up a depth-wise separable convolution. The input and output channels, as well as the spatial dimension of the feature maps, are all affected by traditional convolution. Each input channel is individually transferred to a single convolution in a depth-wise convolution. As a result, the number of output channels and input channels are equal. MobileNet also offers two more parameters that may be used to reduce the number of activities even further: The width multiplier (which can be anywhere between 0 and 1) lowers the number of channels.

The sole difference between MobileNetV2 and the original MobileNet is that it exclusively employs inverted residual blocks with bottlenecking characteristics. It has a significantly fewer number of parameters than the original MobileNet. Any image size bigger than 32 $\times$ 32 is supported by MobileNets, with larger image sizes providing better performance.

5.3.14 DenseNet

DenseNet was created primarily to address the effect of disappearing gradients on the accuracy of high-level neural networks. Simply said, information evaporates before it reaches its destination due to the longer travel between the input and output levels.

5.3.15 Experimental Results Analysis

The design has been experimentally examined. In this section, the following results explain how the desired path can be selected and how the obstacle detection system is working. Finally, the results show how the control system works and experiment with it. Also, the experiment represents the recognition of four different obstacles (i.e., vehicles, doors, pillars, and stairs) for visually impaired people. Four different performance metrics are used to evaluate the proposed method. These performance metrics are accuracy, $F1$-score, precision, and recall and are given in the following equations:

$$\text{Accuracy} = \frac{\text{TN} + \text{TP}}{\text{TN} + \text{TP} + \text{FN} + \text{FP}} \tag{5.1}$$

$$\text{Recall} = \frac{\text{TP}}{(\text{TP} + \text{FN})} \tag{5.2}$$

$$\text{Precision} = \frac{\text{TP}}{(\text{TP} + \text{FP})} \tag{5.3}$$

$$F1 - \text{Score} = 2\frac{\text{Precision} * \text{Recall}}{\text{Precision} + \text{Recall}} \tag{5.4}$$

The suggested system's accuracy, precision, recall, and $F1$-score were computed using the various measurement parameters. Table 5.2 shows that the proposed method achieved the best performance with an accuracy of 99.63%, a precision of 99.59%, a recall of 99.31%, and an $F1$-score of 99.56%.

The proposed model is trained on a pre-trained Caffe Alex-Net model which has the best validation accuracy. The validation accuracy of the model during training is 99.5139% and has a loss validation of 0.0201% as shown in the below graph. Based on the results, the Alex-Net model is chosen and a computer-aided detection (CAD) system is developed using Arduino UNO, ultrasonic sensors, and camera. The research proposed is compared with the present research and is shown in Table 5.3.

The confusion matrices for the classification of deep learning models were calculated for the activities of VIP. For each activity, a comparison is made with performance matrices such as preciseness, precision, recalls, and $F1$ score.

The experimental outcomes of the suggested VIP methods can be illustrated in Figure 5.6. The four types of obstacles (obstacles1, obstacles2, obstacles3, and obstacles4) are recognized from the pre-trained Alex-Net architecture.

Table 5.2 Performance results obtained from pre-trained CNN models (Alex-Net)

Categories	Accuracy	Precision	Recall	$F1$-score
Obstacles1	99.61%	99.04%	100%	99.51%
Obstacles2	99.70%	99.35%	100%	99.67%
Obstacles3	99.23%	100%	98.23%	99.10%
Obstacles4	100%	100%	100%	100%
Total	99.63%	99.59%	99.31%	99.56%

Table 5.3 Performance results comparison with previous work

Categories	Accuracy	False alarm rate	Precision	Recall	$F1$-score
[13]	97%	0.016%	NA	NA	NA
Assistive algorithm (proposed)	**99.56%**	**0.0201%**	**99.59%**	**99.31%**	**99.5%**

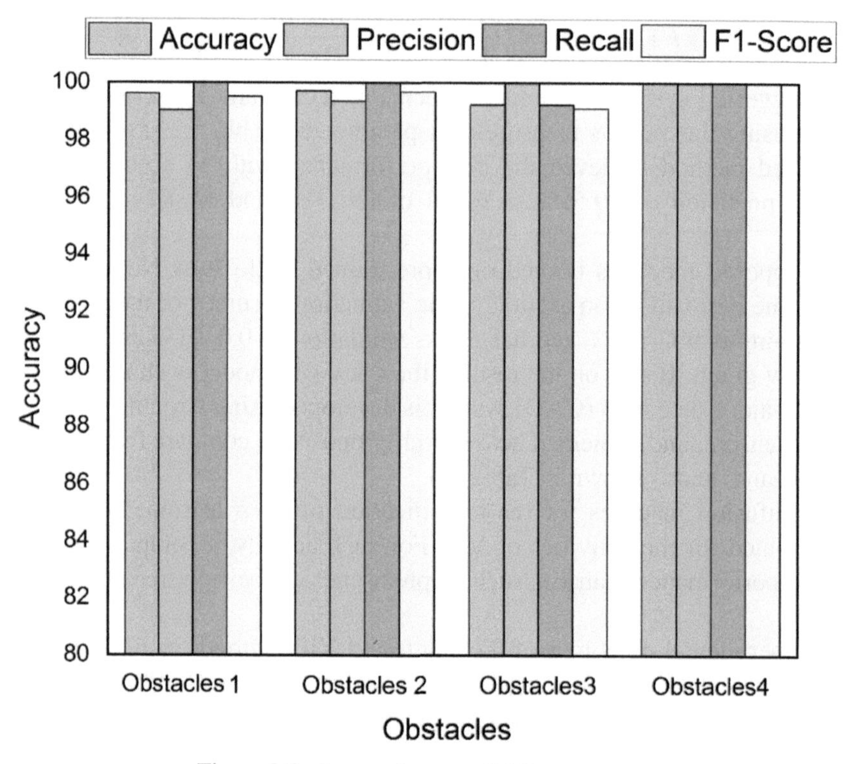

Figure 5.6 Loss and cross-validation accuracy.

Confusion Matrix				
403 23.3%	0 0.0%	0 0.0%	0 0.0%	100% 0.0%
8 0.5%	462 23.3%	0 0.0%	0 0.0%	98.3% 1.7%
10 0.6%	6 0.3%	389 22.5%	0 0.0%	96.0% 4.0%
0 0.0%	0 0.0%	0 0.0%	453 26.2%	100% 0.0%
95.7% 4.3%	98.7% 1.3%	100% 0.0%	100% 0.0%	98.6% 1.4%

Figure 5.7 Confusion matrix.

5.4 Conclusion and Future Directions

This paper proposes a high-level framework for interpreting semantic items in the physical environment utilizing real-time localization methods. With the development of such an intelligent system, we introduce the operating concept that allows users to travel on their own. Our system's capability can be seen both indoors and outdoors without the aid of the guidance of a sighted person. For this purpose, the items are detected in three directions using three sensors (i.e., front, left, and from right) and one is used to detect pothole. The visual sensor is used for the recognition of obstacles in the way of smart cane users. Overall, the proposed design system provides advantages instead of traditional cane. Improvements could be done to make sure that the system more efficient and effective as compared to the currently proposed design. In the future, we aim to install GPS which helps the visually impaired person in an outdoor location that helps their relatives to find them easily and provide a guideline. Also, the proposed system presented thus far provides accurate detection but to provide intelligent directing in terms of obstacle avoidance; it is strongly advised that a newly created neuro-fuzzy control algorithm be programmed into the microcontroller. Another suggested technique is to use excellent outdoor navigation guiding system; the created system might be used with RFID. Last but most import recommendation is to include battery monitoring circuit in the system due to the high-power consumption of the designed system. The precision of obstacle detection will be hampered by a lack of current supply.

Acknowledgements

To do research is a tedious and tough job which cannot be completed in mere isolation. The help and support of friends and colleagues and the guidance of teachers are of great importance in this process. I express my deepest gratitude to those who helped me in this painstaking journey. I would also like to thank Prof. Xin Chen, Ph.D. Associate Dean, School of Automation, China University of Geosciences, Wuhan, Hubei Province, China, who agreed and extended his kind supervision to me for conducting my research work. Finally, I would like to thank my entire family members who have supported me financial for doing the master's degree in computer science.

References

[1] "World report on vision," Who.int, 2020. [Online]. Available: https://www.who.int/publications-detail/world-report-on-vision.

[2] M. M. D'souza, A. Agrawal, V. Tina, V. HR, and T. Navya, "Autonomous walking with guiding stick for the blind using echolocation and image processing," *Methodology*, vol. 7, no. 05, 2019.

[3] P. H. Batavia and S. Singh, "Obstacle detection using adaptive color segmentation and color stereo homography," *in Proceedings of the 2001 ICRA. IEEE International Conference on Robotics and Automation (Cat. No. 01CH37164)*, vol. 1, IEEE, 2001, pp. 705–710.

[4] J. A. Hesch and S. I. Roumeliotis, "Design and analysis of a portable indoor localization aid for the visually impaired," *The International Journal of Robotics Research*, vol. 29, no. 11, pp. 1400–1415, 2010.

[5] V. Pradeep, G. Medioni, and J. Weiland, "Robot vision for the visually impaired," *in Proceedings of the 2010 IEEE Computer Society Conference on Computer Vision and Pattern Recognition-Workshops, IEEE*, 2010, pp. 15–22.

[6] A. Rodríguez, L.M. Bergasa, P.F. Alcantarilla, J. Yebes, and A. Cela, "Obstacle avoidance system for assisting visually impaired people," *in Proceedings of the IEEE Intelligent Vehicles Symposium Workshops*, vol. 35, Madrid, Spain, 2012, p. 16.

[7] F. Shahdib, M. W. U. Bhuiyan, M. K. Hasan, and H. Mahmud, "Obstacle detection and object size measurement for autonomous mobile robot using a sensor," *International Journal of Computer Applications*, vol. 66, no. 9, 2013.

[8] S. Maidenbaum *et al.*, "The "EyeCane", new electronic travel aid for the blind: Technology, behavior & swift learning," *Restorative neurology and neuroscience*, vol. 32, no. 6, pp. 813–824, 2014.

[9] T.-S. Leung and G. Medioni, "Visual navigation aid for the blind in dynamic environments," *in Proceedings of the IEEE Conference on Computer Vision and Pattern Recognition Workshops*, 2014, pp. 565–572.

[10] J. Xiao, S. Joseph, X. Zhang, B. Li, X. Li, and J. Zhang, "An assistive navigation framework for the visually impaired," *IEEE Transactions on Human-Machine Systems*, vol. 45, no. 5, pp. 635–640, 2015. Available: 10.1109/thms.2014.2382570.

[11] S. Suba Raja, M. Vivekanandan, and S. Usha Kiruthika, "Design and implementation of facial recognition system for visually impaired using image processing," *International Journal of Recent Technology and Engineering*, vol. 8, no. 4, pp. 4803–4807, 2019. Available: 10.35940/ijrte.d7676.118419.

[12] M. Vlaminck, Q. L. Hiep, H. Vu, P. Veelaert, and W. Philips, "Indoor assistance for visually impaired people using an RGB-D camera," *in Proceedings of the 2016 IEEE Southwest Symposium on Image Analysis and Interpretation (SSIAI), IEEE*, 2016, pp. 161–164.

[13] E. Ko and E. Yi Kim, "A vision-based wayfinding system for visually impaired people using situation awareness and activity-based instructions," *Sensors*, vol. 17, no. 8, pp. 1882, 2017. Available: 10.3390/s170 81882.

[14] A. Ramadhan, "Wearable smart system for visually impaired people," *Sensors*, vol. 18, no. 3, pp. 843, 2018. Available: 10.3390/s18030843.

[15] A. Nivedita, M. Sindhuja, G. Asha, R. Subasree, and S. Monisha, "Smart cane navigation for visually impaired," *International Journal of Innovative Technology and Exploring Engineering (IJITEE) ISSN*, pp. 2278–3075, 2019.

[16] M. M. D'souza, A. Agrawal, V. Tina, V. HR, and T. Navya, "Autonomous walking with guiding stick for the blind using echolocation and image processing," *Methodology*, vol. 7, no. 05, 2019.

[17] W. Elmannai and K. Elleithy, "Sensor-based assistive devices for visually-impaired people: current status, challenges, and future directions," *Sensors*, vol. 17, no. 3, pp. 565, 2017. Available: 10.3390/s170 30565.

[18] C. Ntakolia, G. Dimas, and D. Iakovidis, "User-centered system design for assisted navigation of visually impaired individuals in outdoor cultural environments," *Universal Access in the Information Society*, 2020. Available: 10.1007/s10209-020-00764-1.

[19] Y. Abdeen Abdalsabour and M. Mohammed Abdulwahab, "Design of intelligent system for visually impaired to access the required location," *Journal of Telecommunication, Electronic and Computer Engineering*, vol. 12, no. 4, pp. 15–19, 2021.

[20] A. Krizhevsky, I. Sutskever, and G.E. Hinton, "Imagenet classification with deep convolutional neural networks," *in Proceedings of the Advances in Neural Information Processing Systems*, 2012, pp. 1097–1105.

[21] K. Simonyan and A. Zisserman, "Very deep convolutional networks for large-scale image recognition," 2014, arXiv preprint arXiv:1409.1556.

[22] K. He, X. Zhang, S. Ren, and J. Sun, "Deep residual learning for image recognition," *in Proceedings of the IEEE Conference on Computer Vision and Pattern Recognition*, 2016, pp. 770–778.

[23] C. Szegedy, V. Vanhoucke, S. Ioffe, J. Shlens, and Z. Wojna, "Rethinking the inception architecture for computer vision," *in Proceedings of the IEEE Conference on Computer Vision and Pattern Recognition*, 2016, pp. 2818–2826.

[24] S. Xie, R. Girshick, P. Doll´ar, Z. Tu, and K. He, "Aggregated residual transformations for deep neural networks," *in Proceedings of the IEEE Conference on Computer Vision and Pattern Recognition*, 2017, pp. 1492–1500.

[25] M. Imad et al., "Pakistani Currency Recognition to Assist Blind Person Based on Convolutional Neural Network", Journal of Computer Science and Technology Studies: Vol. 2 No. 2, 2020

[26] M. Abul Hassan, S. Irfan Ullah, A. Salam, A. Wajid Ullah, M. Imad and F. Ullah, "Energy efficient hierarchical based fish eye state routing protocol for flying Ad-hoc networks", Indonesian Journal of Electrical Engineering and Computer Science, vol. 21, no. 1, p. 465, 2021. Available: 10.11591/ijeecs.v21.i1.pp465-471.

[27] Faiza, S. Irfan ullah, A. Salam, F. Ullah, M. Imad and M. Abul Hassan, "Diagnosing of Dermoscopic Images using Machine Learning approaches for Melanoma Detection," 2020 IEEE 23rd International Multitopic Conference (INMIC), 2020, pp. 1-5, doi: 10.1109/INMIC50486.2020.9318114.

[28] F. Ullah et al., "Navigation System for Autonomous Vehicle: A Survey", Journal of Computer Science and Technology Studies, vol. 2, no. 2, pp. 20-35, 2020.

[29] F. Ullah et al., "Real Time Road Blocker Detection and Distance Calculation for Autonomous Vehicle Based on Camera Vision", Asian Journal of Applied Science and Technology, vol. 04, no. 03, pp. 100-108, 2020. Available: 10.38177/ajast.2020.4314.

[30] F.Ullah et al., "A Vision Based Road Blocker Detection and Distance Calculation for Intelligent Vehicles", International Journal of Computer Science and Information Security (IJCSIS) 18.6 (2020).

[31] F. Ullah et al., "An Efficient Machine Learning Model Based on Improved Features Selections for Early and Accurate Heart Disease Predication", Computational Intelligence and Neuroscience, vol. 2022,

pp. 1-12, 2022. Available: 10.1155/2022/1906466 [Accessed 25 July 2022].

[32] M. Imad et al., "Pakistani Currency Recognition to Assist Blind Person Based on Convolutional Neural Network", Journal of Computer Science and Technology Studies, vol. 2, no. 2, pp. 12-19, 2022.

[33] N. Ullah et al., "Performance Analysis of POX and RYU Based on Dijkstra's Algorithm for Software Defined Networking", European, Asian, Middle Eastern, North African Conference on Management & Information Systems, pp. 24-35, 2020.

[34] I. Ahmad, "Reinforcement Learning-Based Coexistence Interference Management in Wireless Body Area Networks", International Journal of Computer and Systems Engineering, vol. 14, no. 11, pp. 446-453, 2020.

[35] M. Imad, "Automatic Detection of Bullet in Human Body Based on X-Ray Images Using Machine Learning Techniques", International Journal of Computer Science and Information Security (IJCSIS), vol. 18, no. 6, 2020.

6

Application of IoT-Enabled CNN for Natural Language Processing

Ashish Kumar[1], Rishab Mamgai[2], and Rachna Jain[3]

[1]School of Computer Science Engineering and Technology, Bennett University, Greater Noida, India
[2]Bharati Vidyapeeth's College of Engineering, India
[3]Bhagwan Parshuram Institute of Technology, India

Abstract

Internet of Things (IoT) based systems are used to define communication systems based on machine-to-machine interaction. IoT when integrated with convolution neural network (CNN) can provide a system that can communicate with surroundings using human speech. Natural language processing (NLP) can interact with IoT-based deep learning systems to provide development in the automation field. IoT can connect a network of specific devices and exploit deep learning for feature extraction, namely sensor features, radio frequency features, and speech features. IoT with NLP can develop speech-based recognition systems for home automation systems. Smart home applications can be integrated with voice-command-based IoT devices to communicate specific commands to the devices. In addition, NLP-based IoT devices can help disabled people to perform their daily activities. These devices can monitor their health and provide voice-based security alerts. Also, NLP-enabled IoT devices can be helpful for automating environmental data collections which include geographical activities. However, NLP-based IoT implementation has certain limitations, namely language understanding, change in accent, and change in voice. These challenges restrict the efficient and quick utilization of NLP-based IoT devices. The deep learning technology with a big vocabulary database has provided numerous opportunities to train the voice and command recognition system in IoT. IoT-enabled CNN devices for voice recognition act as a boon to society.

Keywords: IoT, deep learning, NLP, home automation.

6.1 Introduction

Internet of Things (IoT) based devices are used to automate configuration systems based on human speech interaction [1, 2]. The interaction can be either with the help of body gestures or voice commands. Emerging trends in deep learning technologies and cloud computing have improved the performance of IoT devices integrated with either voice recognition system or gesture control systems. With the use of these advanced IoT devices, users are now able to perform their daily computing tasks more easily and efficiently. Generally, IoT devices include sensors, control devices, and routing devices. Also, IoT devices contain software focusing on networking, embedded systems, and middleware. All these devices when configured together then contribute a lot in automating the task with the support of deep learning technologies. The integration of natural language processing (NLP) in IoT devices can improve the ability and efficiency of the overall configuration system. Now, the devices can be trained to execute the real-time commands rather than to execute specific task based on some pre-fed commands.

In recent years, IoT has given many applications ranging from development of modern infrastructure and smart cities to different sectors like agriculture, healthcare, home automation, transportation, education, etc. [3]. The main power to these applications is provided by the smart learning mechanism for prediction, pattern recognition, data extraction, or data analytics. There are many machine learning approaches and algorithms which can be employed for the intelligent learning mechanism, but, in recent years, deep learning is the one which is mainly employed for many IoT applications. The imperative reason behind this dominance of deep learning in the IoT devices may be due to the emerging need of analytics which cannot be fulfilled by the traditional machine learning approaches. Instead, there is a need for a variety of modern methods of data analysis, artificial intelligence, and NLP for handling big data from modern day IoT systems. Figure 6.1 illustrates the details of various application areas of NLP-enabled deep-learning-based IoT devices.

With advancement in technology, IoT will make a significant annual economic impact of about $2.7 to $6.2 trillion by 2025 by the McKinsey's report on the global economic impact of IoT [4]. This study predicted that the

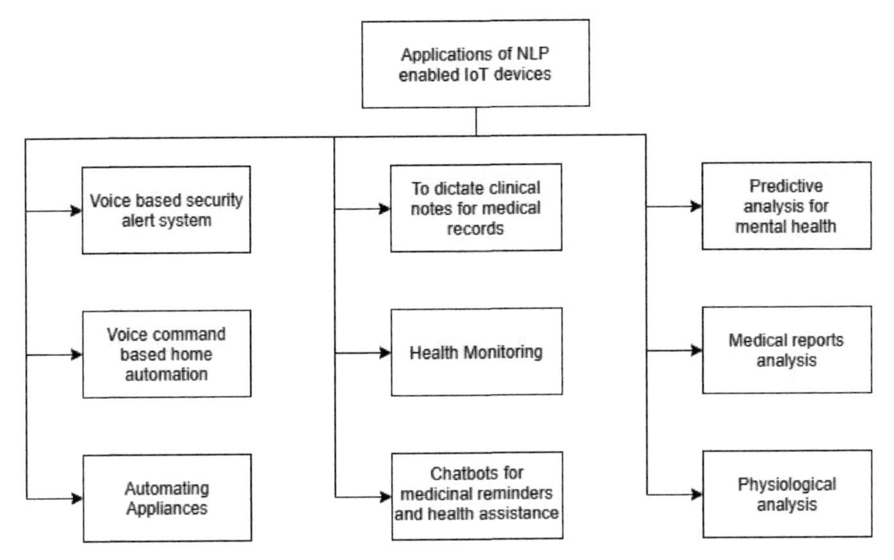

Figure 6.1 Details of various applications of deep-learning-based NLP-enabled IoT devices.

healthcare sector will be a major contributor toward it, followed by industry and energy. On the other hand, transportation, security, urban infrastructure, agriculture, and retail will be the minor contributors but altogether contributing a good chunk to the annual economic impact. Also, the report defines impact of machine learning under automation of knowledge work. The report states that advances in ML techniques, such as deep learning and neural networks, are major contributors in automation. Others like speech and gesture recognition will also be highly benefitting ML technologies. In the report, an estimate of $5.2 trillion to $6.7 trillion worth of potential economic impact of knowledge work automation has been made for the year 2025 for IoT systems.

IoT systems with the help of deep learning technologies and NLP devices can provide a system that can communicate with surroundings using human speech. NLP can interact with IoT-based deep learning systems to provide development in the automation field. IoT can connect a network of specific devices and exploit deep learning for feature extraction, namely sensor features, radio frequency features, and speech features. IoT with NLP can develop speech-based recognition systems for home automation systems. Smart home applications can be integrated with voice-command-based IoT devices to communicate specific commands to the devices [5]. In addition, NLP-based IoT devices can help disabled people to perform their daily activities. These devices can monitor their health and provide voice-based security alerts. Also, NLP-enabled IoT devices can be helpful for automating environmental data collections which include geographical activities. However, NLP-based IoT implementation has certain limitations, namely language understanding, change in accent, and change in voice. These challenges restrict the efficient and quick utilization of NLP-based IoT devices. The deep learning technology with a big vocabulary database has provided numerous opportunities to train the voice and command recognition system in IoT. In sum, IoT-enabled CNN devices for voice recognition act as a boon to society. These latest technologies when embedded in IoT devices can not only improve the efficiency of the existing system but also provide benefits to the users. Real-time voice-command-based and gesture-control-based IoT devices provide world class experience to the people.

6.2 Related Work

In the past few years, a lot of work has been proposed in IoT and deep learning to find out the possible applications of combined implementation of both the fields. Table 6.1 tabulates the representative work using deep

Table 6.1 Representative work under various gesture recognition systems with the performance evaluation

Authors	Gesture	Performance metrics	Summary
Wang *et al.* [10]	Lip motion	91% accuracy	WiHear, a system to hear people talk, using Wi-Fi signals.
Abdelnasser *et al.* [11]	Hand gestures	90% accuracy	WiGest, uses Wi-Fi RSSI to detect hand gestures.
Liu *et al.* [12]	Keystrokes	77.43% accuracy, 30 training samples per key And 93.47% accuracy, 80 training samples per key	This system is based on Wi-Fi signals.
Qian *et al.* [13]	Motion direction	92% accuracy	WiDance, Wi-Fi CSI-based motion direction sensing.
Virmani *et al.* [14]	Activity recognition	91.4% accuracy	WiAg, Wi-Fi based configuration recognition system and virtual samples generated for all possible gestures using translation function.

learning techniques based on gesture recognition. The authors have proposed novel architectures for implementing deep-learning-based IoT systems. In this direction, Gladence *et al.* [6] proposed a significant progress to realize intelligent environments which are capable of providing needed services automatically for user comfort by detecting user's actions and gestures. A system which automatically turns on the AC when a person enters and sits in a room has been proposed. The basic idea behind this system was to automatically turn appliances on/off based on user's activity. In this way, IoT automation can help in saving energy along with keeping up with the needs of humans. Mohammadi *et al.* [7] performed research primarily in deep learning applications in IoT for big data and streaming analytics. The authors discussed about different deep learning techniques with an overview of application of these techniques in the field of IoT [7, 8]. Various frameworks have been proposed by exploiting deep learning with other machine learning approaches such as reinforcement learning, online learning, transfer learning, etc. In addition, applications of deep learning in IoT were discussed for different sectors such as smart city, energy, intelligent transportation system, etc. Jiang *et al.* [9] studied Wi-Fi sensing applications in health monitoring and gesture recognition. The authors provided a brief review of Wi-Fi enabled gesture recognition, where a comparison between various gesture recognition systems has been discussed along with their accuracies. The research predicted that gestures like lip motion, hand gestures, motion direction, and keystrokes can be recognized with the help of Wi-Fi based gesture recognition system. The system achieved an exceptional accuracy of around 90%.

Hussain *et al.* [3] reviewed the applications of machine learning techniques in IoT security challenges and threats. In this, reinforcement learning, deep learning, and their applications in different security problems in IoT network were studied. Machine learning based solutions for IoT security have been provided to address the challenges and threats. Mandula *et al.* [15] analyzed the utilization of IoT in home automation using an Android mobile app and micro-controller-based Arduino board. The authors proposed two models for home automation, one of which used Bluetooth and the other one used ethernet for network communication. Due to limitations of Bluetooth connectivity range, the first model was utilized for indoor domain, whereas ethernet-based model was capable of handling from outside situations. Xiao *et al.* [16] reviewed the ML-based methods for data privacy and security in the IoT. In their study, they also identified several challenges that have to be addressed in ML-based security techniques in practical IoT systems such as partial state observation, computation and communication overhead, backup, and security solutions.

In sum, a lot of work has been proposed using machine-learning-based technique for home automation, addressing the security-related challenges and threats. In addition, gestures-based systems were also utilized to make the experience simpler and useful. The next section will review the deep learning work inspired by voice automation system using IoT.

6.3 IoT-Enabled CNN for NLP

IoT is a system in which various devices having capability to operate over Internet are connected together for performing one or multiple operations such as processing, signaling, sensing, recognition, etc. [17]. As Internet has evolved a lot in past years, it has become simpler and faster to communicate between devices and computers which have led to a major development in the field of IoT, providing with the faster and effective way of information retrieval and automation. There has been a lot of development in the field of machine learning, especially in deep learning where the CNNs have been applied for pattern recognition, image processing, NLP, object tracking, etc. Research works in the past few years have helped a lot in making smart systems, where the devices are now able to detect objects and people, patterns in images, and voice. So, the researches in both fields have opened new ways of automation where we can automate our day-to-day tasks with the help of smart devices using machine learning algorithms for executing tasks along with the IoT devices in physical surroundings. Here, further deep learning with IoT devices, more specifically, applications of CNN-based IoT devices, has been tabulated in Table 6.2.

It has been evident that CNN is the most versatile in comparison to DNN model when applied to the IoT domains [27, 28]. CNN has the capability to recognize even smaller patterns in the data provided which helps it to outperform other deep learning models not just in IoT domain but also in many of the other applications.

6.4 Applications of IoT-Enabled CNN for NLP

This section presents a review of applications of IoT-enabled CNN for NLP. IoT is a major contributor to home automation. Many technologies like Bluetooth [31], GSM [32–34], Zigbee, etc., are used in home automation techniques. GSM is an ideal remote communication technology where

Table 6.2 Details of various deep learning models and the application description

Application	CNN	LSTM	RBM	RNN
Image recognition	CNN for plant disease Recognition. Overall, 96.3% accuracy [18].	Model was based on two-layered LSTM [19].	–	Model achieved 66% top-1 accuracy and 80% top-3 accuracy [20].
Physiological detection	Used a model based on five convolution layers and three pooling layers [21].	Model combines CNN and LSTM for activity recognitions [22].	–	RNN improved the score of the proposed architecture [23].
Localization	Used faster RCNN for integrating feature extraction and classification into one network [24].	–	RBM-based indoor fingerprinting scheme using CSI [25].	–
Smart home	Gesture controlled home automation system achieved an accuracy of 98.12% [26].	–	–	–
Healthcare	CNN with ten convolution layers and two fully connected layers was used to detect cardiovascular disease from mammograms [27].	–	RBM-based posture analysis in fall detection and classification [28].	–
Sports	AlexNet-based hierarchical model for group activity recognition on volleyball dataset [29].	–	–	RNN model for classifying NBA offensive plays [20].

(Continued)

Table 6.2 Continued.

Application	CNN	LSTM	RBM	RNN
IoT infrastructure	CNN for identifying wireless network interference [30].	–	–	–

Internet is not available. Wi-Fi technology due to its communication speed and availability in almost every part of the world is preferred for NLP [35, 36] and other deep learning applications. The advancements in Internet and today's technologies like cloud computing, artificial intelligence, and wireless networking have opened more fields for IoT to contribute. IoT along with deep learning has provided with many applications in healthcare domains such as medical diagnosis, disease prediction, and personal healthcare applications. Major application in healthcare is in physiological domain where the user's health data can be collected through smartphone sensors or other wearable devices with the help of compact IoT devices by implementing motion recognition/detection techniques with the help of deep learning. The great causes for home automation are to assist shape lives of humans with extreme disabilities [37]. NLP-enabled home automation systems can help assist people who are physically challenged but can speak and listen, hence making them dependent on other people. Figure 6.2 illustrates the architecture of basic home automation system based on speech recognition. Further, in this section, we will focus on home automation and how it can be implemented to help disabled people with NLP-based IoT systems.

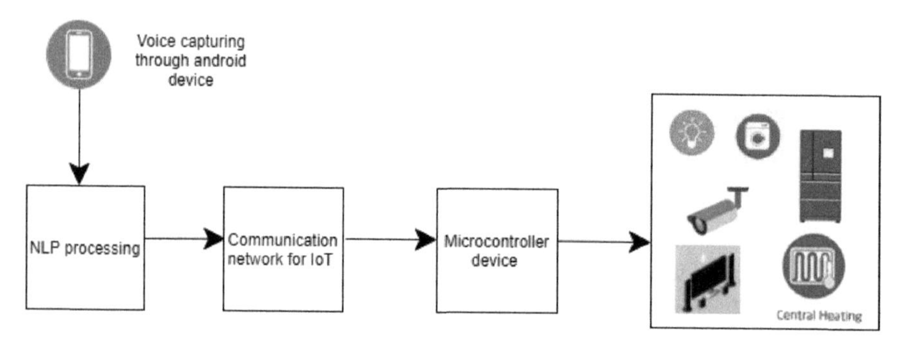

Figure 6.2 Basic architecture for speech-recognition-based IoT automation. The basic flow of information with the details of hardware devices has been shown.

Table 6.3 Details of various deep learning models with the description in IoT applications

Deep learning model	Type of model	IoT applications	Summary
RNN	Supervised	Identify movement pattern Behavior detection [38]	Useful for time-dependent data
CNN	Supervised	Traffic sign detection [39]	Visual tasks such as pattern recognition
LSTM	Supervised	Human activity recognition [40]	Applicable on problems with time-series data, where there is long time lag in data
RBM	Unsupervised, Supervised	Indoor localization [41]	Feature extraction and dimensionality reduction
Auto-encoders	Unsupervised	Fault detection [42]	Training data with noise; noise is removed by encoding data into a complex function for refinement

6.4.1 Home Automation

IoT has a major application in smart home automation. IoT-based systems are used in home automation mainly realized in day-to-day appliances such as fridge, TV, air conditioners, and heating systems for ease of controlling. IoT has provided us with a remote access to our home appliances, such as remotely maintaining the home temperature and monitoring surrounding changes. Indoor localization is a popular topic in home automation [39, 40]. Indoor localization is an alternative technique for GPS. Satellite technologies such as GPS lack precision in places like airports, parking garages, buildings, and underground locations. Indoor localization uses a network of devices to locate people and objects in small places where the technologies like GPS lose precision or fail to track [41, 42]. Indoor localization enables many services like intrusion detection, person monitoring in a localized place, etc. Traditional indoor positioning systems use K-nearest neighbors, Bayesian model, or SVMs.

Paramvir and Venkata [43] proposed an RF-based user location and tracking system. They collected signal strength information for 70 distinct physical locations in all four directions on the floor. For basic analysis, only single nearest neighbor in signal space was selected. Further KNN was applied to signal space for finding the location of the user. The system offers better accuracy for smaller values of k in comparison to its large values. This may be due to deviation of true location due to averaging of signals over the signal space. As the data keeps evolving and is getting bigger, this poses a problem with the traditional machine learning approach in indoor localization. So, the researchers have turned toward deep learning. In this direction, Gu *et al.* [44] proposed a model for Wi-Fi based indoor localization which used deep learning to improve feature classification. To ensure fast learning speed, semi-supervised extreme learning machine (ELM) was also introduced. Gesture-based recognition system has been proposed for home automation with IoT. Mainly, NLP-based smart home automation systems use a normal sound receiving system like through a microphone. The sounds or voice instructions can be collected through a microphone of a smart phone or through a microphone mounted on a micro-controller of an IoT device in the network. Digital personal assistants such as Apple's Siri, Google Assistant, Amazon Alexa, Microsoft's Cortana, etc., are all based on deep learning and have shown great success [45]. They operate on voice commands given by the user. Assistants like these can be successfully implemented within the IoT networks for taking voice commands to perform a series of tasks. Modern day NLP is not just limited to recognizing the instructions given through text or speech, but along with speech recognition, it is also capable of recognizing the person whose voice is being used to give the instructions, hence providing a way of keeping the IoT-enabled devices secured and also reducing misuse of the devices by an unauthorized person. So, this can be seen as an advantage of NLP-based IoT systems as they will be able to take commands and recognize if the user is authorized to use the device in the network; only then the given command will be taken as input by the IoT device to initiate the required action.

Hence, integration of NLP with IoT could be a further great development in the field of home automation where NLP can provide us with a way to use short messages or voice notes to interact with home appliances such as AC and refrigerators. CNN-based IoT architecture can be majorly implemented for home security where CNN models integrated with IoT systems can be used for tracking activity inside and outside the house. For example, an automated face detection model can be integrated with IoT system for automatically opening the doors for the people authorized for entering the premises. As the mobile technology is advancing quickly, indoor localization has become a popular research topic. Also, the technology is not limited to home automation systems, but a lot of work has been proposed to make disabled life easier. The next section will detail about the existing work which helps to improve the life experience to disabled people.

6.4.2 Boon for Disabled People

Home automation is not just limited to automating appliances. Physically challenged people are much more dependent on others for their daily activities. NLP-enabled IoT solutions can surely help ease of their lives. In the past years, monitoring health of a patient or a disabled person was a difficult task for doctors and family members. The advancements in IoT have enabled us to develop micro-devices for health monitoring which has improved the quality of services for a patient or disabled person. Many researchers have proposed and developed android-based models which use IoT devices in physical environment for tracking health information of disabled people such as heart-rate, oxygen level, body temperature, etc. [46–48]. These devices use Bluetooth to share information over close proximity and have been integrated with speech-based SMS system for sharing messages and connecting with the care taker as shown in Figure 6.3. Techniques like dictionary search are

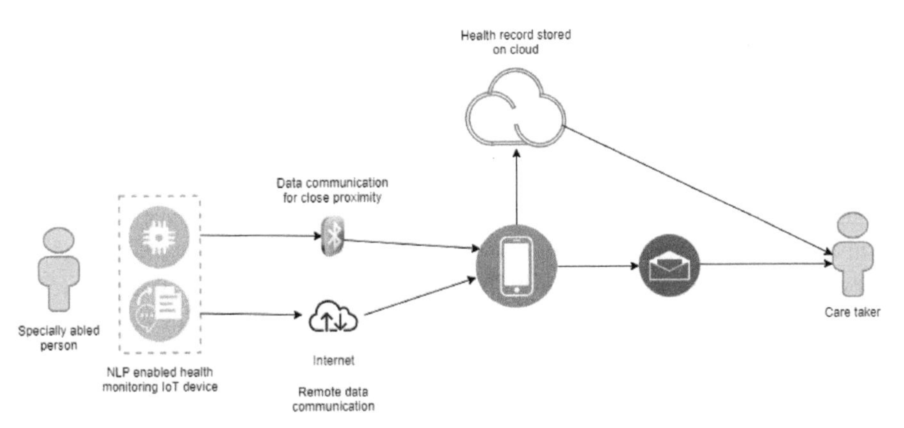

Figure 6.3 IoT-based model for health monitoring. NLP-enabled health monitoring devices have been integrated with connecting services like Internet or Bluetooth. Details of health record are saved on cloud for remote access by health organizations.

also being integrated with the speech recognition system for personalizing the health assistant for better results in remote communication between a disabled person and the care taker [49].

IoT devices can be integrated with deep learning for supporting physically challenged people for the ease of their life, making them independent for daily and frequent activities such as opening doors or windows, switching appliances on/off based on voice commands, etc. Gonzalez *et al.* [50] proposed a device for enhancing the quality of life of visually disabled people. However, the device was capable of performing many tasks such as email reading, medication reminder, music player, and meditation reminder. But the major functionality of this device was recognizing people and detecting any obstacle in the way. The device uses ultrasonic sensors for obstacle detection, which sends and receives wave signals for processing by the Raspberry PI module for detecting objects. For recognizing people, the device is fitted with a video camera which sends images captured to the computer vision module for face detection and recognition which are done with the help of Haar Wavelet and Fisher Faces, respectively. Another important research with this is proposed by Bhargava *et al.* [51]. The main aim of this research was to develop an image to text to speech system for the betterment of visually impaired. The author proposed an IoT system integrated with image processing and NLP module. The system operates on Raspberry PI with a camera module for image acquisition. The function of the system was to detect the text from the images using image processing techniques and convert it to text using OCR. The detected text is then saved for future reference or to be processed by the NLP module for dictating the information present in the image to the visually impaired people.

Automated doors can be implemented in home and other places for assistance of disabled people. For example, implementation of deep learning with IoT devices can help automating opening and closing doors based on voice commands or by detecting motion around them by providing an easy access to disabled people for entering the house without any help. Security systems and devices can be built with the help of deep-learning-based IoT devices for object detection and voice detection which can be installed for security surveillance around or inside the house for providing alerts of intrusion to the respected authority for the safety of old people and physically challenged people. In this direction, Wadhwani *et al.* designed an Arduino-based system to send alerts to the owner in case of any trespassing [52, 53]. IoT-enabled deep learning models can help us to develop devices for this kind of people to make their lives easy and independent of others.

6.5 Applications of IoT-Enabled CNN

IoT has impacted many fields and provided with real-time solutions for the problems. These include farming, infrastructure, and many more.

6.5.1 Smart Farming

Agriculture is an important aspect of sustaining human life by breeding plants and animals to meet the food demand. Before the technological advancement in Internet, traditional farming relied on intensive farming practices. However, with the rising demand of food due to increasing population and certain environmental factors such as rising climate change, extreme weather conditions, etc., there is a paradigm shift from intensive farming practices toward smart farming. In a span of the past 10 years, many researchers have performed research on IoT-based smart farming which includes IoT-based irrigation systems, IoT modules for monitoring soil moisture, soil temperature, and other such important characteristics of soil and environment required for a good yield farming. With the recent developments in deep learning, researchers have started to look forward toward using CNN for much better automation through IoT modules or systems. Since then, many researchers have proposed such CNN-based IoT systems and have also succeeded in developing these kinds of automated systems for smart and efficient farming.

Shekhar *et al.* [54] have proposed a system based on the IoT-based irrigation system. The system was fully automated and monitored the soil temperature and moisture through a sensor network. These important characteristics captured by the sensors were used to prepare a dataset for predicting the requirement of soil irrigation. These predictions were made by a rather simpler machine learning algorithm, *K*-nearest neighbors or KNN. The system includes Arduino Uno and Raspberry Pi3 as key components for machine-to-machine communication for controlling the whole irrigation process. Another important part of this system was that the prepared dataset and anticipated dataset both were made available to the farmer through cloud server.

Figure 6.5 illustrates a basic architecture for ML-based IoT controlled irrigation system. A similar one has been proposed by the authors of [55]. The data acquisition module consists of sensor networks for collecting data related to environmental conditions such as moisture level, surrounding temperature, etc. This data is stored on regular basis for training purpose or directly transmitted to the ML module for analysis and irrigation requirement predictions.

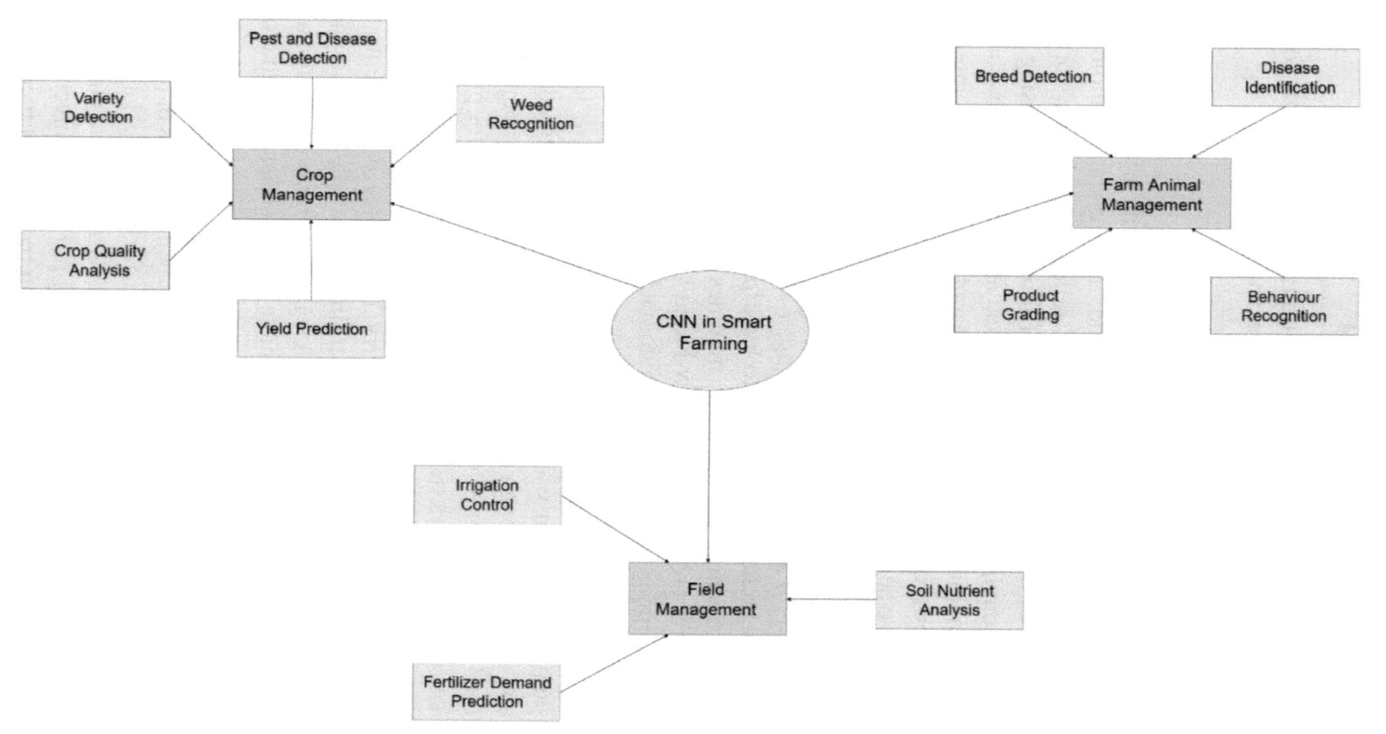

Figure 6.4 Description of various domains of smart farming in which IoT is integrated with CNN.

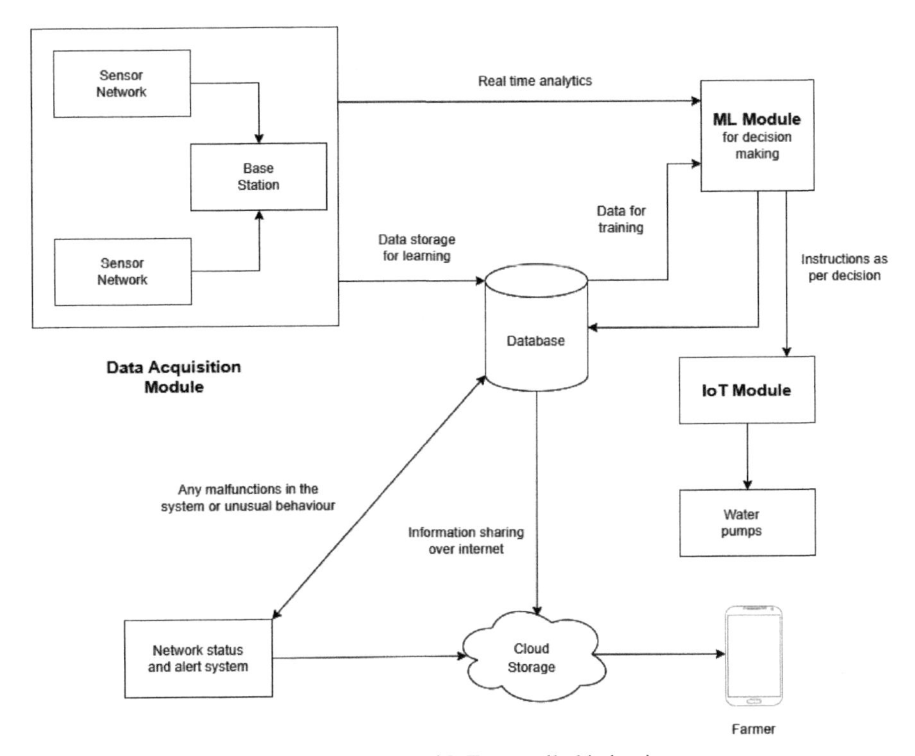

Figure 6.5 ML automated IoT controlled irrigation system.

A further improvement to this system could be a two-way communication between cloud storage and database. This could help in better predictions by considering other factors such as soil composition, nutrition levels, moisture requirement of a certain crop, and other related parameters which change according to the type of crop being cultivated. This is due to the fact that these factors can only be analyzed by certain laboratory tests, which, in turn, need to be updated regularly by the end user.

Luigi *et al.* [62] proposed a CNN-based model for plant disease detection and diagnosis by analyzing the leaves. The model was trained with about 87,800 plant images of 25 different plants categorized into 58 different plant-disease combinations. The best performance was achieved with a testing dataset of about 17,550 images with an accuracy of 99.53% in identifying plant-disease combinations. The model used was VGGCNN. This research infers the success of using CNN in real time for analyzing different plant images so as to detect plant diseases. Indhu Mathi *et al.* [63] proposed an

Table 6.4 Details of various proposed deep learning models based on available deep learning frameworks for the application in smart farming

Smart farming applications	**Framework**	**Performance metrics**	**Description**
Recognize plant species	AlexNet	99.5% accuracy	Authors [56] prepared two datasets from the original one using data augmentation techniques and concluded that venation structure is an important feature for recognizing plant species.
Identify crop species and diseases	AlexNet + GoogLeNet	0.9934 $F1$-score and 99.35% accuracy	The authors [57] experimented with these two deep learning architectures following two training mechanisms, transfer learning and training from scratch with three types of datasets colored, grayscale, and segmented. As expected, transfer learning technique gave better results which were found with GoogLeNet.
Leaf disease detection	CaffeNet	96.3% accuracy	A new dataset was generated from available datasets by using image augmentation techniques. An important aspect of this research [58] is that the dataset includes background images which are classified with an accuracy of 98.21% which gives good separation of plant leaves and surroundings.
Classify weeds from crop	VGGNET	86.20% classification accuracy	The authors [59] used six different datasets which included 22 different species to prepare a new dataset through image augmentation techniques. A 50% dropout rate was applied before two fully connected layers. A mini-batch of 200 images was selected for training the model initialized with weights from VGG16 network.

Table 6.4 Continued.

Smart farming applications	**Framework**	**Performance metrics**	**Description**
Fruit counting	Inception-ResNet	70%–100% with an average accuracy of 91.03%	Authors [60] came up with a new CNN model which was integrated with four layers of a modified Inception-ResNet-A and a modified inception network. The training was done over 24,000 images with Adam optimizer and initial weights were taken from Xavier initializer.
Crop disease classification for early disease symptoms	Deep ResNet	Balanced accuracy of 0.84 in real testing conditions	The integral part of the system [61] was image processing module which was based on superpixel-based tile extraction which was achieved using SLIC superpixel extraction algorithm. The significance of this part was to avoid degrading of visual features of early signs of diseases by avoiding the downscaling of part of image where the disease could be present.

automated diagnosis and classification system. Two clusters, healthy leaf areas and infected areas, were formed for each plant image. On the basis of selected cluster, features were extracted from that cluster to estimate the percentage of infected leaves. The algorithm comprises mainly three parts: first is K-means for clustering, then GLCM or gray level co-occurrence matrix for feature selection, and then, finally, SVM or support vector machine for classification so as to identify the plant disease.

Santosh *et al.* [64] proposed a system based on standard YOLO model for detecting tomato plant diseases. The authors used a dataset consisting of image samples of leaves of three types of tomato plant diseases namely Gray Spot, Late Blight, and Bacterial Canker, further adding 275 image samples of healthy tomato plant leaves. The system's computing unit was implemented on Raspberry PI with a graphical user interface (GUI) for collecting images and capturing videos. The main purpose of image processing module of the system is image acquisition through IoT system and performing image augmentation and feature extraction on the acquired images or video. The average loss of the trained model is 0.0634 and the mean average precision (MAP) is 0.76. An overall accuracy of 89% is found over plant village dataset.

6.5.2 Smart Infrastructure

Nowadays, we are quite familiar with the term "smart cities," which means a city with a proper management and functioning so as to promote the economic growth. With the modern day changing world and increasing population, the cities are getting much more populated and the management is becoming a crucial and difficult task. The solution to this problem is a smart information and communication network which can possibly allow us to improve and automate the management and functioning tasks. As we are familiar with the thing that IoT can provide us with the smart monitoring systems through a network of sensors and receivers. We have also seen a significant research growth in developing machine learning systems which can help in decision making. Upon CNN integration with IoT, we can look toward more smarter and intelligent transportation systems [65], smart lighting, smart parking management, etc. Some of the researches based on intelligent traffic management and monitoring system are discussed in Table 6.5.

Machine learning or artificial intelligence based IoT systems can be implemented for many different functionalities such as for monitoring vibrations in materials of bridges and buildings to check for some unusual sounds which can help in identifying air pockets and unusual gaps in the structure and further help in analyzing structural strength to prevent any possible collapse. A major part of these kinds of systems requires sensor networks, radio frequency identification, and video surveillance devices in order to collect the required information for computation by the machine learning module for decision making to implement the appropriate actions by the IoT module. Such machine-learning-based IoT systems can contribute a lot toward developing and managing modern day infrastructures in much smarter and automated ways.

6.6 Challenges in NLP-Based IoT Devices and Solutions

Security and privacy are two big challenges in implementations of the IoT devices in commercial environments. IoT is becoming a major part of modern infrastructure dealing with billions of objects and humans. Hence, security in

Table 6.5 Intelligent traffic control and monitoring systems based on AI-integrated IoT systems

Authors	System	Description
Ying *et al.* [66]	Intelligent traffic control system	The system was an artificial-intelligence-based IoT system for handling traffic lights. It was a distributed multi-agent Q learning system which could be deployed for controlling local traffic lights for vehicles as well as pedestrians. Surveillance cameras were used to check the queue lengths which were considered as the number of vehicles and pedestrians waiting at the intersection. Different agents were deployed at different intersections with a computation and control module.
Hasan Omar *et al.* [67]	Intelligent traffic information system	The system is able to acquire real-time traffic information and monitor it. The system was based on IoT and wireless communication. The architecture uses wireless sensors and active radio-frequency identification (RFID) for collecting real-time traffic data and monitoring the traffic flow which is done by the multi-agent-based system. These agents are autonomous and intelligent. Hence, they can interact in a useful way with their environment without much human interference. This system allows automatic representation, tracking, and querying of tagged traffic objects.
Danping Wang *et al.* [68]	Urban traffic guidance system	This was conducted on employing IoT devices for urban traffic control. The main aim of developing this kind of system was to deal with the problem of traffic jam in urban cities. A large amount of data was collected through integrated devices on connected vehicles and roadside units. The traffic guidance information module acquires data related to road transport infrastructure and real-time transportation information. The optimal path search module implements resistive multi-objective-based database constrained optimal path algorithm.

IoT is a necessary and complex task to assure information security and consistent operation of the devices. IoT is like a smart wireless sensor network, somewhat architecturally inheriting from wireless sensor networks (WSNs). So, it also inherits the same security challenges from WSNs such as network breach, physical tampering, jamming, selective forwarding, wormhole attack, etc. [69]. As the main communication channel for IoT is the Internet, many of the IoT devices are integrated with the Internet for remote access which opens the backdoors for security attacks, such as simple security breaches for stopping the workflow of the device. The security attacks can be classified into two categories, namely "passive attacks" and "active attacks" [70]. Passive attacks are focused on deducing system information by monitoring messages and data transmission. The main motive is to gather information from the system rather than affecting system resources and operations. These kinds of attacks can go undetected because the aim is to deduce system information instead of altering any data or operation. Active attack is when the motive of attack is to alter system resources and affect its operation. Figure 6.6 shows an illustration of passive and active attacks.

Implementation of NLP also poses a major challenge of data stealing in the case of security breach where training data, such as facial recognition images or voice patterns, can be manipulated. A major limitation of the IoT devices is due to their environment of operation, adding more security challenges to the operation of devices. To ensure data security, techniques like cryptography can be used [70]. Modern day cryptography technique is implemented in one of the two forms: "private-key cryptography" or "public-key cryptography." Private-key method using symmetric key algorithm provides both the sender and receiver with a secret key for their communication. Whereas, public-key method includes asymmetric key algorithms, where a public key is provided to all the communication parties and its private key is kept secret [71, 72].

Confidentiality of data is preventing it from being accessible to any unauthorized user. As modern day IoT systems are getting smarter, a larger amount of data is being generated. This analytics data needs to be handled carefully, as if this data is available to wrong hands, and then it could lead to user privacy violation or tampering of data which could lead to stalling or malfunctioning of IoT devices in the network.

Integrity is maintaining the completeness of data being transmitted over the network in a system. As IoT systems include many devices connected together, communicating through certain transmission protocols over the

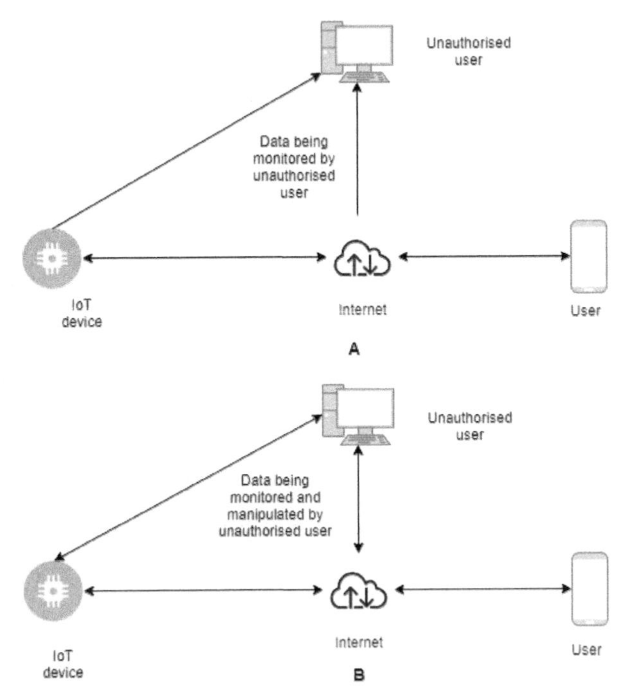

Figure 6.6 Details of various attacks on IoT-based devices. (a) Passive attack. (b) Active attack.

Internet or GSM. So, the chances of distortion or addition of noise in the data being transmitted over the communication channel are higher.

Authentication is the process of identifying user or device. It is used to identify the user so as to provide usage access according to the need. A strong authentication is a must in an IoT system to prevent unauthorized users from making control commands or having access to the system data which could lead to system abuse.

Authorization is providing users access rights to an IoT system. Users could include humans, sensors, or other IoT devices in the network. The main challenge is how to provide successful access to all kinds of user entities in such an environment [56]. In addition, the data and control should be made available to only authorized user.

Availability is a fundamental feature which needs to be ensured for successful deployment, as authorized user entities should always have access to the IoT system. Threats such as DoS, jamming, and node capture can lead to unavailability of the services by IoT devices.

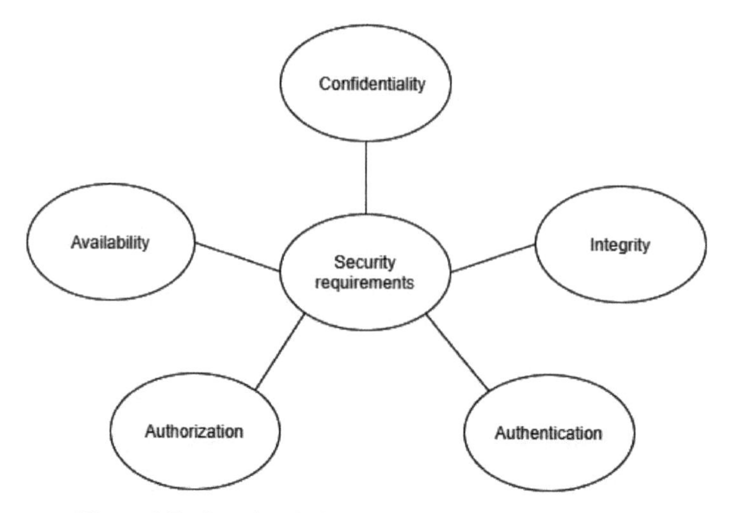

Figure 6.7 Security challenges and hazards in IoT network.

The above five security properties as shown in Figure 6.7 should be considered very well for ensuring security of IoT systems. Exploitation of any of these properties can put system at risk of several security threats which could lead to several issues.

6.7 Conclusion

In this chapter, we have reviewed and analyzed the work in the domain of IoT. IoT is an emerging domain with the capability to automate tasks and can improve the life experience for the disabled too. However, IoT has limited capability to recognize patterns, voice, or anomalies by learning from data. Application of CNN-based models in IoT has shown great development for automating tasks by addressing those issues. In addition, automation tasks can be made simpler and accessible by embedding voice commands/speech recognition through NLP. These tasks not only provide the customer a different experience but also make disabled people more independent for their day-to-day tasks. Also, we have discussed various challenges and security issues related to CNN-based IoT systems. In order to ensure the privacy and confidentiality of the user data, it is necessary to prevent the IoT-based devices from cyberattacks. Various cyberattacks in IoT have been reviewed and the details are tabulated with the solutions.

References

[1] S. Kumar, S. Benedict and S. Ajith, "Application of natural language processing and IoT cloud in smart homes", *in Proceedings of the 2019 2nd International Conference on Intelligent Communication and Computational Techniques (ICCT)*, pp. 20-25, 2019.

[2] A. Al-Fuqaha, M. Guizani, M. Mohammadi, M. Aledhari and M. Ayyash, "Internet of Things: A survey on enabling technologies, protocols, and applications", *IEEE Communications Surveys & Tutorials*, vol. 17, no. 4, pp. 2347-2376, Fourthquarter 2015.

[3] F. Hussain, R. Hussain, S. A. Hassan and E. Hossain, "Machine learning in IoT security: Current solutions and future challenges", *IEEE Communications Surveys & Tutorials*, vol. 22, no. 3, pp. 1686-1721, Thirdquarter 2020.

[4] J. Manyika, M. Chui, J. Bughin, R. Dobbs, P. Bisson, and A. Marrs, *Disruptive Technologies: Advances that will Transform Life, Business, and the Global Economy*. McKinsey Global Institute, San Francisco, CA, USA, vol. 180, 2013.

[5] P. J. Rani, J. Bakthakumar, B. P. Kumaar, U. P. Kumaar and S. Kumar, "Voice controlled home automation system using natural language processing (NLP) and Internet of Things (IoT)", *in Proceedings of the 2017 3rd International Conference on Science Technology Engineering Management (ICONSTEM)*, pp. 368-373, March 2017.

[6] L. M. Gladence, M. Anu, R. Rathna and B. Prince, "Recommender system for home automation using IoT and artificial intelligence", *Journal of Ambient Intelligence and Humanized Computing*, 2020.

[7] M. Mohammadi, A. Al-Fuqaha, S. Sorour and M. Guizani, "Deep learning for IoT big data and streaming analytics: A survey", *IEEE Communications Surveys & Tutorials*, vol. 20, no. 4, pp. 2923-2960, Fourthquarter 2018.

[8] X. Ma *et al.*, "A survey on deep learning empowered IoT applications", *IEEE Access*, vol. 7, pp. 181721-181732, 2019.

[9] H. Jiang, C. Cai, X. Ma, Y. Yang and J. Liu, "Smart home based on WiFi sensing: A survey", *IEEE Access*, vol. 6, pp. 13317-13325, 2018.

[10] G. Wang, Y. Zou, Z. Zhou, K. Wu, and L. M. Ni, "We can hear you with wifi!", *in Proceedings of the 20th Annual International Conference on Mobile Computing and Networking (MobiCom)*, 2014.

[11] H. Abdelnasser, M. Youssef and K. A. Harras, "Wigest: A ubiquitous wifi-based gesture recognition system", *in Proceedings of the 34th*

Annual Joint Conference of the IEEE Computer and Communications Societies (INFOCOM), 2015.

[12] K. Ali, A. X. Liu, W. Wang and M. Shahzad, “Keystroke recognition using wifi signals”, *in Proceedings of the 21st Annual International Conference on Mobile Computing and Networking (MobiCom)*, 2015.

[13] K. Qian, C. Wu, Z. Zhou, Y. Zheng, Z. Yang and Y. Liu, “Inferring motion direction using commodity wi-fi for interactive exergames”, *in Proceedings of the 2017 CHI Conference on Human Factors in Computing Systems (CHI)*, 2017.

[14] A. Virmani and M. Shahzad, “Position and orientation agnostic gesture recognition using wifi”, *in Proceedings of the 15th Annual International.*

[15] K. Mandula, R. Parupalli, C. A. S. Murty, E. Magesh and R. Lunagariya, “Mobile based home automation using Internet of Things (IoT)”, *in Proceedings of the 2015 International Conference on Control, Instrumentation, Communication and Computational Technologies (ICCICCT)*, pp. 340-343, 2015.

[16] L. Xiao, X. Wan, X. Lu, Y. Zhang and D. Wu, “IoT security techniques based on machine learning: How do IoT devices use AI to enhance security?”, *IEEE Signal Processing Magazine*, vol. 35, no. 5, pp. 41-49, Sep. 2018.

[17] P. P. Ray, “A survey on Internet of Things architectures”, *EAI Endorsed Transactions on Internet of Things*, vol. 2, 2016.

[18] S. Sladojevic, M. Arsenovic, A. Anderla, D. Culibrk and D. Stefanovic, “Deep neural networks based recognition of plant diseases by leaf image classification,” *Computational Intelligence and Neuroscience*, pp. 1-11, 2016.

[19] R. Shah and R. Romijnders, “Applying deep learning to basketball trajectories”, *in Proceedings of the ACM KDD’16*, 2016.

[20] K.-C. Wang and R. Zemel, “Classifying NBA offensive plays using neural networks”, *in Proceedings of the MIT SLOAN Sports Analytics Conference*, 2016.

[21] A. Toshev and C. Szegedy, “Deeppose: Human pose estimation via deep neural networks”, *in Proceedings of the IEEE Conference on Computer Vision and Pattern Recognition*, pp. 1653-1660, 2014.

[22] F. J. Ordo´nez and D. Roggen, “Deep convolutional and LSTM recurrent ˜neural networks for multimodal wearable activity recognition”, *Sensors*, vol. 16, no. 1, p. 115, 2016.

[23] L. Pigou, A. Van Den Oord, S. Dieleman, M. Van Herreweghe and J. Dambre, "Beyond temporal pooling: Recurrence and temporal convolutions for gesture recognition in video", *International Journal of Computer Vision*, pp. 1–10, 2015.

[24] A. Xiao, R. Chen, D. Li, Y. Chen and D. Wu, "An indoor positioning system based on static objects in large indoor scenes by using smartphone cameras", *Sensors*, vol. 18, 2018.

[25] X. Wang, L. Gao, S. Mao and S. Pandey, "DeepFi: Deep learning for indoor fingerprinting using channel state information", *in Proceedings of the 2015 IEEE Wireless Communications and Networking Conference (WCNC)*, pp. 1666-1671, 2015.

[26] N. Kheratkar, S. Bhavani, A. Jarali, A. Pathak and S. Kumbhar, "Gesture controlled home automation using CNN", *in Proceedings of the 2020 4th International Conference on Intelligent Computing and Control Systems (ICICCS)*, pp. 620-626, 2020.

[27] J. Wang *et al.*, "Detecting cardiovascular disease from mammograms with deep learning", *IEEE Transactions on Medical Imaging*, vol. 36, no. 5, pp. 1172-1181, May 2017.

[28] P. Feng, M. Yu, S. M. Naqvi and J. Chambers, "Deep learning for posture analysis in fall detection", *in Proceedings of the International Conference on Digital Signal Processing, DSP*, 2014, pp. 12-17, 2014.

[29] M. S. Ibrahim, S. Muralidharan, Z. Deng, A. Vahdat and G. Mori, "A hierarchical deep temporal model for group activity recognition", *in Proceedings of the 2016 IEEE Conference on Computer Vision and Pattern Recognition (CVPR)*, pp. 1971-1980, 2016.

[30] M. Schmidt, D. Block and U. Meier, "Wireless interference identification with convolutional neural networks", pp. 180-185, 2017.

[31] R. Piyare and M. Tazil, "Bluetooth based home automation system using cell phone", *in Proceedings of the 2011 IEEE 15th International Symposium on Consumer Electronics (ISCE)*, pp. 192–195, Jun. 2011.

[32] B. Yuksekkaya, A. A. Kayalar, M. B. Tosun, M. K. Ozcan and A. Z. Alkar, "A GSM, internet and speech controlled wireless interactive home automation system", *IEEE Transactions on Consumer Electronics*, vol. 52, no. 3, pp. 837–843, Aug. 2006.

[33] A. Alheraish, "Design and implementation of home automation system", *IEEE Transactions on Consumer Electronics*, vol. 50, no. 4, pp. 1087-1092, Nov. 2004.

[34] R. Teymourzadeh, S. A. Ahmed, K. W. Chan and M. V. Hoong, "Smart GSM based home automation system", *in Proceedings of the 2013 IEEE*

Conference on Systems, Process Control (ICSPC), pp. 306-309, Dec. 2013.

[35] P. S. N. Reddy, K. T. K. Reddy, P. A. K. Reddy, G. N. K. Ramaiah and S. N. Kishor, "An IoT based home automation using android application", *in Proceedings of the 2016 International Conference on Signal Processing, Communication, Power and Embedded System (SCOPES)*, pp. 285-290, Oct. 2016.

[36] H. Ren, Y. Song, S. Yang and F. Situ, "Secure smart home: A voiceprint and internet based authentication system for remote accessing", *in Proceedings of the 2016 11th International Conference on Computer Science Education (ICCSE)*, pp. 247-251, Aug. 2016.

[37] P. J. Rani, J. Bakthakumar, B. P. Kumaar, U. P. Kumaar and S. Kumar, "Voice controlled home automation system using natural language processing (NLP) and internet of things (IoT)", *in Proceedings of the 2017 3rd International Conference on Science Technology Engineering Management (ICONSTEM)*, pp. 368-373, Mar. 2017.

[38] D. Arifoglu and A. Bouchachia, "Activity recognition and abnormal behaviour detection with recurrent neural networks", *Procedia Computer Science*, vol. 110, pp. 86-93, 2017.

[39] K. S. Boujemaa, I. Berrada, A. Bouhoute and K. Boubouh, "Traffic sign recognition using convolutional neural networks", *in Proceedings of the 2017 International Conference on Wireless Networks and Mobile Communications (WINCOM)*, pp. 1-6, 2017.

[40] S. W. Pienaar and R. Malekian, "Human activity recognition using LSTM-RNN deep neural network architecture", *in Proceedings of the 2019 IEEE 2nd Wireless Africa Conference (WAC)*, pp. 1-5, 2019.

[41] X. Wang, L. Gao, S. Mao and S. Pandey, "CSI-based fingerprinting for indoor localization: A deep learning approach", *IEEE Transactions on Vehicular Technology*, 2016.

[42] S. M. H. Maasoum, A. Mostafavi and A. Sadighi, "An autoencoder-based algorithm for fault detection of rotating machines, suitable for online learning and standalone applications", *in Proceedings of the 2020 6th Iranian Conference on Signal Processing and Intelligent Systems (ICSPIS)*, pp. 1-6, 2020.

[43] P. Bahl and V. N. Padmanabhan, "RADAR: An in-building RF-based user location and tracking system", *in Proceedings of the 19th Annual Joint Conference of the IEEE Computer and Communications Society (INFOCOM)*, pp. 775-784, Mar. 2000.

[44] Y. Gu, Y. Chen, J. Liu and X. Jiang, "Semi-supervised deep extreme learning machine for Wi-Fi based localization", *Neurocomputing*, vol. 166, pp. 282–293, Oct. 2015.

[45] L. Deng, "Artificial intelligence in the rising wave of deep learning: The historical path and future outlook [perspectives]", *IEEE Signal Processing Magazine*, vol. 35, no. 1, pp. 177-180, Jan. 2018.

[46] M. A. Kumar and Y. RaviSekar, "Android based health care monitoring system", *in Proceedings of the IEEE Sponsored 2nd International Conference on Innovations in Information Embedded and Communication Systems ICIIECS'15*, Mar. 2015.

[47] B. Sneha, V. Bhavana, S. Brunda, T. S. Murali, S. Puneeth and B. A. Ravikiran, "A wireless based patient monitoring system using Android technology", *in Proceedings of the 2015 International Conference on Applied and Theoretical Computing and Communication Technology (iCATccT)*, Oct. 2015.

[48] G. Dharmale, V. Thakre and D. D. Patil, "Intelligent hands free speech based SMS system on Android", *in Proceedings of the International Conference on Advances in Human Machine Interaction (HMI - 2016)*, Mar. 03-05, 2016.

[49] I. McGraw, R. Prabhavalkar, R. Alvarez, M. G. Arenas, K. Rao, D. Rybach, Q. Alsharif, H. Sak, A. Gruenstien, F. Beaufays and C. Parada, "Personalized speech recognition on mobile devices", *in Proceedings of the International Conference on Acoustics*, Speech and Signal Processing (ICASSP), pp. 5955-5959, 2016.

[50] L. González-Delgado, L. Serpa-Andrade, K. Calle-Urgiléz, A. Guzhñay-Lucero, V. Robles-Bykbaev and M. Mena-Salcedo, "A low-cost wearable support system for visually disabled people", *in Proceedings of the 2016 IEEE International Autumn Meeting on Power, Electronics and Computing (ROPEC)*, pp. 1-5, 2016.

[51] A. Bhargava, K. V. Nath, P. Sachdeva and M. Samel, "Reading assistant for the visually impaired", *International Journal of Current Engineering and Technology*, vol. 5, no. 2, 2015.

[52] S. Wadhwani, U. Singh, P. Singh and S. Dwivedi, "Smart home automation and security system using Arduino and IoT", *International Research Journal of Engineering and Technology*, vol. 5, no. 2, pp. 1357-1359, 2018.

[53] M. Ejigu and J. Santhosh, "IoT based comprehensive autonomous home automation and security system using M2M communication", *Recent Advances in Computer Science and Communications*, vol. 13, 2020.

[54] Y. Shekhar, E. Dagur, S. Mishra, R. J. Tom, M. Veeramanikandan and S. Sankaranarayanan, "Intelligent IoT based automated irrigation system", *International Journal of Applied Engineering Research*, vol. 12, pp. 7306-7320, 2017.

[55] S. Sicari, A. Rizzardi, L. A. Grieco, and A. Coen-Porisini, "Security, privacy and trust in Internet of Things: The road ahead", *Computer Networks*, vol. 76, pp. 146-164, 2015.

[56] S. H. Lee, C. S. Chan, P. Wilkin and P. Remagnino, "Deep-plant: Plant identification with convolutional neural networks", *in Proceedings of the IEEE International Conference on Image Processing (ICIP)*, pp. 452–456, 2015.

[57] S. P. Mohanty, D. P. Hughes and M. Salathé, "Using deep learning for image-based plant disease detection", *Frontiers in Plant Science*, vol. 7, pp. 1419, 2016.

[58] S. Sladojevic, M. Arsenovic, A. Anderla and D. Stefanović, "Deep neural networks based recognition of plant diseases by leaf image classification", *Computational Intelligence and Neuroscience*, pp. 1-11, 2016.

[59] M. Dyrmann, H. Karstoft and H. Midtiby, "Plant species classification using deep convolutional neural network", *Biosystems Engineering*, vol. 151, pp. 72-80, 2016.

[60] M. Rahnemoonfar and C. Sheppard, "Deep count: Fruit counting based on deep simulated learning", *Sensors (Basel, Switzerland)*, vol. 17, 2017.

[61] A. Picon, A. Alvarez-Gila, M. Seitz, A. B. Ortiz, J. Echazarra and A. Johannes, "Deep convolutional neural networks for mobile capture device-based crop disease classification in the wild", *Computers and Electronics in Agriculture*, vol. 161, 2018.

[62] Y. Ampatzidis, L. De Bellis and A. Luvisi, "iPathology: Robotic applications and management of plants and plant diseases", *Sustainability*, vol. 9, 2017.

[63] K. Indhu Mathi, V. B. Kavya, P. Meera and K. Sankar Ganesh, "Lesion characterisation for plant disease diagnosis", *International Research Journal in Advanced Engineering and Technology (IRJAET)*, vol. 4, no. 2, pp. 2959-2963, 2018.

[64] S. Adhikari, B. Shrestha, B. Baiju and K. C. Saban, "Tomato plant diseases detection system using image processing", *in Proceedings of the 1st KEC Conference on Engineering and Technology*, 2018.

[65] R. Jain, "A congestion control system based on VANET for small length roads", *Annals of Emerging Technologies in Computing*, vol. 2, 2018.

[66] Y. Liu, L. Liu and W.-P. Chen, "Intelligent traffic light control using distributed multi-agent Q learning", pp. 1-8, 2017.

[67] H. Omar, "Intelligent traffic information system based on integration of Internet of Things and agent technology", *International Journal of Advanced Computer Science and Applications*, vol. 6, 2015.

[68] D. Wang and K. Hu, "Research on path optimization of urban traffic guidance system", *Applied Mechanics and Materials*, vol. 624, pp. 520-523, 2014.

[69] M.-L. Messai, "Classification of attacks in wireless sensor networks", 2014.

[70] N. Komninos, E. Philippou and A. Pitsillides, "Survey in smart grid and smart home security: Issues, challenges and countermeasures", *IEEE Communications Surveys & Tutorials*, vol. 16, no. 4, pp. 1933-1954, Fourthquarter 2014.

[71] J. Benoit, "An introduction to cryptography as applied to the smart grid", *Cooper Power Systems*, Feb. 2011.

[72] S. Seo, X. Ding and E. Bertino, "Encryption key management for secure communication in smart advanced metering infrastructures", *in Proceedings of the 2013 IEEE International Conference on Smart Grid Communications (SmartGridComm)*, pp. 498-503, 2013.

7

Classification of Myocardial Infarction in ECG Signals Using Enhanced Deep Neural Network Technique

K. Manimekalai[1] and A. Kavitha[2]

[1]Department of Computer Applications,
Sri GVG Visalakshi College for Women, India
[2]Department of Computer Science,
Kongunadu Arts and Science College, India
E-mail: gvgmanimekalai@gmail.com; manimekalai@gvgvc.ac.in; akavitha_cs@kongunaducollege.ac.in

Abstract

The categorization of ECG signals is critical in therapeutic treatment. Traditional approaches have hit their performance ceiling, yet there is always room for improvement. The objective of this research is to use ECG signals to accurately locate and identify myocardial infarction (MI). In the proposed algorithm enhanced deep neural network (EDN), deep learning methods like convolutional neural network (CNN) and long short-term memory (LSTM) algorithms were applied. Vector operations such as matrix multiplication and gradient descent were conducted in parallel with GPU support on large matrices of data. Parallelism in EDN decreases the time it takes for a procedure to run. For the PTB database, the proposed model EDN has a higher accuracy of 87.80% according to the confusion matrices of the algorithms. Based on criteria such as precision, recall, *F*1 measure, and Cohen kappa coefficient, the suggested model demonstrates performance improvisation. These enhancements to EDN's performance would aid in the saving of human lives.

Keywords: Classification, CNN, DNN, EDN, LSTM, MI.

7.1 Introduction

The most prevalent cause of cardiac muscle loss induced by persistent ischemia is myocardial infarction (MI), sometimes known as "heart attack." Pain radiating from the chest to the shoulder, arm, and neck is the most common symptom of a myocardial infarction. Atherosclerosis is the most common cause of myocardial infarction, as evidenced by autopsystudies on the causes of MI. Atherosclerosis is the process of hardening blood arteries in the body, which is characterized by atheroma or lesions that appear in the blood vessels. The center of an atheromatous plaque is formed predominantly of cholesterol esters and cholesterol, and it is surrounded by white fibrous caps made mostly of smooth muscle cells.

Myocardial infarctions are heart muscle deaths. It is a portion of the spectrum that arises in acute coronary syndrome and is an implication of coronary artery disease (CAD), which might result in suctioned blood flow to the heart. Regardless of other causes of acute coronary syndromes (ACS), such as unstable angina (UA), a myocardial infarction (MI) occurs when cells die, as determined by a trooping or CK-MB blood test of cardiac enzymes (Liu *et al.*, 2018). When an MI is suspected, it is called an ST segment elevation myocardial infarction (STEMI) or a non-ST segment elevation myocardial infarction (NSTEMI), and it is discovered on ECG readings.

The principal clinical occurrence in patients with atherosclerosis discovered on the coronary arteries is myocardial infarction (MI), often known as an acute coronary syndrome (ACS); however, the actual phrase is based on the existing definition under which its many presentations are generated. The formation of thrombus is a rare occurrence during the growth of a coronary artery blockage. As a result, coronary heart disease (CHD), also known as coronary artery disease (CAD) or ischemic heart disease (IHD), would be unlikely to be fatal in the absence of thrombus development.

Myocardial infarction is one of the major challenges facing the healthcare system, with a global death rate of 265 per lakh population and around 224 in the Mediterranean region. According to estimates, deaths from cardiovascular diseases will increase by 15% in affluent countries with a rapid economic growth, such as the United States (Wang *et al.*, 2019), 77% in China, and an astonishing 16% in other Asian countries. Finland conducted analytical investigations and determined the largest rate of myocardial infarction, while Japan projected the lowest rate. The British Heart Foundation predicts 0.6%

yearly rate for men and 0.1% for women for acute myocardial infarction in people aged 30–69 years.

Specifically, the World Health Organization and all 194 member countries have decided to work on a general platform inducing global mechanisms to reduce CVD risk by 25% by 2025 by enhancing the "global action strategy that are essential for the prevention and measurement of NCDs 2013–2020" by focusing on the both prevention and controlling measures needed to detect the cardiovascular diseases.

Cardiovascular disease is a potentially fatal ailment that affects more than 17 million individuals worldwide each year. According to Abdulrazaq *et al.*, an electrocardiogram (ECG) is a recording of the electrical function of the human heartbeat. MI is described as an abnormality or interruption in the myocardium's regular activation process.

PTB database is used to gather MI signals. Because traditional ECG analysis approaches are time-consuming and tedious, various automated solutions have arisen in recent years (Benjamin *et al.*, 2017). With the advent of contemporary signal processing, data mining, and machine learning techniques, the ECG's diagnostic power has exploded.

To detect and classify the ECG signal, numerous algorithms have been proposed. For displaying experimental findings, some methods use time domain, while others use frequency domain. Various unique qualities in ECG classification and recognizing the different pathological classes based on the different beats obtained from the ECG signal were defined based on the algorithms.

7.2 Related Work

According to the findings, ECG signal categorization is an important component that is predominantly used in the clinical diagnosis of heart illness. The biggest problem with utilizing ECG to diagnose heart illness is that a typical ECG varies from individual to individual, and it has been seen that one disease might exhibit distinct symptoms in ECG signals for different patients. According to the researchers, the two disorders may have similar impacts on normal ECG signals. Because of these issues, identifying heart disease is challenging. As a result, by using the pattern classifier method, it is possible to advance the patient's ECG for MI diagnosis. This study provided a frugal solution. This study suggested a framework for ECG classification based on a binary-class classification problem with two classes: normal and myocardial infarction (Rajkumar *et al.*, 2019).

The two aspects of ECG data are the classification of the ECG signal and the classification of individual ECG beats. Many waves, including the P, Q, R, S, T, and U waves, are found to be present in a single cardiac cycle, which are frequently used to depict one ECG pulse. A single ECG signal is formed by combining thousands of ECG beats.

Pre-processing, feature extraction method, and classification methods are all important elements in the ECG classification process. ECG signals can cause many types of noise (for example, baseline wander noise), which is the key factor that influences feature extraction and classification.

The ECG aids in the provision of meaningful data that reveals both the occurrence of MI and its location. MI characteristics include ST-segment elevation, T-wave inversion, and abnormal Q-wave appearance. These are mostly inferred by the feature vector classification method. Variability among ECG features, a lack of standardization of ECG features, individuality of ECG features, a lack of optimal classification rules for ECG classification, variability in patients' ECG waveforms, and, finally, the selection of the most appropriate classifier are all major issues in ECG classification. These descriptions of the issues would be beneficial to the new research. According to Shweta H. Jambukia (2015), these descriptions of the issues will assist new researchers in recognizing the risks associated with ECG classification.

1. Slight variation of ECG elements: The issue concerns electrocardiogram pattern margins; the amplitude domain is not found to be standard, fixed, or heuristically on time. The feature extraction approach partially chooses ECG elements, and its reliability is primarily dependent on these discovered features. On big datasets, a small deviation in these crucial properties can lead to misclassification.
2. Diversity of ECG elements: The heart rate of a person will change depending on biological and physiological variables. An increase in heart rate might be caused by stress, exercise, enthusiasm, or other work-related activities. The RR interval, QT interval, and PR interval all alter as a result of changes in heart rate. These qualities must be rigorously modified, and the influence of the variable heart rate must be removed.
3. Uniqueness of ECG sequences: The potential for intraclass similarity and interclass diversity of testing patterns acquired from ECG data is denoted by the uniqueness of ECG patterns. It is useful to demonstrate how far the ECG patterns can be scaled in a large enough dataset.
4. The absence of optimal categorization means that ECG classification does not exist.

5. Patients' ECG waveforms may differ: Diverse patients' ECGs may have various slops of signal, amplitude, and timing, resulting in a shift in the ECG waveforms. When processing the classification, it is vital to treat and classify the ECG signal with caution.
6. Changes in heartbeat in a single ECG: Thousands of beats can be found in a single ECG, yet these beats are classified as distinct types (i.e., myocardial infarction, arrhythmia, etc.). The classification model should be trained in such a way that just a few minor errors are discovered on the test dataset.
7. Looking for the best classifier that can categorize MI in real time, which can be difficult because classification accuracy is affected by a variety of factors.

The best characteristic that fabricates a flawless model for clustering P-QRS-T waves contained in ECG signals that detect the myocardial infarction was proposed in this work.

The framework of the ECG signal classification system consists of pre-processing, feature extraction, feature selection, and classification.

7.3 The Normal ECG Signal

A normal ECG is made up of the ECG signal in its natural state waves, segments, intervals, and complex, which are addressed in the following sections, with a graphical representation in Figure 7.1.

- Wave: A wave is considered as a +ve or –ve deflection beginning at the baseline and representing a precise electrical activity. The ECG waves are the P wave, R wave, Q wave, T wave, S wave, and ultimately the U wave.
- Interval: The span of time that elapses between two separate ECG events.

On an electrocardiogram, the PR interval, QRS interval (also known as QRS duration), RR interval, and QT interval are all routinely determined intervals.

√ Segment: The distance between two separate ECG regions with the same amplitude level (not –ve or +ve). On an ECG, there are three segments: the PR segment, the TP segment, and the ST segment.

√ Complex: A collection of numerous waves that have been clustered together. The QRS complex is the sole important complex shown on an ECG.

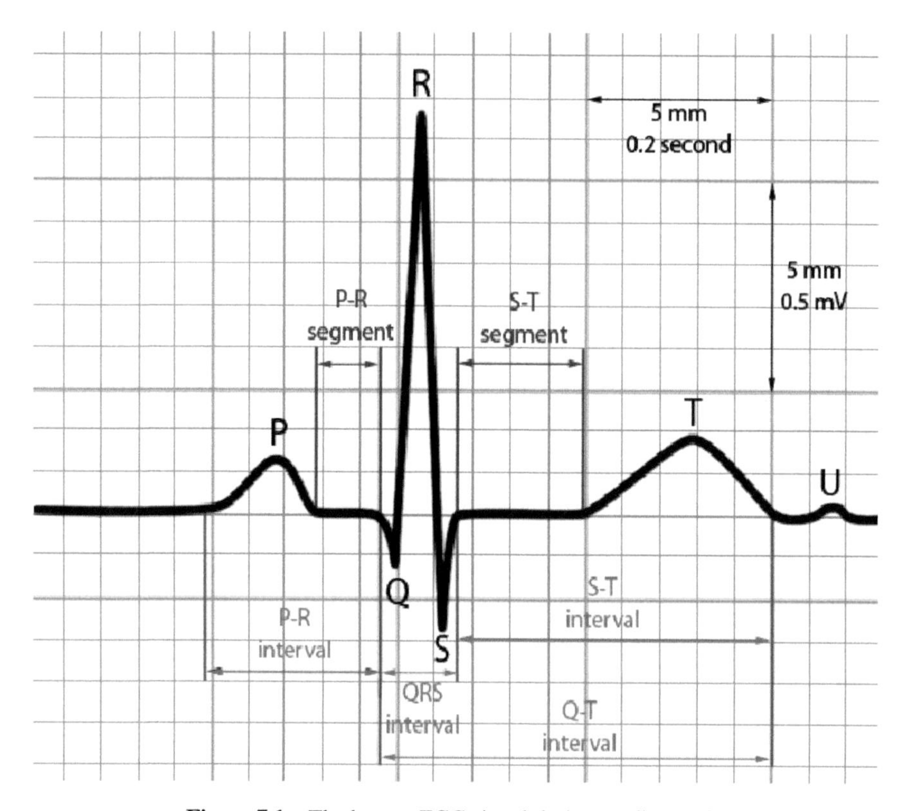

Figure 7.1 The human ECG signal during cardiac cycle.

√ Point: On an ECG, there is just one point that indicates the J point, which is when the QRS interval stops and the ST interval begins.

√ The depolarization of the atria is depicted by the P wave. During normal atrial depolarization, the electrical vector passes through the SA node on its route to the AV node. After rising from the right atrium, it moves from the right to the left atrium. Finally, when seen on an ECG, it is designated as a P wave. It has an 80-ms period.

The spreading depolarization is seen in the QRS complex's right and left ventricles. The ventricles are described as having significant muscle mass related to the atria, and the QRS complex has significantly bigger amplitude than the P wave. A typical QRS complex lasts 80–100 ms. According to convention, any set of these waves is referred to as a QRS complex. The T wave, on the other hand, signals ventricular repolarization.

- The absolute refractory period is the time interval between the beginning of the QRS complex and the peak of the T wave. The second half of the T wave is defined as the relative refractory period. The T wave is typically characterized as positive, with a period of 150–350 ms.
- The period between an R wave and the following R wave is denoted by the RR interval; the usual resting heart rate is between 60 and 100 beats per minute. It lasts from 0.6 and 1.2 seconds on average.
- The PR interval is limited between the performance of the P wave and the onset of the effective QRS complex. The PR interlude represents the time it takes an electrical impulse to go from the sinus node to the AV node and then to the ventricles.
- This could be the most accurate evaluation of the AV node process. It has a duration of 120–200 ms.
- The PR segment connects the P wave to the other QRS complex waves.
- From the AV node through the Bundle divisions, and eventually to the fiber walls, a different sort of impulse vector is raised. This form of electrical action does not immediately cause a decrease and simply moves toward the ventricle portions. This depicts the flat on the ECG and its duration, which is approximately 50–120 ms.
- J point is measured from the point at which the QRS complex terminates its process and the ST segment begins. It is commonly used to assess the degree of ST progression or depression.
- The ST segment connects the QRS complex and the crucial T wave. The ST segment is denoted here to highlight the period during which the ventricles are depolarized. The deviation at the J point with regard and its relationship to the isoelectric line are used to determine or calculate the ST amplitude (PQ or TP segment). From a clinical aspect, it may result in positive or negative deviations belonging to the ST segment that are greater than 1–2 mm, indicating the presence of myocardial ischemia.
- ST interval is estimated from the J point to the T wave's end point. It has a time interval of about 320 ms.
- QT interval is computed from the start of the QRS complex to the end of the T wave. A prolonged QT interval is a risk factor for ventricular arrhythmia, which can lead to sudden death. It varies according to heart rate. It requires a change for clinical studies, giving the QT = (QT interval/sqrt (R-R interval)). It can last up to 420 ms for a heart rate of 60 beats per minute.

7.3.1 ECG Features

The ECG wave has a variety of characteristics. Each one is quite reliable for identifying and documenting the current state or condition of the heart. Starting from lead II of a conventional subject, a separate ECG wave signal. The ECG has a distinct morphology consisting of three waves: the P wave, the QRS complex wave, and the T wave. Waves, complexes, each segment, and intervals are all components of the standard ECG signal, which can be examined by comparing the voltage on the vertical axis to the time on the horizontal axis. It is possible to witness a single ECG waveform that begins and ends at the isoelectric line.

A segment is a line that is flat, straight, and isoelectric. There are two waveforms that are collectively referred to be complex. Interval refers to a waveform or complex that is linked to a segment. Positive deflections are defined as ECG tracings that are presented or positioned above the isoelectric line, whilst negative deflections are defined as tracings that are presented or located below the isoelectric line, as stated by Maharaj *et al.* (2014).

7.3.2 12-Lead ECG System

A typical ECG has 12 leads, according to Rajesh *et al.* (2021). Because the leads are linked to the individual's arms and/or legs, they are referred to as "limb leads." Because they are positioned on the torso, the remaining six leads are referred to as "pre-cordial leads" (precordium). As a result, each of the six limbs has its own lead, which is labeled lead I, II, III, aVL, aVR, and ultimately aVF. The letter "a" stands for "augmented" because these leads are a combination of leads I, II, and III. The letters V1, V2, V3, V4, V5, and V6 represent the six remaining pre-cordial leads. This is a normal 12-lead ECG trace.

A lead is a pair of electrodes that are placed on the human body at precise anatomical places in order to record each node of electrical activity or heart function. Each lead is equipped with two electrodes: a negative (–) and a positive (+). The conventional 12-lead ECG system includes three bipolar leads, three improved unipolar leads, and six chests known as pre-cordial leads, as shown in Figure 1.5 and Table 7.1. The polarity of electrodes can be changed using the "lead selector" on the ECG equipment. This helps in the creation of distinct lead selections with merits without the need to physically move these groups of lead wires or electrodes. The following is a summary of ECG features as defined by Xue *et al.* (2020).

Table 7.1 Standard 12-lead ECG system description

Standard leads	Limb Leads	Chest leads
Bipolar Leads	Unipolar Leads	Unipolar leads
Lead I Lead II Lead III	aVR aVL Avf	V1 V2 V3 V4 V5 V6

Following are the detailed descriptions of the 12-lead systems:

√ Bipolar leads capture variations in both the positive and negative pole potentials.
√ Unipolar leads have a single probing electrode for measuring electrical potential.
√ Unipolar limb leads that record potentials between the aVR, aVL, and aVF.
√ Einthoven seizes the initiative.
√ Lead III is used to record potentials between the left arm and left leg.
√ Lead II, which catches potentials between the right and left arms.
√ Lead I tracks potentials in both the left and right arms.

Chest leads are made up of the following components:

- V6: Those shown in the midaxillary line.
- V5: The V5 rib is positioned on the anterior axillary line.
- V4: In the fifth intercostal gap, in the midclavicular line.
- V3: A voltage increases between the second and fourth electrodes.
- V2 is on the left sternal margin of the fourth intercostal gap.
- V1: Located on the 4th intercostal space's right sternal edge.

Only a few electrodes are required to study the heart rhythm; typically, ten electrodes are utilized for acquisition when multiple waveforms and morphological information are required.

7.4 Proposed Methodology

The proposed ECG classification framework is based on deep learning techniques like native CNN, LSTM, and the proposed EDN algorithms, as shown in the overall architecture of Figure 7.2.

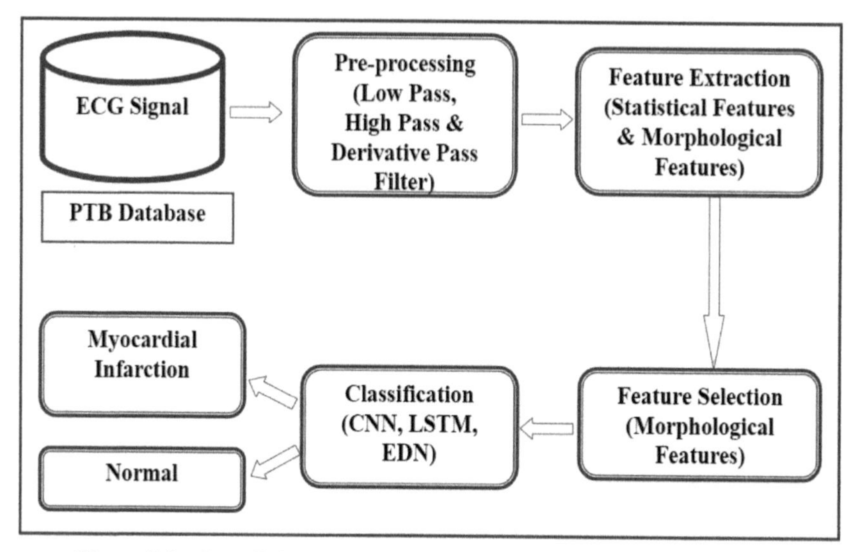

Figure 7.2 Overall framework of the proposed ECG classification system.

When there is noise in the signal, it becomes distorted. As a result, the accuracy of the signal's feature selection and classification degrades.

7.4.1 Phase I: Pre-Processing

The presence of noise has an effect on the signal. As a result, the signal's accuracy in feature selection and classification suffers. It is required to remove noise from the ECG wave signal during the pre-processing stage.

ECG Signal Filtering

Filters on ECGs are necessary for removing artifacts; however, their incorrect use may result in misdiagnosis.

A. Low-Pass Filters

Low-pass filters in ECG signals minimize high-frequency noise. In the noisy signal, the smoothed ECG signal is blended with high-frequency noise [9].

B. High-Pass Filter

In ECG signals, high-pass filters reduce low-frequency noise.

C. Base Derivative Filters

With slope computations, derivative filters are frequently utilized. The derivatives of the received signals represent significant quantities.

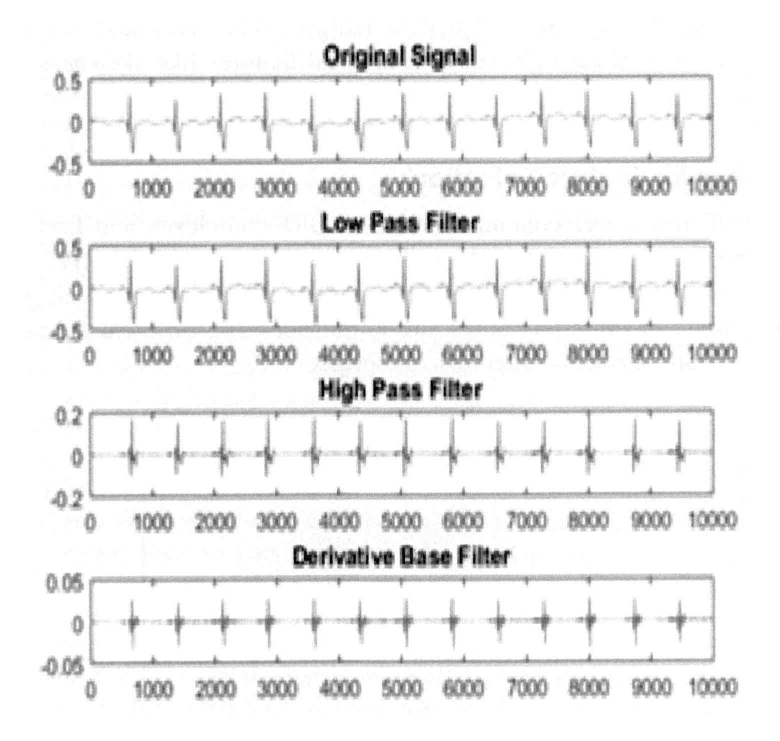

Figure 7.3 Comparison of original signal and filtered signal.

7.4.2 Phase II: Feature Extraction

The frequency domain and temporal domain of the ECG signal have distinct characteristics. To extract features, first-order and higher-order statistics are

Feature extraction for ECG signal			
Mean	Variance	Skewness	Kurtosis
72.763190	2841.590151	5.441525	36.595744

successfully used. First-order statistical features like mean and variance are retrieved, as well as higher-order statistical features like skewness and kurtosis.

7.4.3 Phase III: Feature Selection

Multiple ECG beats, each containing P waves, QRS complexes, and T waves, make up an ECG signal. Intervals such as "PR," "RR," "QRS," "ST," and "QT," as well as peaks such as P, Q, R, S, and T of ECG signals with their usual amplitude or duration values, are eliminated. The different forms of ECG features are segments, intervals, and peaks.

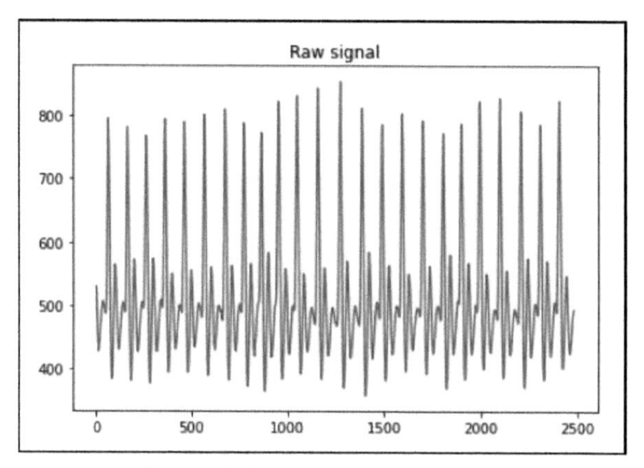

Figure 7.4 PQRST wave format.

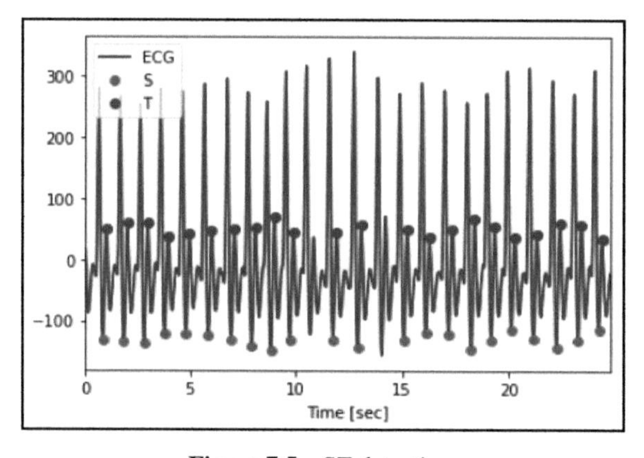

Figure 7.5 ST detection.

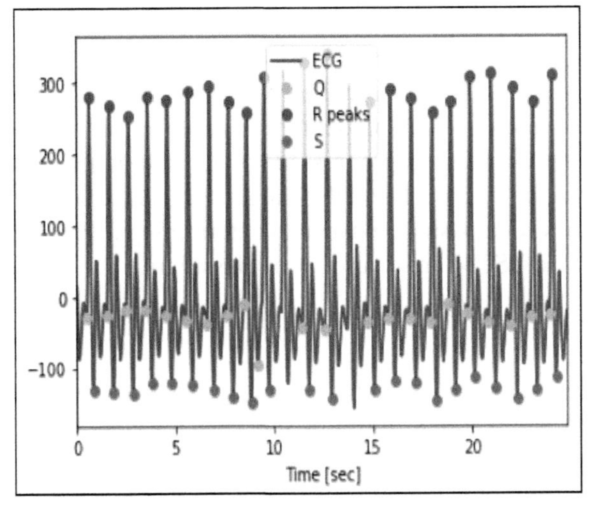

Figure 7.6 QRS detection.

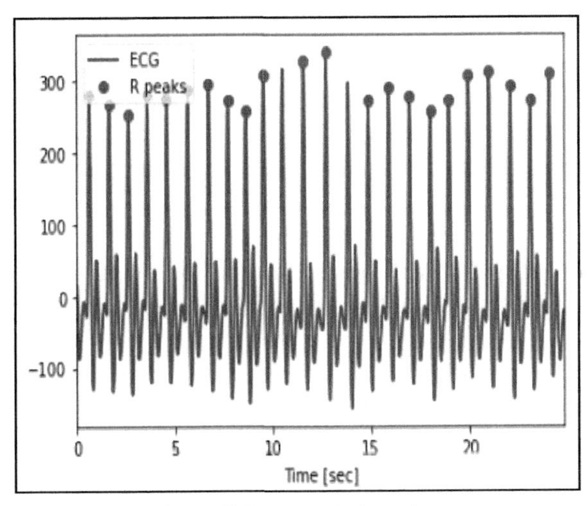

Figure 7.7 R peak detection.

7.5 ECG Classification Using Deep Learning Techniques

The major purpose of this study is to enhance the features of an automated model that may be used on mobile healthcare devices to diagnose MI using ECG data. When using the machine learning method, the features were frequently extracted from raw ECG data using traditional methods or by assuming the features using advanced machine learning models. According to Chen *et al.* (2020), deep learning methods are used to extract hidden information from raw data, which will be useful for data classification. This research effort provides an EDN model that is ideal for this purpose.

Deep neural networks have progressed to the point where they can now accurately analyze a signal. The DNN algorithm and the deep learning approach are useful for classifying data and determining if the data has MI or is healthy. CNN and LSTM are examples of classifiers.

7.5.1 CNN

CNN is called feature learners and have a large capacity for automatically extracting appropriate features from raw input data. CNNs are composed of two main components, each of which performs a distinct purpose. According to Baloglu *et al.* (2019), the feature extractor is the first component in CNNs, and it is generally in charge of ensuring that the features are extracted automatically. The classifier, also known as a fully linked network and a multi-layer perceptron (MLP), is discussed in the second part.

The fully linked segment is in charge of classifying data based on the learned features collected from the first phase. Two common layers are included in the feature extraction section: a convolutional and a pooling. The convolutional layer can extract feature maps from the previous layer. The convolutional layer is composed of several convolution kernels, also known as filters, which are added by bias and then employed in the activation function to extract a feature map that will be used in the following layer. One of the most common sub-layers encountered in the feature extraction section is the pooling layer.

The pooling layer is divided into three types: max, min, and average pooling. The feature maps' resolution is reduced as a result of the pooling layer. The suggested model also includes a max-pooling function, which aids in determining the largest value discovered in a set of close inputs. However, when considering in terms of complexity, there is a distinction between the two. When compared to the completely connected layers, the feature extraction section, which is called the primary layer, helps to function more calculations that comprise feature extraction and feature selection procedures.

Several comparable properties between CNN structure and ANN structure with input layer, hidden layer, and output layer can be discovered in Liu *et al.* (2018). CNN is sometimes known as the developed form of the ANN since, unlike NN, it is translational and shift invariant. CNNs are frequently made up of a variety of layers, including input, average pooling, drop layer, convolution, max-pooling, and softmax layer, among others, and play a vital part in the model. The role of feature extraction is critical for automatically obtaining useful characteristics from ECG signals, while the classification part is responsible for accurately classifying signals using the extracted features.

The fundamental components for feature extraction, as stated above, primarily comprise two layers: convolutional layers and downsampling layers. Convolutional layers, commonly known as C-layers, are a type of fuzzy filter that helps to differentiate the characteristics of original signals while simultaneously reducing noise. Convolution is performed between the higher layer's feature vectors and the current layer's convolution kernel. Finally, according to Li *et al.* (2017), the activation function contributes to the output of convolution calculation results. Eqn (7.1) describes the result of the convolutional layer:

$$x_j^i = f(\sum_{i \in M_j} x_i^{l-1} * W + b_j^l) \tag{7.1}$$

where l_j x is the characteristic vector corresponding to the *j*th convolution kernel of the *l*th layer, and "*M*" is the current neuron's perceptive component, and l_{ij} and *W* are the bias coefficients corresponding to the *j*th convolution kernel of the *l*th layer, and *f* is a nonlinear function (2018).

Concurrently, pooling technology is used to store features classified using three functions: displacement, scaling, and invariance. The downsampling layer contains the function for further feature extraction; however, the spatial resolution found between the hidden layer is diminishing, and its formula is defined in the following equation:

$$X_j^l = f(\beta_j^l \text{down}\left(X_j^{l-1}\right) + b_j^l) \tag{7.2}$$

Here, down() denotes the downsampling function, l_j denotes the weighting coefficient, and l_j b denotes the bias coefficient. There are a number of input and output layers in addition to the C-layers and S-layers. The ECG signal is split as input data before training the network model, and the target's output vector is also mentioned. Eqn (7.3) is used to identify and analyze the error, which is then compared to the provided goal output vector.

$$E = \frac{1}{2}\sum_{k=0}^{n-1}(\ d_k - y_k\ )^2 \leq \varepsilon \tag{7.3}$$

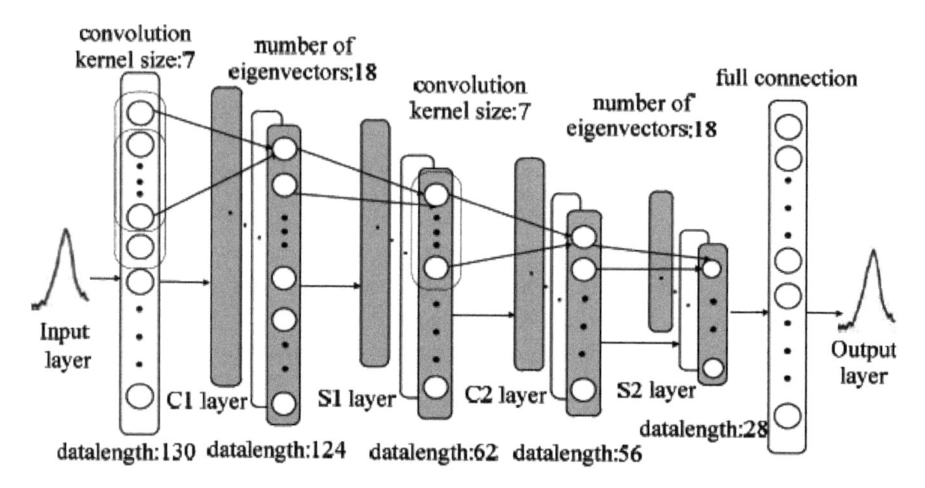

Figure 7.8 CNN architecture for ECG classification.

where E stands for the total error function, k y stands for the output vector, and k d stands for the goal vector. If the training is convincing, it is completed; in the meantime, the weights and thresholds are saved. Each weight is specified to be constant, and the classifier is built. Similarly, if the stage is not reached, the process would be repeated. Convolutional neural networks were designed to deal with two-dimensional data. However, it is mostly used to deal with one-dimensional data. As a result, the CNN model's structure must be managed. Figure 7.8 depicts the CNN model used for ECG categorization.

An input layer, two convolutional layers, a full connection layer, two downsampling layers, and an output layer are the four primary layers in total. According to Li *et al.* (2000), the input of each neuron is closely related to the output of the previous layer, which is primarily utilized to retrieve local information. In the convolutional layer C1, 18 convolution kernels with a length of 7 sampling points are produced, and the input is a 130-sample-point segment of ECG data. It also generates 18 feature vectors and 124 sampling points.

The C1 layer's feature vectors are pooled in the S1 layer, yielding 62 sample points; the C2 layer includes 18 convolution kernels with a length of 7 sampling points, and its output is said to contain 324 feature vectors with 56 sampling points. When the feature vectors are pooled again with the S2 layer, their length is distributed into 28 sampling points, which are then sent to the output layer for classification results evaluation.

7.5.2 LSTM

RNN is important for assessing the underlying representation of a time-varying signal, and it is one of the network topologies developed to address the sequential problem, also known as sequence classification.

Hochreiter and Schmidhuber introduced a significant upgrade to RNN in 1997, which they dubbed the long short-term memory (LSTM). As a result, numerous LSTM applications have lately been improved.

LSTMs are the most widely used and are one of the types of RNN architecture. As a result, LSTM networks (units) are an important structure that contains memory blocks and memory cells, as well as gate units. The use of multiplicative input gate units can assist in defending against the negative effects caused by unrelated inputs (Zhang *et al.*, 2019). According to Kumar *et al.* (2021), the input gate controls data flow into the memory cell, while the output gate controls data flow from the memory cell to other LSTM blocks.

The data flow in an LSTM cell is controlled by the read gate (denoted as "i_t"), forget gate (denoted as "f_t"), and write gate (denoted as "o_t"). As a result, the hidden variables and cell state of the LSTM at time-stamp t are denoted as h_t and c_t, respectively. The letter h_t represents the output of an LSTM cell. x_t represents the input vector to an LSTM cell at time stamp t. Vanilla LSTM use $x_t + L$ as the last input to an LSTM cell, where L denotes the number of segments in an ECG sequence used for recognition. This is depicted in Figure 7.9.

However, it was subsequently found that the final time-stamp output may not precisely describe all of the previously acquired data. Furthermore, an error arises during the back-propagation function due to the first time-stamp,

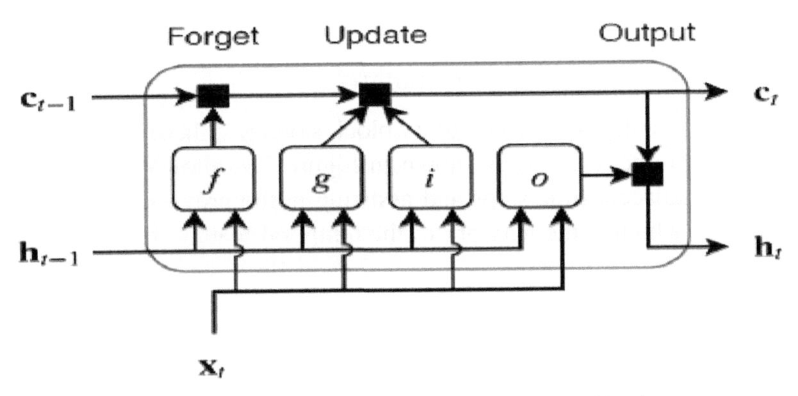

Figure 7.9 Architecture of LSTM for ECG classification.

which may become unimportant (Shadmand *et al.*, 2016). Rather than using the hidden variable as the output, we have used this work to characterize the output at each time-stamp. At each time-stamp, a dropout layer captures and passes the output of the LSTM cell. The output of the dropout layer is then routed through a fully linked layer.

A softmax function is employed in the final portion to calculate the likelihood of an ECG sequence. Each time interval's values are stored in the cell unit. The rest unit gates control the flow of data into and out of the unit. In the memory block structure, the forget gate is controlled by an efficient network known as a simple one-layer neural network. The following equation can be used to calculate the functions of this gate:

$$f_t = \delta(W[X_t, h_{t-1}, C_{t-1}] + b_f) \tag{7.4}$$

in which x_t is the input sequence, c_{t--1} is the previous LSTM block memory, and h_{t--1} is the earlier block output The bias vector is denoted by b_f, the fundamental logistic sigmoid function by, and the various weight vectors used for each input by Darmawahyni *et al.* (2019). An input gate is a unit that constructs new memory based on the previous memory block effect by using a simple NN with an activation function denoted as tanh. As a result, eqn (7.5) is employed in order to evaluate these operations [see eqn (7.6)]

$$i_t = \delta(W [X_t, h_{t-1}, C_{t-1}] + b_i) \tag{7.5}$$

$$C_t = f_t. \mathrm{C_{t-1}{+}i_t}. \tanh(W [X_t, h_{t-1}, C_{t-1}] + b_i) \tag{7.6}$$

The output gate serves as an output of the present LSTM block and can be created using the following equations:

$$\delta_t = \delta(W [X_t, h_{t-1}, C_t] + b_o) \tag{7.7}$$

$$h_t = \delta_t .\tanh(C_t) \tag{7.8}$$

where b_i and b_i are the previous memory block's outputs; these denoted units are connected to one another, as shown in Figure 7.9, allowing the data to cycle between adjacent time steps and assisting in the provision of an inner feedback state, which is the network to the temporal feature in the provided data.

The forget gate in the memory block structure is controlled by a simple one-layer neural network, where x_t represents the input sequence, h_{t-1} represents the previous block output, C_{t-1} represents the previous LSTM block memory, and b_f represents the bias vector. The letter "W" signifies the independent weight vectors for each input in the logistic sigmoid function. The sigmoid activation function, which is considered the output of the forget gate, is used for the preceding memory block by implementing element-wise multiplication. As a result, the quantity of memory consumed in the previous memory block will be determined.

If the activation output vector yields values close to zero, the preceding memory will be deleted. The input gate, on the other hand, is a sector where new memory is formed by combining the preceding memory block effect with a basic NN with the tanh activation function. These functions are then assessed, and the resulting output gate is passed into another section, which generates the output of the current LSTM block.

7.5.3 Enhanced Deep Neural Network (EDN)

MI is distinguished by irregular and unpredictable beats that can be single or many. As a result, the suggested network should be able to handle signals of different lengths. These characteristics are well addressed by the established proposed EDN framework.

The LSTM unit functions by combining a "memory" cell with a gating mechanism comprised of three nonlinear gates: an input gate, an output gate, and a forget gate. The objective of the gate is to control the flow of signals into and out of the cell in order to ensure successful RNN training and to regulate long-term dependence (Feng *et al.*, 2019). Since its inception, the LSTM unit has undergone numerous modifications to improve performance. The number of components in the LSTM architecture can be increased to improve performance.

As a result, CNN and LSTM approaches are primarily used in the suggested enhanced deep neural network model. Incorporating extra components into the LSTM design will improve performance. This model's input layer now contains both healthy and MI data. They are let in via hierarchically organized EDN and dropout layers, which are subsequently translated into various-sized feature maps. Class prediction can only be coded using the thick layer. The dropout method prevents the model from overfitting during training.

According to Manimekalai *et al.* (2020), the model inspects the entire training dataset at each epoch. If the epoch number is large enough, a model can remember the training data. Vector operations such as matrix multiplication and gradient descent are advantageous for huge matrices of data that are processed in parallel using GPU. The bias of h is frequently included to the cell state vector in this manner, which helps to increase performance efficiency. Because the output gate is regarded as less important than the forget gate and the input gate, the recommended approach aids in the adjustment of the hidden state vector by implying point-wise Hadamard multiplication of the prior output gate parameter and previous cell state vector. Because of the parallelism, EDN has a faster progressive performance time. Data that helps each vector in the hidden state can be implicitly stored on a previous cell state unit. Convolutional neural networks (CNN) are used to create an effective neighborhood identification process.

The CNN and LSTM algorithms are used in the proposed enhanced deep neural network (EDN) framework. The number of components in the LSTM architecture may be considered to improve performance. The input layer of this model receives both healthy and MI information. They are transformed into feature maps of varied sizes after passing through EDN and dropout layers in a hierarchical order. The dense layer then provides class prediction automatically. During training, the dropout method protects the model from overfitting. At each epoch, the model examines the complete training dataset. If the epoch number is large enough, a model can memorize the training data. Vector operations on large data matrices, such as matrix multiplication and gradient descent, are performed in parallel with GPU support. In this approach, the bias of h is added to the cell state vector to improve performance. The proposed EDN framework is depicted in Figure 7.10.

Because the output gate is less critical than the input and forget gates, the proposed technique updated the hidden state vector by adding point-wise Hadamard multiplication among the prior output gate parameter and previous cell state vector. Because of parallelism, EDN minimizes the time it takes for a process to run. It has the capability of processing data in such a way that each vector in the hidden state is implicitly dependent on a preceding cell state unit. The proposed model employs CNN to improve the efficiency of the neighborhood identification process. The EDN model implementation reduces training time when compared to an LSTM model.

The EDN model application reduces training time when compared to an LSTM model (Fu *et al.*, 2020). The stateful LSTM units in the model are handled by a fully connected softmax layer, resulting in a possible distribution over system call integers. The functional LSTM and CNN models were both considerably more successful on ECG signals.

To increase the performance of the suggested methodology, the bias of h was added to the cell state vector. Because it was less important than the input and forget gates, the output gate was deleted. It changed the hidden state

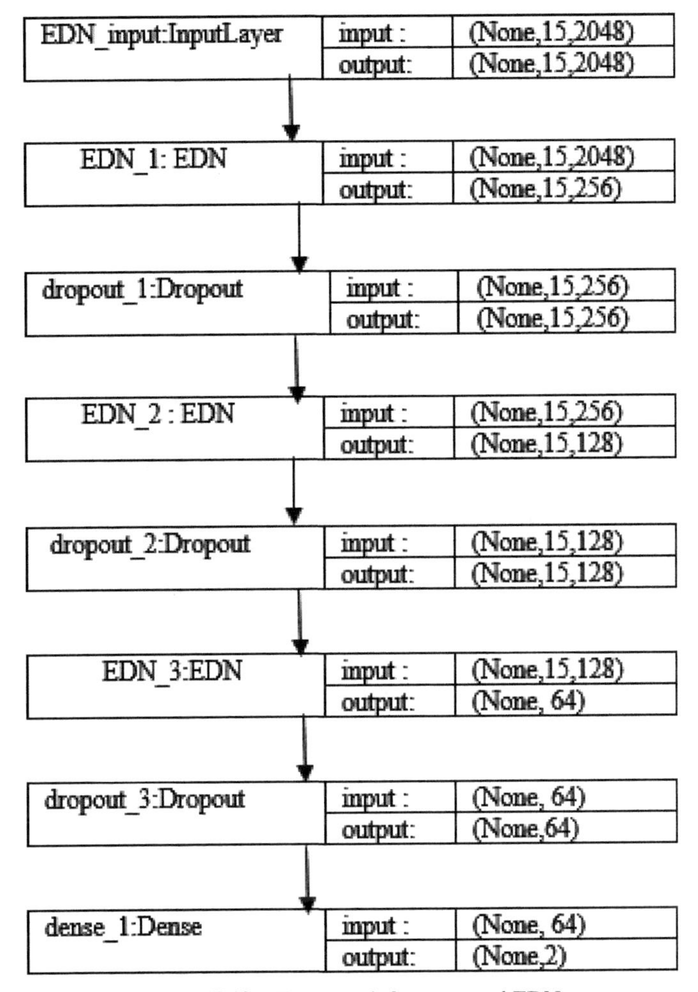

Figure 7.10 Framework for proposed EDN.

vector by combining the prior output gate parameter and the preceding cell state vector using point-wise Hadamard multiplication.

$$x_j^i = f(\sum_{ieM_j} x_i^{l-1} * k_{ij}^l + b_j^l) \tag{7.9}$$

$$i_t = \sigma(W_{xi}x_t + W_{hi}h_{t-1} + W_{ci}c_{t-1} + b_i) \tag{7.10}$$

$$f_t = \sigma(W_{xf}x_t + W_{hf}h_{t-1} + W_{cf}c_{t-1} + b_f) \tag{7.11}$$

$$o_t = \sigma(W_{xo}x_t + W_{ho}h_{t-1} + W_{co}c_{t-1} + b_o) \quad (7.12)$$
$$C_t = \tanh(W_c x_t + r_t(W_c h_{t-1} + b_{ch}) + b_h) \quad (7.13)$$
$$h_t = (1 - o_t * C_t + o_t * C_{t-1}) \quad (7.14)$$

where x_{ji} is the m-dimensional input vector, i_t is the input gate at time t, and f_t is the forget gate at time t, vectors utilizing the sigmoid function of point-wise multiplication vectors. O_t is the output gate at time t. These vectors for the input gate, forget gate, and output gate are all n-dimensional. C_t is the cell state vector, which concatenates vectors using tanh activation. It is the n-dimensional activation unit for cell state. h_t is the hidden state vector, which employs the point-wise Hadamard multiplication operator. Each was determined using an equation ranging from (7.9) to (7.14).

It may process input sequentially, which allows each vector in the hidden state to be implicitly dependent on the previous cell state unit. EDN employs convolutional neural networks to improve the efficiency of the neighborhood identification process.

As a result, it is argued that no extraction from hand-crafted features is required in deep learning models, making them very simple to use. As a result, the properties of these two algorithms were implemented in this research effort in order to function the diagnosis of myocardial infarctions.

Each convolution operation is carried out by moving the kernel across a section of the input vector one sample at a period, multiplying and then

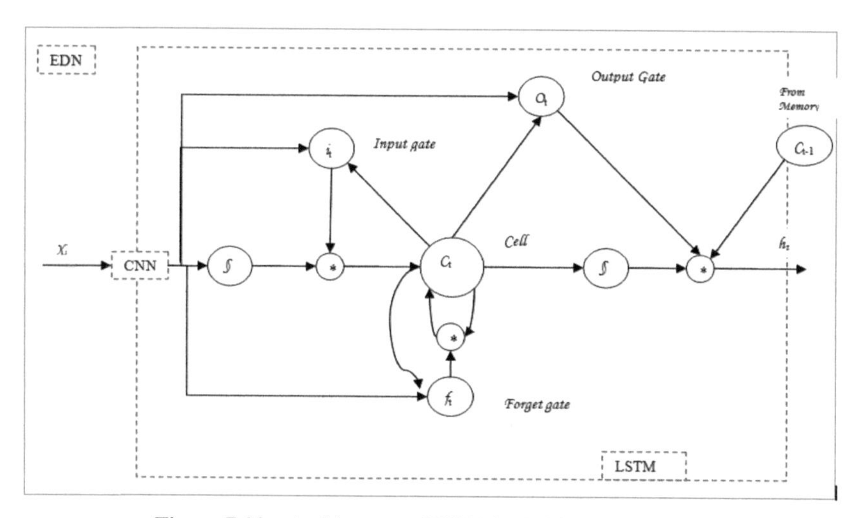

Figure 7.11 Architecture of EDN for ECG classification.

Algorithm EDN: Enhanced Deep Neural Network

Input : $x_i^l = (x_1^l + x_2^l + x_3^l + \cdots + x_i^l)$ - a Chain of Independent Variables
D — Denotes the number of memory blocks.
Sj – the number of cells in Block j.
Step 1: Read the data, then calculate the standard deviation and separate the data.
Step 2: Calculate the total number of independent variables.
Fix(log3(data size)) Mapsize = fix(log3(data size)) -1
Step 3: Create CNN layers by combining Input Layers and Subsampling Layers.

$$\partial^L = (W^{L+l})^T \partial^{L+l} f(u^L)$$

$$x_j^L = f\left(\sum_{i \in M_j} x_i^{l-1} * k_{ij}^l + b_j^l\right)$$

Step 4: Send the CNN output layers units as Input vectors to the LSTM unit's Input gate, Forget Gate, and Output Gate.
Step 5: For each memory block, compute the Input, Forget, and Output gates for j=1 to D.
Evaluate the Input Gate: $i_t = \sigma(W_{xi}x_t + W_{hi}h_{t-1} + W_{ci}c_{t-1} + b_i)$
Evaluate the Forget Gate: $f_t = \sigma(W_{xf}x_t + W_{hf}h_{t-1} + W_{cf}c_{t-1} + b_f)$
Evaluate the Forget Gate: $O_t = \sigma(W_{xo}x_t + W_{ho}h_{t-1} + W_{co}c_{t-1} + b_o)$
for V =1 to S_j do
$C_t = \tanh(W_c x_t + r_t(W_c h_{t-1} + b_{ch}) + b_h)$
Finally, compute to update the hidden state.
Assess the Hidden State: $h_t = (1 - o_t) * C_t + o_t * C_{t-1}$
End for
Step 6: EDN Layers should be returned.

---

adding the values of the superimposed matrices (Shu *et al.*, 2018). In order to achieve the relevant spatial information identified in the provided data, the weights of kernel *k* must be regularly updated by using the network during the training stage. As a result, rather than using a suitable convolution approach, the proposed work employs full convolution characteristics.

This structure was recommended because the entire convolution is made up of shorter length segments that are already padded with zeros. Furthermore, no bias is used or added during the convolution method function to ensure the integrity of the zero padding. Because it is derived from the hypothesized convolutional layer, the output of 10 is believed to be a zero padded sequence. Use the size 2 max-pooling filter technique with non-overlapping stride to cut the size of the input description in half, as seen in the feature maps after each convolutional layer is applied.

In the following part, the necessary LSTM layer is inferred to extract temporal data from these feature maps. The features extracted using the two most commonly used methods, convolution and pooling, are then divided into sequential components and fed into the commonly used LSTM units for temporal analysis. However, in the entirely linked layer that predicts myocardial infarction, just the output from the LSTM's last phase, as stated above, is used.

In comparison to other conventional investigations, the proposed EDN model has delivered efficient findings. By removing the recurrent connection network and dense connections from the network, high classification accuracy of 87.80% can be achieved. The CNN is considered to be better at obtaining spatial features, whereas the LSTM is said to be effective in learning temporal features and to be more efficient at doing so.

By combining these two modalities, the technology was able to increase not only diagnosis accuracy but also the features of the model used in classifying cardiac signals with various sequence lengths. While the introduced deep learning network has trained from start to finish utilizing noisy ECG signals, pre-processing methods such as noise reduction and feature extraction are not necessary.

Assuming that the test segment contains just one type of myocardial infarction model, the suggested system can classify ECG signals of varying lengths. However, this is not always true because the ECG signal seen in the real world may comprise a range of distinct forms of myocardial infarction. In the future, this research might be enhanced to include the use of the featured auto-encoder network on these ECG data to examine element-wise analysis by linking or comparing each pixel to a class label.

This will be able to parallelize the beat detection function while also including the categorization technique as a result of this. High-end graphics cards are necessary to accelerate the model's training process. Instead of using weighted loss for training, data augmentation 21 can be utilized to increase training data variability and alleviate the increasing class imbalance problem.

7.6 Experimental Results

The PTB database was employed in this study, with 80% of the data being imposed for training purposes. Table 7.3 lists the enduring 20% of data that were used for testing purposes. PTB dataset has utilized to identify 148 MI patients' records and 52 normal patients' records for this study, with 118 MI recordings and 41 normal recordings used for training and 30 MI recordings and 11 normal recordings used for testing.

The three deep learning models, CNN, LSTM, and EDN, were implemented in the Google Open Source Research Laboratory (COLAB) using Python code. It is an Internet tool that is free to use. The healthy signal

Table 7.2 PTB dataset used for training and testing purpose

ECG recordings in PTB dataset		
	MI	**Normal**
Total	148	52
Training	118	41
Testing	30	11

and MI signal were classified from the input ECG signals using an ECG signal classification system that was created, trained, and tested, and the experimental findings were presented here.

Table 7.4 shows the ECG classification system's confusion matrix based on the test dataset. Beginning with the raw signal and ending with the final stage, which included the use of a combination of filtering approaches, feature extraction and feature selection, classification methods like CNN, LSTM, and EDN were used in a series of steps. Precision, sensitivity, specificity, and accuracy were determined using the classification approaches' confusion matrix to calculate the performance of the ECG classification system.

Deep learning system contains its own interconnected neurons which help to send and receive messages between each other. These interconnections between neurons are assigned with weights, which signify a network state and are reorganized during the learning process. A feed-forward neural network consists of 10 hidden layers that are used for the classification of myocardial infarction in this research work. As a result, this deep neural network was implemented on a Python notebook in Google Co-laboratory, and the number of neurons present in each hidden layer was limited to 50, requiring this network to be trained utilizing a GPU-based system. An activation works upon the rectified linear unit (ReLU) that is implied for the hidden layers and a sigmoid function was implied at the output layer. Back-propagation which contains stochastic gradient decay is useful for generating the network weights. Thus, the learning rate was enhanced by implying grid search for accuracy and to reduce overfitting.

7.6.1 Performance Evaluation

The performance level of the ECG signal classification system depended on numerous important factors including dataset used for experimental purpose, filtering process for noise removal, identifying and selecting important features that are present in the data, and specific classification methods that suit as well as give better results.

Table 7.3 Confusion matrix for the ECG classification system

			Actual class	
			Abnormal	Normal
Raw signal + CNN	Predicted class	Abnormal	23	6
		Normal	7	5
			Actual class	
			Abnormal	Normal
Raw signal + LSTM	Predicted class	Abnormal	23	7
		Normal	7	4
			Actual class	
			Abnormal	Normal
Raw signal + EDN	Predicted class	Abnormal	24	7
		Normal	5	5
			Actual class	
			Abnormal	Normal
Filtered signal + CNN	Predicted class	Abnormal	25	5
		Normal	4	6
			Actual class	
			Abnormal	Normal
Filtered signal + LSTM	Predicted class	Abnormal	24	4
		Normal	5	7
			Actual class	
			Abnormal	Normal
Filtered signal + EDN	Predicted class	Abnormal	26	5
		Normal	4	6
			Actual class	
			Abnormal	Normal
Filtered signal with selected features + CNN	Predicted class	Abnormal	28	3
		Normal	4	6
			Actual class	
			Abnormal	Normal
Filtered signal with selected features + LSTM	Predicted class	Abnormal	27	3
		Normal	3	8
			Actual class	
			Abnormal	Normal
Filtered signal with selected features + EDN	Predicted class	Abnormal	29	4
		Normal	1	7

7.6.2 Evaluation Metrics

The estimate metrics were used in the experiments to estimate the performance of ECG classification approaches; the performance of the ECG classification system will be evaluated based on the results of these experiments. Precision, recall, $F1$ measure, Cohen kappa coefficient, and accuracy were the metrics used to measure the system's performance in this research work.

Basically, the confusion matrix is used to measure the performance of an algorithm. Confusion matrix contains these four values from the actual data and the predicted data.

- True negative (TN) means number of patients without disease, who shows a negative test result with the assay in question.
- True positive (TP) means number of patients with the disease, who shows a positive test result with the assay in question.
- False negative (FN) means number of patients with the disease, who shows a negative test result with the assay in question.
- False positive (FP) means number of patients not affected by the disease, who shows a positive test result with the assay in question.

Precision defines number of the proportion of patients with a positive test result which describes the patients do have the disease. It is denoted in eqn (7.15).

$$\text{Precision} = \frac{\text{TP}}{(\text{TP+FP})} \tag{7.15}$$

Recall defines number of the proportion of patients affected by the disease who get a positive test result with the assay in question. It is denoted in eqn (7.16).

$$\text{Recall} = \frac{\text{TP}}{(\text{TP+FN})} \tag{7.16}$$

$F1$ measure defines twice the number of the proportion of patients affected by a disease and the number of the proportion of patients with a positive test result with the assay in question. It is denoted in eqn (7.17).

$$F1\ \text{measure} = \frac{2\text{TP}}{2\text{TP} + \text{FP} + \text{FN}} \tag{7.17}$$

Cohen kappa coefficient defines the probability of actual class minus the probability of predicted class divided by 1 minus the probability of predicted class. It is denoted in eqn (7.18).

$$\text{CKC} = \frac{P_0 - P_e}{1 - P_e} \tag{7.18}$$

where P_0 is the probability of the actual class and P_e is the probability of the predicted class.

Accuracy defines the number of correctly classified data such as normal and abnormal, to the total number of classified results. Accuracy describes the correctness of the measurement that predicts the correct value. It is denoted in eqn (7.19).

$$\text{Accuracy} = \frac{\text{TP+TN}}{(\text{TP+TN+FP+FN})} \tag{7.19}$$

7.7 Results and Discussion

The three deep learning models, CNN, LSTM, and EDN, were implemented in the Google Open Source Research Laboratory, or COLAB, using Python code. It is an open source online tool. ECG signal classification system was generated, trained, and tested to classify the healthy signal and the MI signal from the input ECG signals and the experimental results were discussed here.

Table 7.5 clearly compares ECG signal categorization methods based on performance metrics precision. Based on the resultant data obtained in Table 7.5, it is clear that the filtered signal with selected features along with EDN performance is significant while comparing with combinations of CNN and LSTM algorithms.

Figure 7.12 illustrates a graphical picture of the comparison of ECG classification methods using CNN, LSTM, and EDN, beginning with the combination of raw ECG signals and progressing to pre-processed and selected features to classify the ECG signals. By comparing the performance of the ECG classification system in terms of precision, it is clear that the filtered signal with selected features along with EDN yields 96.67% precision while comparing with CNN and LSTM, which yields 90.32% and 90%, respectively.

The comparison of ECG signal categorization models based on the performance metrics accuracy is clearly seen in Table 7.6. When comparing the filtered signal with selected features along with EDN performance to

Table 7.4 Precision-based comparison of ECG classification methods

Methods	**Precision (%)**
Raw signal + CNN	79.31
Raw signal + LSTM	76.67
Raw signal + EDN	77.42
Filtered signal + CNN	83.33
Filtered signal + LSTM	85.71
Filtered signal + EDN	83.87
Filtered signal with selected features + CNN	90.32
Filtered signal with selected features + LSTM	90.00
Filtered signal with selected features + EDN	**96.67**

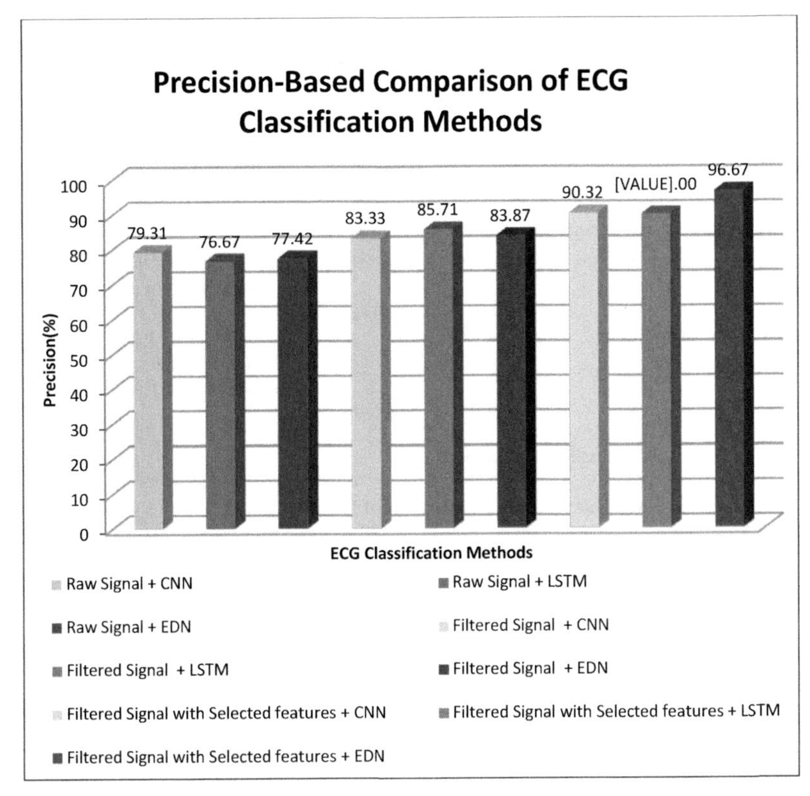

Figure 7.12 Precision-based comparison of ECG classification methods.

Table 7.5 Accuracy-based comparison of ECG classification methods methods

Methods	Accuracy
Raw signal + CNN	68.29
Raw signal + LSTM	65.85
Raw signal + EDN	70.73
Filtered signal + CNN	77.50
Filtered signal + LSTM	77.50
Filtered signal + EDN	78.05
Filtered signal with selected features + CNN	82.93
Filtered signal with selected features + LSTM	85.37
Filtered signal with selected features + EDN	**87.80**

combinations of CNN and LSTM algorithms, it is evident that the filtered signal with selected features along with EDN performance is substantial.

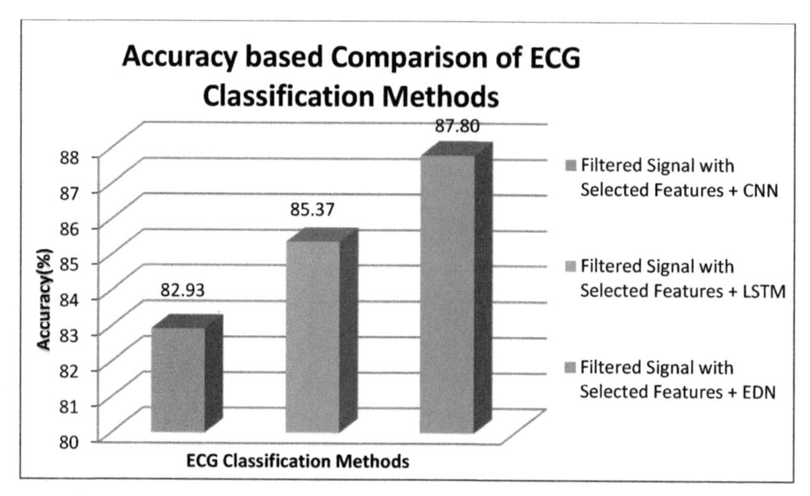

Figure 7.13 Accuracy-based comparison of ECG classification methods.

Table 7.6 Techniques for ECG classification comparison

Method	CNN (%)	LSTM (%)	EDN (%)
Precision	90.32	90.00	**96.67**
Recall	87.50	90.00	**96.67**
*F*1 measure	88.89	90.00	**92.06**
Accuracy	82.93	85.37	**87.80**

Figure 7.13 shows a graphical representation of the classification methods used to classify ECG signals: pre-processed signal with selected features + CNN, pre-processed signal with selected features + LSTM, and pre-processed signal with selected features + EDN.

When comparing the accuracy of the ECG classification method, pre-processed signals with selected features + EDN scores 87.80%, whereas CNN and LSTM score 82.93% and 85.37%, respectively. The comparison of ECG categorization techniques is shown in Table 7.7.

As shown in Table 7.7, the classification methods of pre-processed signal with selected features + CNN, pre-processed signal with selected features + LSTM, and pre-processed signal with selected features + EDN were used to classify the ECG signals in terms of precision, recall, *F*1 measure, and accuracy. When compared to CNN and LSTM, EDN outperforms them on all criteria Fu *et al.* (2020).

Finally, the proposed algorithm is demonstrated in execution Fu *et al.* (2020). The EDN model's loss and accuracy for training and testing data

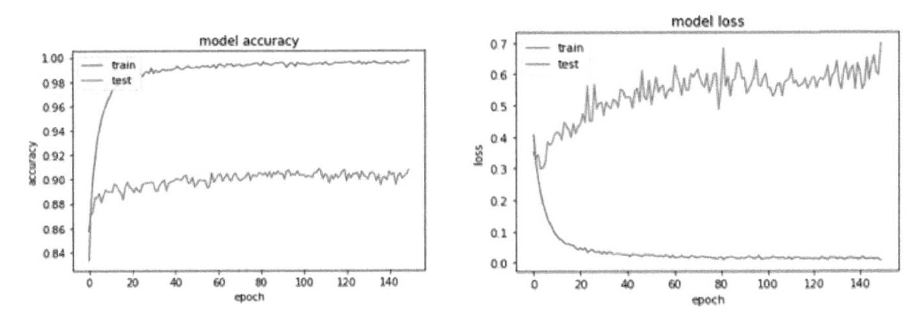

Figure 7.14 Loss and accuracy for training and testing data for EDN algorithm.

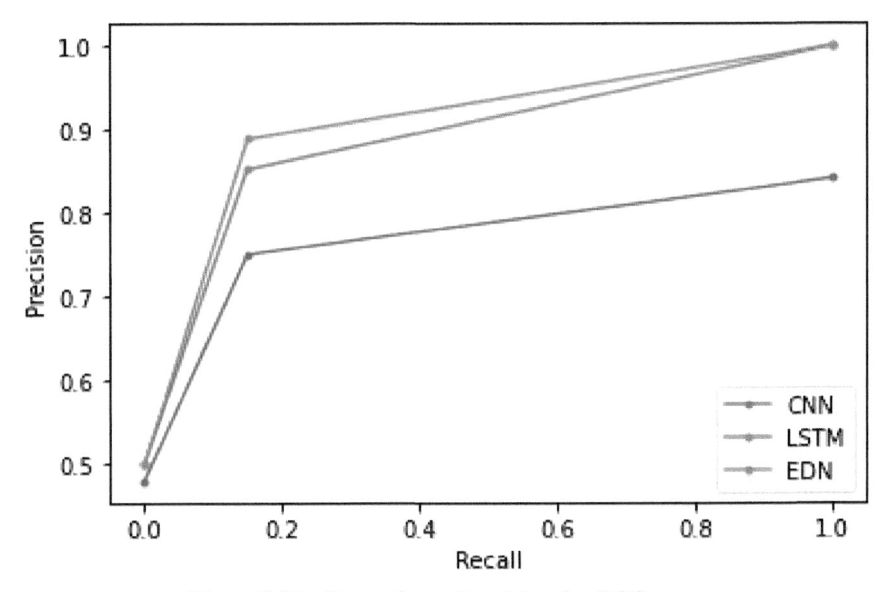

Figure 7.15 Comparison of models using ROC curve.

are shown in Figure 7.14. The ROC curve represents the performance of the classification algorithm. Figure 7.15 depicts the precision–recall curve.

7.8 Conclusion

This chapter offers an outline of the classification strategies used to categorize ECG data as well as an explanation of why the recommended approach for classifying the ECG signal is necessary. Convolution neural network, LSTM, and EDN algorithms were used in the classification procedure. When comparing ECG classification system accuracy, pre-processed signals with selected features + EDN obtain 87.80%, whereas CNN and LSTM achieve 82.93% and 85.37%, respectively. The overall findings of the performance metrics reveal that the suggested EDN algorithm overtakes deep learning techniques such as CNN and LSTM when compared to the experimental results. According to the confusion matrices of the algorithms, the proposed model EDN has a higher accuracy of 87.80% for the PTB database. The recommended model displays performance improvisation based on metrics such as precision, recall, *F*1 measure, and Cohen kappa coefficient. These improvements to EDN's performance would help to save human lives.

References

Baloglu UB, Talo M, Yildirim O, Tan RS, Acharya UR. (2019). Classification of Myocardial Infarction With Multi-Lead ECG Signals and Deep CNN. Pattern Recognition Letters. Vol:122, pp:23–30.

Benjamin EJ, Blaha MJ, Chiuve SE, Cushman M, Das SR, Deo R. (2017). Heart Disease and Stroke Statistics, Update: A Report from the American Heart Association.

Chen C, Zhengchun H, Ruiqi Z, Guangyuan L, Wanhui W. (2020). Automated Arrthythmia Classification Based on a Combination Network of CNN and LSTM. Biomedical Signal Processing and Control.

Darmawahyuni A, Nurmaini S, Sukemi, Caesarendra W, Bhayyu V, M Naufal Rachmatullah, Firdaus. (2019). Deep Learning with a Recurrent Network Structure in the Sequence Modeling of Imbalanced Data for ECG-Rhythm Classifier. Algorithms. pp:118–118.

Debasish J, Samarendra D. (2020). An LSTM Based Model for Person Identification using ECG Signal. IEEE Sensors Letters.

Dan L, Jianxin Z, Qiang Z, Xiaopeng W. (2017). Classification of ECG Signals Based on 1D Convolution Neural Network, 2017 IEEE 19th International Conference on e-Health Networking, Applications and Services (Healthcom).

Feng K, Pi X, Liu H, Sun K. (2019). Myocardial Infarction Classification based on Convolutional Neural Network and Recurrent Neural Network. Applied Sciences. Vol:9, Is:9, pp:1869–1879.

Fu L, Lu B, Nie B, Peng Z, Liu H, Pi X. (2020). Hybrid Network with Attention Mechanism for Detection and Location of Myocardial Infarction Based on 12-Lead Electrocardiogram Signals. Sensors. Vol:20, Is:4, pp:1020–1020.

Lana AA, Muzhir SA-A. CNN-LSTM Based for ECG Arrhythmias and Myocardial Infarction Classification. Advances in Science, Technology and Engineering Systems Journal.

Liu W, Huang Q, Chang S, Wang H, He J. (2018). Multiple-Feature-Branch Convolutional Neural Network for Myocardial Infarction Diagnosis using Electrocardiogram. Biomedical Signal Processing and Control. Vol:45, pp:22–32.

Maharaj, Alonso AM. (2014). Discriminant Analysis of Multivariate Time Series: Application to Diagnosis Based on ECG Signals. Computational Statistics & Data Analysis. Vol:70, pp:67–87.

Manimekalai K, Kavitha A. (2020). Deep Learning Methods in Classification of Myocardial Infarction by Employing ECG Signals. Indian Journal of Science and Technology. Vol:13, Is:28, pp:2823–2832.

Manimekalai K, Kavitha A. (2019). Methodology of Filtering the Noise and Utilization of STEMI in the Myocardial Infarction. Vol:9, Is:1, pp:2249–8958.

Praveen K, Priyanka S, Aghijeet S, Ankush P *et al.* (2021). Prediction of Real-World Slope Movements via Recurrent and Non-recurrent Neural Network Algorithms: A Case Study of the Tangni Landslide. Indian Geotechnical Journal.

Rajesh N, Ramachandra AC, Prathibaha. (2021). Detection and Identification of Irregularities in Human Heart Rate, International Conference on Intelligent Technologies (CONIT).

PTB Diagnostic ECG Database. (2004). Available from: https://www.physionet.org/ content/ptbdb/1.0.0/.

Rajkumar A, Ganesan M, Lavanya R. (2019). Arrhythmia Classification on ECG Using Deep Learning, 2019 5th International Conference on Advanced Computing & Communication Systems (ICACCS), pp. 365-369. IEEE.

Shadmand S, Mashoufi B. (2016). A New Personalized ECG Signal Classification Algorithm Using Block-Based Neural Network and Particle Swarm Optimization. Biomedical Signal Processing and Control. Vol:25, pp:12–23.

Shu LO, Eddie YKN, Ru ST, Rajendra AU. (2018). Automated Diagnosis of Arrhythmia Using Combination of CNN and LSTM Techniques with Variable Length Heart Beats. Computers in Biology and Medicine.

Shweta HJ, Vipul KD, Harshadkumar BP. (2015). Classification of ECG Signals Using Machine Learning Techniques: A Survey, 2015 International Conference on Advances in Computer Engineering and Applications.

Wang HM, Zhao W, Jia DY, Hu J. (2019). Myocardial Infarction Detection Based on Multi-Lead Ensemble Neural Network, 2019 41st Annual International Conference of the IEEE Engineering in Medicine & Biology Society (EMBC).

Xue X, Sohyun J, Jianqiang L. (2020). Interpretation of Electrocardiogram (ECG) Rhythm by Combined CNN and biLSTM. IEEE Access.

Zhang X, Li R, Hu Q, Zhou B, Wang Z. (2019). A New Automatic Approach to Distinguish Myocardial Infarction Based on LSTM. In: and others, editor. 2019 8th International Symposium on Next Generation Electronics (ISNE). IEEE.

8

Automation Algorithm for Labeling of Oil Spill Images using Pre-trained Deep Learning Model

V. Sudha[1] and Anna Saro Vijendran[2]

[1]Department of Computer Science, Sri Ramakrishna College of Arts and Science, India
[2]Sri Ramakrishna College of Arts and Science, India
E-mail: sudhavaiyapuri@gmail.com; annasarovijendran@srcas.ac.in

Abstract

Image annotation is a tough venture in deep learning. Without the annotation, it is hard to perceive objects for machines. This chapter's main goal is to improve a concept of an automatic annotation system that includes a pre-trained semantic segmentation model and MATLAB Image Labeler Tool. In this chapter, automatic annotation of synthetic aperture radar images of the oil spills is carried out using pixel-wise semantic segmentation that is perhaps the most mainstream errand in computer vision. Presently, deep-learning-based convolutional neural networks redriving significant advances in semantic segmentation due to their incredible feature representation. This proposed method includes a pre-trained DeepLabv3+ along with ResNet18 as a backbone model to create an automation algorithm. DeepLabv3+ assigns different categories to each pixel in an input image. Image Labeler is used to create an automation algorithm for automatic labeling of oil spill images. The novelty of the article is due to adapting pre-trained DeepLabV3+ uses ResNet18 as the backbone for image annotation using the Image Labeler feature to improve the generalization ability of the system. Broad analyses on proposed semantic image segmentation division using ResNet18 as backbone are performed over oil spill SAR dataset, and results are accurate when compared to Xception and MobileNet backbone model.

Keywords: Synthetic aperture radar, deep learning, semantic segmentation, ResNet18, Xception, MobileNet computer vision.

8.1 Introduction

Image annotation is a complicated activity of detecting and classifying every object in a dataset. Although the process is very essential in some cases, the complexity in manual annotation limits the use of annotation and object detection in their tasks. So, a new automation algorithm is developed with the intention to do the most effective object detection and classification. Image annotation helps machines to recognize the varied objects through computer vision and understand like humans. It labels the data using the annotation algorithm to create a massive training dataset for machine learning. Without the annotation, it is difficult for machines to recognize the objects. Automatic image annotation (AIA) is one among the image retrieval techniques in which the images are annotated with semantic keywords automatically and then it will be easier to retrieve the images. Image annotation is the process of labeling photographs with a set of pre-set descriptions based on the image attribute. This can assist in bridging the gap between low-level visual features and the high-level meanings derived from the image. The core idea behind picture annotation is to extract semantic notions from a large number of sample photographs and apply them to new images automatically. The photos have already been labeled, allowing them to be found quickly using keywords. Due to diverse imaging settings, mind-boggling and difficult to depict objects, highly textured background, and occlusions, automatic picture annotation is a difficult process [23].

Problems and challenges of image annotation are mentioned below.

- Manual annotation has traditionally been employed for databases with vast quantities of photos. Manual annotation, on the other hand, is an exceptionally tedious and exorbitant interaction for an enormous number of datasets.
- Segmenting oil spills from an image having different classes is a crucial task. Usually, classes and class labels will be different from one class to another.
- Low dimension images from the Internet are very difficult to process in an automation algorithm. The semantic gap occurs when low-level features are obtained from image annotation.

- Annotation of new images will be possible only after training and learning of the model. The task of object recognition for semantic prediction is a challenging work [3].

Pixel-based semantic segmentation allows the classification of objects making computer vision localize the images and predicts the object more precisely. It coordinates the various articles in various regions as a similar class and uses them in preparing the model. It names every pixel of an image for understanding the features in images and solves the computer vision problem into deep-learning-based segmentation. It upgrades the precision in locating multiple objects using computer vision. The satellite images like monitoring of oil spills are used with semantic segmentation model to gather the information.

The deep learning method employs a convolution neural network; it uses the concept of machine learning along with stacking of depth and width of layer architecture. The extraction of discriminative features or image representations from the input data determines the performance of a pre-trained network. Machine learning algorithms perform a better feature extraction compared to traditional models. Many hidden layers in neural networks [18] are capable of deriving great levels of abstraction from raw data. Convolutional neural networks (CNNs) [21] are used to learn image representations and can be utilized to solve a variety of computer vision challenges. Deep CNNs, in particular, are made up of numerous layers of linear and non-linear processes that are all learned at the same time. The settings of these layers are learned over numerous iterations to solve a specific task. In recent years, CNN-based approaches for feature extraction from pictures and video data have gained popularity. A CNN is made up of convolutional and pooling layers that alternate in appearance. Convolutional layers are made up of stacks of predefined-size filters that are convolved with the layer's input. The depth of the CNN can be raised by making the pooling layer's output the next convolutional layer's input. The convolution layer effectively learns visual characteristics. It takes an input image and creates an output with the same dimensions and classes. Convolution, activation or ReLU, and pooling are the most popular layers.

- Convolution processes the incoming images through a series of convolutional filters, each of which activates different aspects of the images.
- By mapping negative values to zero and preserving positive values, the rectified linear unit (ReLU) enables faster and more successful training.

Because only the activated characteristics are carried on into the next layer, this is frequently referred to as activation.

- Pooling reduces the number of parameters that the network must learn by conducting non-linear downsampling on the output.

The performance of CNNs with smaller filter sizes (3 × 3) and deeper architectures has improved. The ResNet18 [19] is one such example (shown in Figure 8.3.

8.2 Related Work

8.2.1 Image Annotation Algorithm

Image annotation the usage of CCA-KNN, that's a brand-new version primarily based totally on combining the functions including CNN and phrase embedding vector. The foremost goal of this technique is to keep away from more than one function computing and additionally beneficial in lots of real-global applications. The findings are presented for all three versions of the CCA designs, with the linear, kernel, and *k*-nearest neighbor clustering. CCA-KNN, which beats previous results and achieves similar results on all the datasets [24], has been used to obtain outstanding results. Image annotation at pixel level is executed through guided filter network (GFN). GFN facilitates in developing labels and optimizing iteratively to label the very last photograph. Comparing the conventional weakly supervised segmentation methods, semantic segmentation performs well [26]. From the multi-label dataset, decided on labels are used primarily based totally on rating function. The annotation set of rules consists of the fusion of CNN functions and VGG16 spine community alongside ideal thresholding. This thresholding idea avoids the downsampling of photos and predicts the right label masks. The gain of this technique is the stepped forward parameters and downside is because of shallow CNN; better stage functions cannot be anticipated accurately. It may step forward the usage of deepening the layers [9]. The convex deep mastering fashions including tensor deep stacking network and kernel deep convex network are used to annotate the photograph and it, in particular, makes use of CNN functions. It takes much less time to teach the photograph [14].

Faster RCNNs with pre-skilled fashions VGG-16 and RFCN with ResNet101 are fine-tuned to categorize items into both foreground and background. It has been found that the proposed automated annotation method could be very green in detecting any unknown items even in an unknown environment. Robustness of the version to any new item has been confirmed with foreground detection outcomes while examined on completely new item sets. The version is likewise tested to be strong to photos of any digital digicam resolutions and unique light conditions. The proposed annotation method is framed to generate a square ROI round every item, however will now no longer be capable of generating a segmented item location the usage of the given architecture. In order to get the precise contour of an item, these paintings may be prolonged by making use of pixel-clever semantic segmentation techniques, like Mask RCNN or PSPNet in the area of faster-RCNN/RFCN [21]. The most important demanding situations get up from the problem of characterizing complicated and ambiguous contents of the satellite TV for PC photos and the excessive human exertions value due to getting ready a huge quantity of education examples with brilliant pixel-stage labels in absolutely supervised annotation methods. This trouble is overcome in a weakly supervised style with a green excessive-stage semantic function moving scheme. The proposed method made complete use of auxiliary satellite TV for PC photograph information set to examine excessive-stage functions primarily based totally at the SDSAE deep mastering technique after which transferred the found-out functions to carry out annotation with handiest a small quantity of education information, which basically decreased the value of labeling education information. Evaluations have proven that our technique can offer aggressive overall performance in comparison with the absolutely supervised methods [27]. To create a multi-label learning issue, use the automatic picture annotation method on deep CNN. When training this model, the images generated can help to reduce overfitting and improve the generalization capabilities of the CNN model [10].

Superpixel segmentation, a novel set of criteria, is combined with a hierarchical Dirichlet technique to analyze objects in images that are represented as a bag of words. Superpixels that appear in a cluster on a regular frequency are stitched together to create false composite images, and the associated labeling is examined using a reverse image search [16]. The combined prediction, post-processing, and adjustment model can be used in annotating images or videos. The aerial images, which have much less salient capabilities for detection. Also, by means of leveraging consumer click on facts and the adjustment version, we are able to enhance the general IOU and expand the framework at some stage in runtime to evolve to new instructions whose classified schooling statistics are not conveniently available [19].

To address the issues faced by diverse objects under varied lighting circumstances, Labeling Tool which was created for image segmentation is more efficient. Superpixels are the fastest for items that are somewhat large and have a lot of contrast with their surroundings. Polygons are useful for large or low-contrast objects, but the brush is useful for small details. The ground truth labels have been shared as mask files in a publicly available dataset for other researchers to use. The image dataset, as well as the photo labeling tool, will be upgraded in the future [20].

8.2.2 Semantic Segmentation

For segmentation, semantic segmentation models such as FCN, U-Net, and DeepLabV3 were utilized. Even though the dataset has one-of-a-kind versions in digital digicam calibration, the pre-skilled deep-learning-primarily based totally segmentation fashions deliver correct effects. Each fashion's overall computing performance on the CPU and GPU is calculated. DeepLabV3 and U-Net are a little slower than the FCN version in terms of inference time. Deep-learning-based segmentation models outperform standard segmentation approaches by a significant margin [8]. The DeepLabV3+ semantic segmentation framework mixed with the Xception version produces inadequate facts withinside the nearby aspect [7].

All the prevailing capabilities presently have the hassle of currently no longer sufficiently describing the images. The goal of the AIA technique is to bridge the gap between low-stage visible photo capabilities collected by machines and high-stage semantic notions perceived by humans. Another annoyance is the simultaneous execution of all annotations. When confronted with huge education datasets, several photo annotation fashions require a lengthy time and computational complexity inside the education phase, making them computationally extensive [1]. The CMGGAN network is utilized to develop a unique technique for semantic segmentation of range sensor data. The suggested technique completely avoids the luxurious and difficult labeling of ranging sensor facts because semantic segmentation networks are pre-trained on the fact sets. Although the experimental outcomes look to be promising, it is critical to train the model on a much larger dataset, which will most likely include more dynamic item detections [12]. On SAR pictures, the one-of-a-kind implementation of the U-Net structure is investigated, with one where U-Net is used from the scratch, while the other is from pre-trained weights. The switch was flipped. U-Net is capable of recognizing minute details inside a picture, such as little rivers and other features [18]. Pixel-level semantic segmentation is improved by combining global context information with local picture attributes. First, in the encoder network, we establish a fusion layer that allows us to merge global and local characteristics. In the decoder network, a stacked pooling block is used to extend the receptive fields of features maps and is needed to contextualize local semantic predictions. This strategy is based on ResNet18, which decreases the number of parameters in our model and allows it to predict better than earlier models [22].

8.3 Proposed Method

Image annotation has major critical worth in recovering and concealing pictures with various classes. This chapter incorporates the annotating of pictures utilizing semantic segmentation. The proposed method has two stages. At first, the semantic segmentation is done utilizing pre-trained DeepLabV3+ model and later the image labeling is done by MATLAB Image Labeler. The proposed work is carried out on utilizing MATLAB Image Labeler and distinctive SAR datasets that incorporate ENVISAT, ALOSPOLSAR, TERRASAR, and SENTINAL.

MATLAB Computer Vision tool kit is more proficient in picture handling than OpenCV. The target of this proposed calculation is to remove classes from a picture and mark them with a reasonable class utilizing DeepLabV3+ (a pre-prepared convolutional neural network). Figure 8.2 depicts an architectural overview of the proposed framework. The basic steps are: image acquisition, pre-processing, semantic segmentation, and automation algorithm, and, finally, annotated image is generated.

The proposed work comprises the accompanying modules.

- Image pre-processing
- Semantic segmentation
- Annotation algorithm

8.3.1 Image Pre-Processing

It is utilized to upgrade pictures that aid in the precision of results. MATLAB has image pre-processing instruments that aid in including investigation and commotion decrease. In this proposed work, forgetting the blunder-free outcomes in semantic division measures, the following pre-processing steps are necessary.

1. Data Acquisition

SAR oil spill datasets are collected from different satellites such as ENVISAT, SENTINEL, ALOSPOLSAR, and TERRASAR. The characteristics of

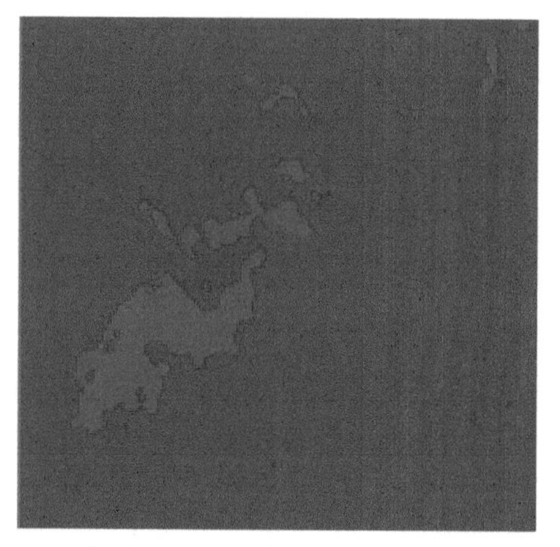

Figure 8.1 Ground truth image.

Table 8.1 Characteristics of SAR satellites

Satellite sensors	Spectral bands	Spatial resolution	Temporal resolution	Swath width	No. of datasets
TERRASAR	X	18.5	11	260	2
ENVISAT	C	30	3	400	12
ALOS POLSAR	L	6	14	350	13
SENTINEL	SWIR	10	5	290	13

satellites are shown in Table 8.1. The oil spill images are obtained from the Gulf of Mexico (2010), the Mediterranean shoreline of Israel (2021), and MV Wakashio oil spill of Mauritius (2020). All pixels within the image had been categorized into classes, specifically oil, and background as shown in Figure 8.1. The labels had been used to create ground truth records for training and validation of the semantic segmentation algorithm. The Image Labeler tool in MATLAB software was used to carry out the labeling technique. Labeling can be done by hand, semi-automatically, or automatically with the use of an automation program. Automated labeling is used in this study to guarantee that the region of interest is appropriately delineated.

2. Image Resizes

The original picture size and pixel picture size for semantic division network actuations ought to be the same for all datasets, to keep away from the DAGNetwork mistake.

3. Data Augmentation

As the training dataset only contains 40 images, overfitting is a serious concern. If we train for a couple of epochs over this small dataset, we would fear that our version will begin becoming the noise on this small dataset, main too bad overall performance on out-of-pattern examples. This trouble may be relatively mitigated by means of records augmentation. It helps in expanding network exactness. DeepLabV3+ is trained using augmented image data. Data augmentation enabling saves you the community from overfitting and memorizing the precise information of the schooling images. Load the pattern data, which includes oil spills.

4. Log Normalization

It is a pre-processing for the machine learning process in MATLAB. It is a method for standardizing the data. It helps to enhance dark features. It applies the log transformation to the pixel values in the image. It is used to increase the range of dark areas while avoiding clipping bright areas.

$$x' = \log(x + 1).$$

The pixel values of the output and input images are x' and x, respectively. To avoid a 0 value in pixels, each pixel value is multiplied by 1. The minimum pixel intensity value should always be at least one.

Log Transformations' Properties:

- The range of gray levels is enlarged for input images with lower amplitudes.
- The range of gray levels is compressed for greater amplitudes of the input image.

5. Hybrid Median Filter

It is the extension of the median filter and preserves the edges better. It smoothens the noise in the image. The steps involved in HMF are as follows.

1. Get median value for horizontal and vertical pixels.
2. Get median value for diagonal pixels.
3. Find the center pixel value.
4. Find median again for the first, second, and third steps and keep replacing for new value.

Figure 8.2 depicts the suggested method's overall architecture. It shows the process of automatic annotation of the dataset. The dataset is used from

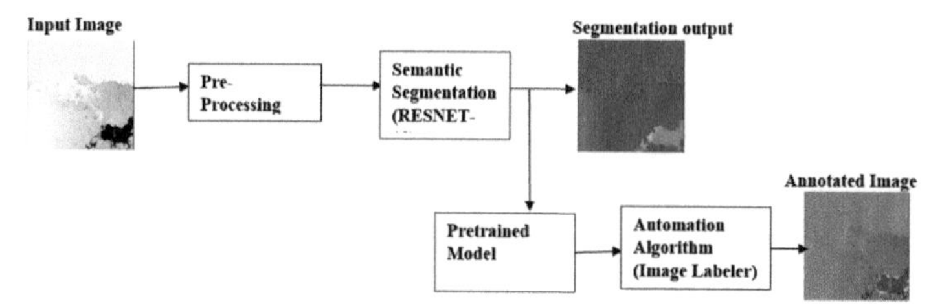

Figure 8.2 An architectural overview of the proposed framework.

the several SAR satellites listed in Table 8.1. The pre-processing is applied to the dataset. DeepLabV3+ semantic segmentation with ResNet18 as the backbone is trained and saved in MATLAB. Load the pre-trained model in custom algorithm creation to annotate the image automatically.

8.3.2 Semantic Segmentation

Semantic segmentation is an advancement to make forecasts from coarse to fine deduction by making thick expectations construing names for each pixel; so every pixel is marked with the class. The process of semantic segmentation includes the following.

1. Label Data or Obtain Labeled Data

For semantic segmentation, a pixel-wise label or ground truth label is applied, which means for each pixel in the training set, there is a label pixel for it. And all pixels have their own label. For example, we have 512 × 512 × 3 image dimensions; so we will have 512 × 512 label data. Every pixel in the image will have one label. So, in this work, there are two objects such as oil and background in the image, 0,1 as the classes label. The label image will have 512 × 512 containing label 0,1 for each pixel.

2. Create a Datastore for Original Images and Labeled Images

The datastore is created for original input and ground truth data. Once the image is labeled using the Image Labeler tool, the labeled images are stored as Pixel Label Data. Both the datasets are further used for training or data augmentation process.

3. Create a Semantic Segmentation Pre-Trained Network

An encoder and a decoder form the foundation of image segmentation architecture. Filters are used by the encoder to find exact features from the image.

The decoder is responsible for generating the final output, which is often a segmentation mask representing the shape of the object. This architecture or a version of it can be found in almost all architectures. Deep lab architecture is used in this study. Convolutions with upsampled filters are employed in this architecture for jobs that need dense prediction. To partition objects at various scales, atrous spatial pyramid pooling is utilized. CNN is employed to improve object boundary localization. Upsampling the filters through the insertion of zeros or sparse sampling of input feature maps is used to produce atrous convolution.

4. Train and Evaluate the Network

The input images used color images with the resolution of 512 pixels × 512 pixels for both DeepLabV3+ and SegNet networks. Due to the different network architectures used, the setup parameters of each architecture were fine-tuned differently. For instance, the batch size of each model was adjusted due to the single CPU. However, the number of epochs in this study was set to be consistent to allow fair comparison and for less training time. To avoid overfitting, the training is terminated once the loss computed on the validation set has worsened for four consecutive epochs.

Training Phase:

DeepLabV3+ was trained using stochastic gradient descent with momentum (SGDM) method with an L2 regularization value of 0.005. The learning rate followed a piecewise schedule that reduced it by a factor of 0.3 every three epochs from an initial value of 0.0002. For every 10 epochs, the network multiplied the starting learning rate of 0.001 with 0.3, and training was completed over 30 epochs. All the models were trained and validated using MATLAB R2020a on a system with a single CPU. Table 8.2 summarizes the hyperparameters used for each network architecture in this study.

Table 8.2 Hyperparameters for network architecture

Model	**Sequential parameters**
Activation function (input)	ReLU
Activation function (output)	SoftMax
Optimizer	SGDM
Loss function	Cross entropy
No. of epochs	3
Batch size	5
Validation split	0.10

Inference Phase:

The inference phase was carried out to determine how well the trained networks performed when new images were being tested. In order to do that, all three models are applied to several images that have not been used during the training phase to determine how general the model was.

DeepLabV3+:

Chen *et al.* [5] created a segmentation model based on deep learning. Chen *et al.* [5] created a deep-learning-based segmentation model recently. Encoder–decoder architecture is commonly used in mannequins, as shown in Figure 8.5. The DeepLabV3 network [10] uses atrous convolution, which allows the convolutional layer to increase the corresponding receptive subject of convolution without lowering the spatial dimension or increasing the number of parameters of this connection, hence strengthening the network's segmentation impact. Simultaneously, V3 enhances the ASPP module and refers to the hybrid dilated convolution (HDC) [9] concept, which is utilized to reduce the "Gidding issue" produced by accelerated convolution and extend the receptive area to mixed world input while keeping the spine [4, 22] The encoder–decoder structure is used in "DeepLabV3+." To improve the object boundaries, DeepLabV3 is utilized to encode the rich contextual information, and an easy yet effective decoder module is used. It is also worth noting that the atrous convolution may be used to extract encoder aspects at any resolution using the available computation resources [11]. The images are segmented using the following typical deep networks.

1. ***AlexNet***: It is an eight-layered CNN composed of five convolutional layers, max-pooling ones, ReLU as non-linearities, three completely convolutional layers, and dropout. The component extraction is done in every one of the layers. The AlexNet with more profound highlights is more abstract [4].
2. ***VGG16:*** It utilizes a heap of convolution layers with little responsive fields and is better than having a huge layer.
3. ***ResNet18:*** It is notable because of its profundity (152 layers) and the presentation of leftover squares. The leftover squares address the issue of preparing a truly profound design by presenting character skip associations so that layers can duplicate their contributions to the following layers shown in Figure 8.3.
4. ***MobileNetV2:*** It is a convolutional neural organization layout that appears to carry out properly on mobile phones. It relies upon a

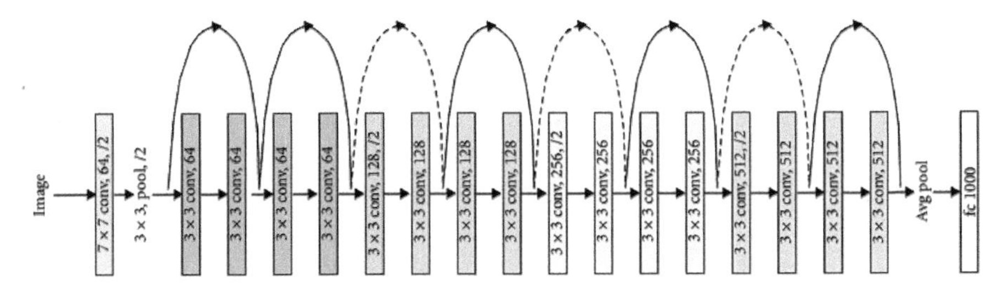

Figure 8.3 ResNet18 architecture.

disappointing leftover layout in which the ultimate connections are among the bottleneck layers. The center extension layer channels highlight as a source of non-linearity by using light weight depth-wise convolutions. Overall, MobileNetV2's structure carries an underlying 32-channel absolutely convolution layer, which is followed by 19 remaining bottleneck layers. Figure 8.5 depicts the mobile net architecture [13].

5. ***Xception:*** The ex-foothold base of the organization is shaped by 36 convolutional layers in the Xception design. With the exception of the first and last modules, the 36 convolutional layers are grouped into 14 modules, all of which have straight lingering linkages around them. So, the Xception engineering is a straight heap of depth-wise divisible convolution layers with remaining connections. It is shown in Figure 8.4 [6].

In this work, ResNet18, MobileNetV2, and Xception backbones are trained and results are compared. ResNet18 which is more accurate is used as the pre-trained model for image annotation. In this proposed work, ResNet18 is used as the backbone and it performs well compared to Xception. The advantages of ResNet18 compared to other backbones are as follows.

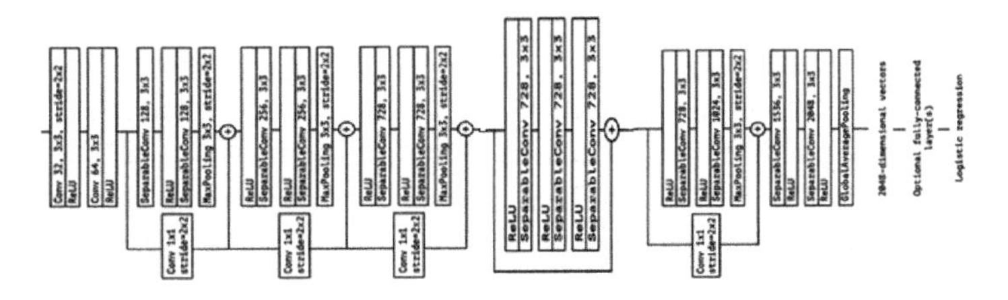

Figure 8.4 Xception architecture.

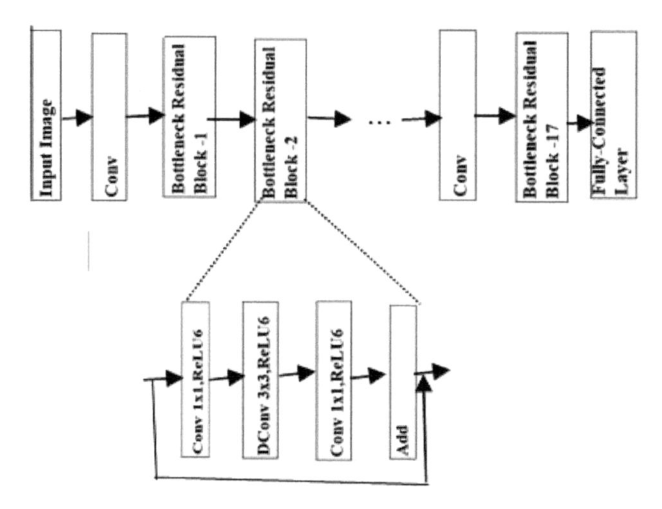

Figure 8.5 MobileNet architecture.

- DeepLabV3+ models using ResNet18 that were optimal in terms of inference time and accuracy.
- Networks with a large number of layers can be easily taught without increasing the percentage of training errors.
- ResNet18 can help with identity mapping to solve the vanishing gradient problem.

8.3.3 Automation Algorithm

By labeling ground truth data in a collection of datasets, the Image Labeler supports semantic segmentation. The Image Labeler tool includes the following features, which are listed below.

- Labeling images has different types of ROI regions of interest labels such as rectangular, polyline, pixel, polygon, and scene labels.
- Use built-in or custom algorithms to label your ground truth data.
- Automation algorithms are evaluated using a visual summary.
- Export the labeled ground truth as an object that is used for training semantic segmentation networks.

All picture file formats are supported by the Image Labeler program. For high image sizes, it employs a blocked image. A blocked image is a huge image that has been broken down into tiny blocks to fit into memory. MATLAB

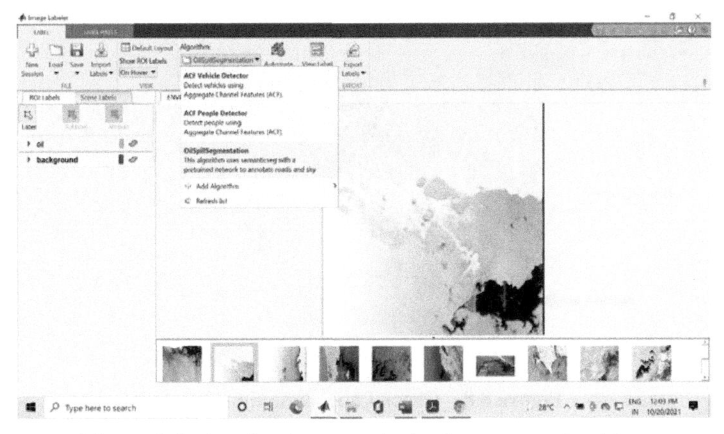

Figure 8.6 Loading image and selecting custom algorithm.

Image Labeler is used for creating algorithms. The steps involved in creating an automation algorithm are given below.

- Open Image Labeler and load the image to be annotated as shown in Figure 8.6.
- Create ROI labels as used in creating semantic segmentation.
- Click the Select Algorithm dropdown to select an existing algorithm named Oil Spill Segmentation as shown in Figure 8.6 or create a new algorithm.
- Click Automate and run the algorithm to annotate images.
- Save the annotated images as shown in Figure 8.7.

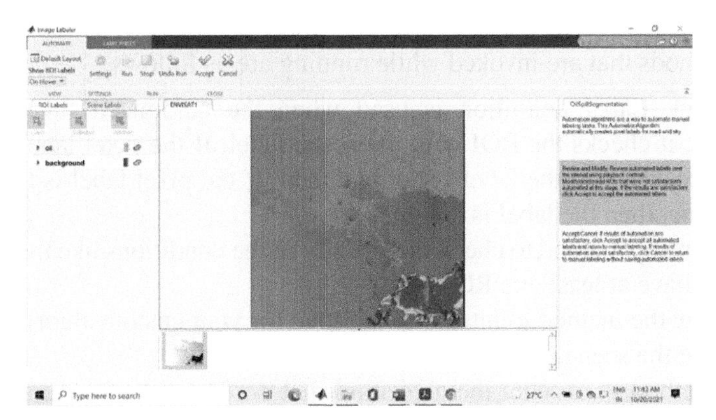

Figure 8.7 Annotated image.

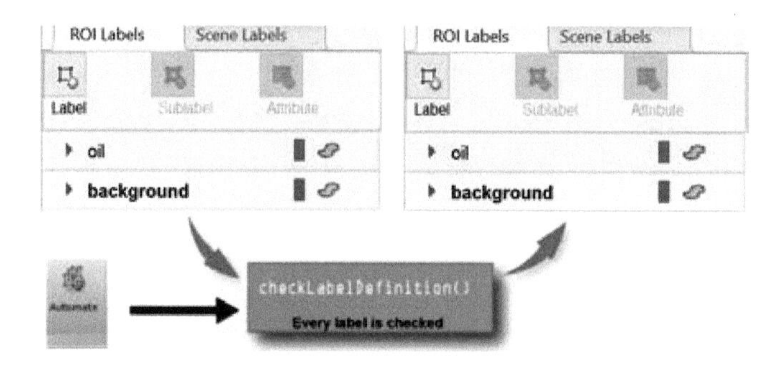

Figure 8.8 Check Label definition

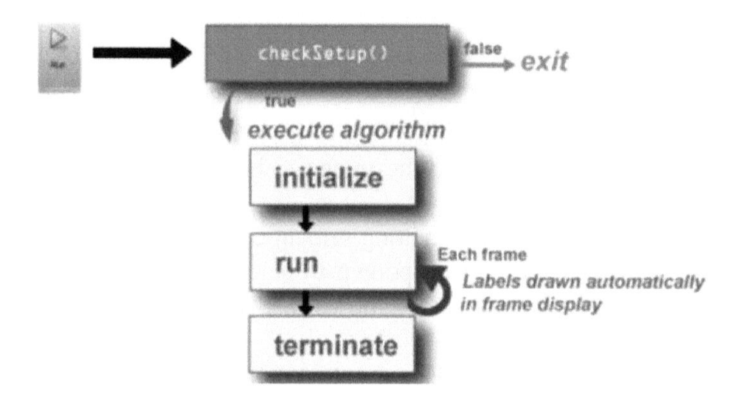

Figure 8.9 Flow of algorithm execution.

The methods that are invoked while running are as follows.

- A Check Label definition is used when the automated function is clicked; it checks the ROI with the scene label. If the pixel label is not in the scene, then the label is not included. If the pixel label is there in the scene, then the label is assigned.
- Check-setup method to check the validity of the conditions like the scene should have at least one ROI label.
- Initialize the method to introduce the state for your custom algorithm by utilizing the scene.
- Run method to execute the algorithm that performs the image annotation. The annotated image is shown in Figure 8.7.

- Terminate method to end the state of the automation algorithm after the algorithm runs.

8.4 Performance Measures

In the proposed work, using oil spill SAR images, the evaluation of DeepLabV3 plus backbones such as ResNet18, MobileNet, and Xception are compared. ResNet18 gives more accurate results than Xception. The comparisons of the backbone of the DeepLabV3+ network are shown in Figure 8.12. Data augmentation improves the outcome. DeepLabV3+ backbones are trained using the "stochastic gradient descent with momentum optimizer" with a batch size of three picture tiles on a single CPU. Le-3 learning rates are fixed in all backbones. The training process contains a dataset of 40 images that are divided into two subsets and trained on the ResNet18 and Xception pre-trained models for three epochs. ResNet18 performance is accurate compared to Xception, and then all the 40 datasets are trained using ResNet18. The training accuracy and loss rate of ResNet18 are shown in Figure 8.15. Some frequently accepted performance measures, such as accuracy, BF score, and intersection over union, are presented in Table 8.3, together with the confusion matrix, enabling quantitative assessment of the accuracy of semantic segmentation outputs. The performance measures of different backbones are shown in Figure 8.11. The confusion matrix of the predicted labels and ground truth labels are shown in Table 8.2. The confusion matrix shows the segmentation correctly classified 990,321 pixels as oil and 8,758,554 pixels as background. The confusion matrix also shows that the segmentation misclassified 146,139 background pixels as oil and 457,267 oil pixels are misclassified as background. Class-specific accuracy measures are used to calculate the proportion of correctly identified pixels from the reference (sensitivity) and the fraction of correctly classified pixels from the output. The training progress of all the three backbones of DeepLabV3+ is shown in Figures 8.13–8.15. Comparatively, ResNet18 performs well and an annotation algorithm is implemented using this algorithm.

8.4.1 Evaluation of Segmentation Models

1. Pixel Accuracy

The pixel accuracy is normally pronounced for every class one after the other in addition to globally throughout all training. The per-class pixel accuracy is evaluated by the use of a binary mask. This statistic might occasionally produce deceptive results when the class illustration is small inside the image because the degree can be skewed in reporting how effectively you detect negative cases when the class illustration is small within the image.

$$\text{Accuracy} = \frac{\mathrm{TP} + \mathrm{TN}}{\mathrm{TP} + \mathrm{TN} + \mathrm{FP} + \mathrm{FN}}$$

where TP = A pixel that has been appropriately identified as belonging to the specified class. FP = A pixel that is incorrectly assigned to the supplied class. TN = A pixel that has been successfully identified as belonging to a class other than the one specified. FN = A pixel that has been incorrectly identified as not belonging to the specified class.

2. Intersection Over Union

The Jaccard index, often known as the intersection over union (IoU) measure, is a strategy for quantifying the percentage of overlap between the masks and the output of our forecast. This metric is closely related to the Dice coefficient, which is often used in training as a loss characteristic. Using the whole range of pixels accessible across each mask, the IoU metric splits the range of pixels at a common point across the target and prediction masks. The IoU is calculated using the formula below.

$$\mathrm{IoU} = \frac{\text{Target} \cap \text{Prediction}}{\text{Target} \cup \text{Prediction}} = .$$

The intersection (AB) is made up of pixels from both the prediction and ground truth masks, whereas the union (AB) is made up of all pixels from either the prediction or the target mask. The IoU is determined one by one for each class and then averaged across the entire of our semantic segmentation prediction.

3. Mean BF Score

To assess if a point on the anticipated boundary matches the ground truth boundary, the BF score is the harmonic mean ($F1$-measure) of the precision and recall values multiplied by a distance error tolerance

$$\text{BFscore} = \frac{2 \times (\text{precision} + \text{recall})}{\text{precision} + \text{recall}}$$

It shows how well each class's anticipated boundary matches the true boundary. In comparison to the IoU metric, the BF score correlates better with human qualitative assessment. When you use a confusion matrix as the function's input, this statistic is not available.

- A class's mean BF score is the average BF score for all photos in that class.
- The average BF score of all classes in a given image is the image's mean BF score.
- The average BF score of all classes in all photos is the mean BF score of a separate dataset.

Confusion Matrix:

A confusion matrix table describes the performance of segmentation on a set of test data for which the true values are known. Each entry in a confusion matrix represents the number of predictions made by the model where the classes were properly or incorrectly classified. Figure 8.10 depicts the confusion matrix. ResNet18's confusion matrix is displayed in Table 8.5. The total loss rate, false positive rate or type I error, and false negative rate or type II error are determined using the confusion matrix of all backbones and are displayed in Table 8.4. ResNet18 has a loss rate of 0.4, which is lower than other backbones. There is a 0.03 false negative rate as a result. ResNet18 has a high level of accuracy. The following is a list of rates derived from the confusion matrix:

1. Accuracy = (TP + TN)/total
2. Misclassification Rate = (FP + FN)/total
3. True Positive Rate = TP/Positive Value
4. False Positive Rate = FP/Negative Value
5. True Negative Rate = TN/Negative Value
6. Precision = TP/Positive Prediction
7. Prevalence = Positive Value/total

The accuracy demonstrates the class's total correct predictions. The entire error rate of the class, which is equal to 1 minus accuracy, is shown by the misclassification rate or error rate. The true positive rate, also known as "sensitivity" or "recall," indicates whether or not a forecast was true. When it is actually incorrect and predicts as positive, the false positive rate indicates that. True negative rate, also known as "specificity," indicates when a prediction is inaccurate and forecasts a positive outcome. It is the same as 1 minus the false positive rate. Precision can tell the difference between a true positive and an accurate result. The term "prevalence" refers to how frequently a positive result appears.

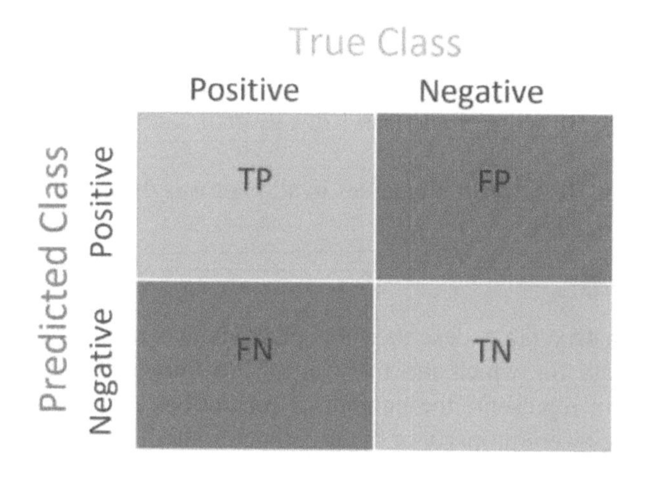

Figure 8.10 Confusion matrix.

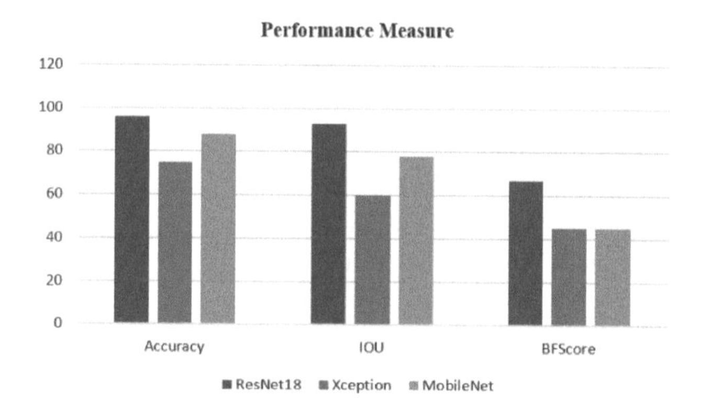

Figure 8.11 Performance measure comparisons of backbones of DeepLabV3+.

Table 8.3 Comparisons of backbones of DeepLabV3+

Method	*Model*	*Accuracy*	*IoU*	*BF score*
DeepLabV3+	ResNet18	96	93	67
	Xception	75	60	45
	MobileNet	88	78	45

Table 8.4 Comparison of loss rate of pre-trained model

Method	***Model***	***Overall loss rate***	***False positive***	***False negative***
DeepLabV3+	ResNet18	0.4	0.12	0.03
	Xception	0.20	0.18	0.04
	MobileNet	0.12	0.12	0.25

Table 8.5 Confusion matrix of ResNet18 pre-trained model

Label	***Oil***	***Background***
Oil	990,321	457,267
Background	146,139	8,758,554

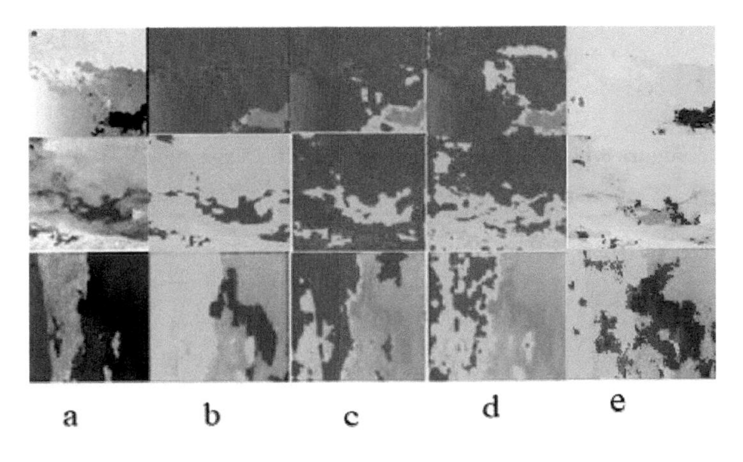

Figure 8.12 Comparisons of ResNet18 and Xception backbone architecture. a) Original image. (b) Ground truth image. (c) ResNet18. (d) Xception. (e) MobileNet.

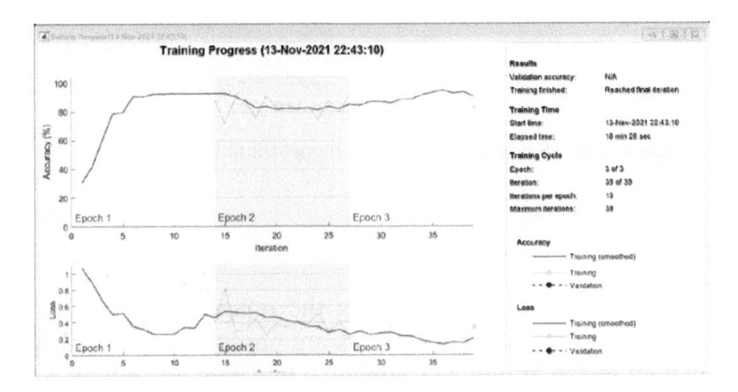

Figure 8.13 MobileNet pre-trained model training progress.

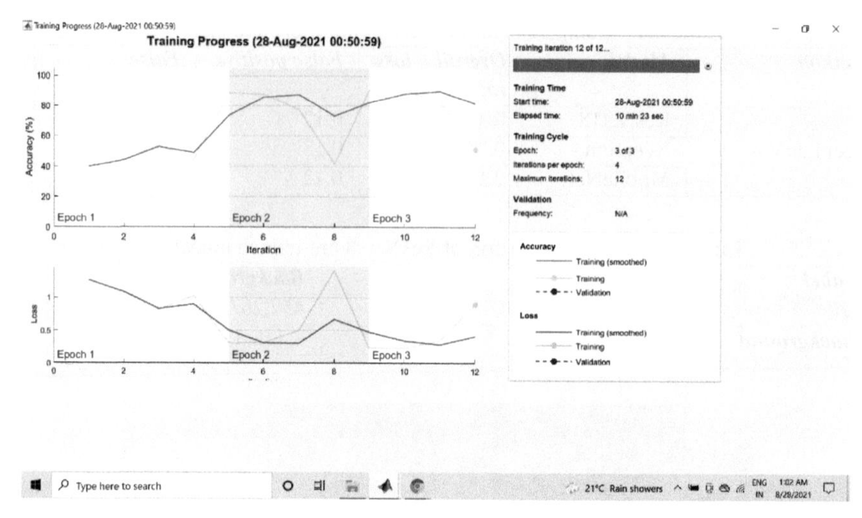

Figure 8.14 Xception pre-trained model training progress.

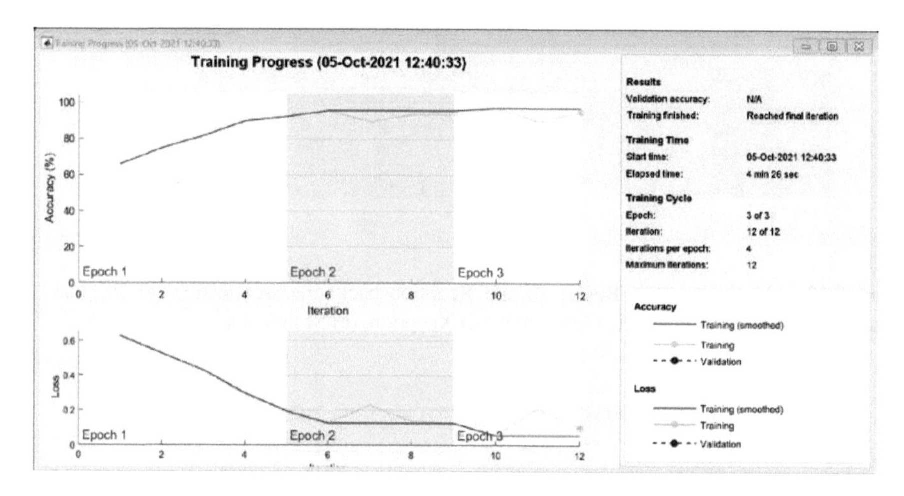

Figure 8.15 ResNet18 pre-trained model training progress.

8.5 Conclusion

The proposed work is used for annotating the images using SAR dataset. For every image in the dataset, annotations are produced and are displayed. When we input an image, it will be segmented to detect oil spill by semantic segmentation based on DeepLabV3+ with ResNet18 backbone, and the input image is compared to the correctness of the objects detected. The proposed method gives us a robust methodology for extracting oil spills in images at low time complexity. We have tested our result with the Xception backbone network, but it takes a lot of time and space to segment the image. Our proposed method takes comparatively less time than the other method to detect an object in the image. Pixels assist in the modeling and annotation of data. Finally, to improve usability in the realm of computer vision, the system can be connected with various retrieval methods. DeepLabV3+ backbones were compared, and ResNet18 was shown to be superior to the others. Both networks could distinguish features of the oil from the background with some misidentification of pixels due to similar features. This work focused on the accuracy factor only without considering the training time and memory usage. When measuring semantic segmentation metrics on the overall test set results, it was found that the DeepLabV3+ network with ResNet18 performed better than other networks by achieving above 96% for overall accuracy and IoU metrics, and approximately 67% for BF score.

In the future, improving the accuracy of the system will have a different ensembles approach. Better and more training images per semantic notion may result in more stable models. Future work can be extended using an extensive labeled dataset and hyperparameter fine-tuning to improve the segmentation accuracy.

References

[1] M. M. Adnan, M. S. M. Rahim, A. Rehman, Z. Mehmood, T. Saba, and R. A. Naqvi. Automatic image annotation based on deep learning models: A systematic review and future challenges. IEEE Access, 1–1, 2021. doi:10.1109/access.2021.3068897.

[2] Andrew G. Howard, Menglong Zhu, Bo Chen, Dmitry Kalenichenko, Weijun Wang, Tobias Weyand, Marco Andreetto, and Hartwig Adam. Mobilenets: Efficient convolutional neural networks for mobile vision applications. CoRR, 2017. abs/1704.04861.

[3] R. Avinash, G. Apoorva, and S. G. Gurushankar, Image annotation framework. May 2017.

[4] Chen Tianhua, Zheng Siqun, and Yu Junchuan. Remote sensing image segmentation using improved deeplab network. Measurement and Control Technology, 37(11), 2018.

[5] L.-C. Chen, Y. Zhu, G. Papandreou, F. Schroff, and H. Adam. Encoder-decoder with atrous separable convolution for semantic image

segmentation. European Conference on Computer Vision (ECCV), 2018, pp. 801–818.

[6] Francois Chollet Google, Inc., Xception: Deep learning with depthwise separable convolutions.

[7] Hongxing Peng, Chao Xue, Yuanyuan Shao, Keyin Chen, Juntao Xiong, Zhihua Xie, and Liuhong Zhang. Semantic segmentation of litchi branches using DeepLabV3+ model. 2020.

[8] Imran Ahmed, Misbah Ahmad, Fakhri Alam Khan, and Muhammad Asif. Comparison of deep-learning-based segmentation models: Using top view person images. 2020.

[9] Jianfang CaoID, Aidi Zhao1, and Zibang Zhang, Automatic image annotation method based on a convolutional neural network with threshold optimization. 2020.

[10] X. Ke, J. Zou, and Y. Niu. End-to-end automatic image annotation based on deep CNN and multi-label data augmentation. IEEE Transactions on Multimedia, 1–1, 2019. doi:10.1109/tmm.2019.2895511.

[11] Liang-Chieh Chen, Yukun Zhu, George Papandreou, Florian Schroff, and Hartwig Adam. Encoder-decoder with atrous separable convolution for semantic image segmentation.

[12] V. Lekic and Z. Babic. Using GANs to enable semantic segmentation of ranging sensor data. 2018 Zooming Innovation in Consumer Technologies Conference (ZINC), 2018. doi:10.1109/zinc.2018.8448963.

[13] Mark Sandler, Andrew Howard, Menglong Zhu, Andrey Zhmoginov, and Liang-Chieh Chen. MobileNetV2: Inverted residuals and linear bottlenecks. 2018 IEEE/CVF Conference on Computer Vision and Pattern Recognition.

[14] Niharjyoti Sarangi and C. Chandra Sekhar. Automatic image annotation using convex deep learning models. 2015.

[15] A. Novozamsky, D. Vit, F. Sroubek J. Franc, M. Krbalek, Z. Bılkova, and B. Zitova. Automated object labeling for CNN-based image segmentation. 2020.

[16] M. Niemeyer and O. Arandjelovic. Automatic semantic labelling of images by their content using non-parametric bayesian machine learning and image search using synthetically generated image collages. 2018 IEEE 5th International Conference on Data Science and Advanced Analytics (DSAA), 2018. doi:10.1109/dsaa.2018.00026.

[17] Philipe A. Dias, Amy Tabb, and Henry Medeiros. Multispecies fruit flower detection using a refined semantic segmentation network. 2018.

[18] M. Pai, V. Mehrotra, S. Aiyar, U. Verma, and R. Pai. Automatic segmentation of river and land in SAR images: A deep learning approach. 2019 IEEE Second International Conference on Artificial Intelligence and Knowledge Engineering (AIKE), 2019. doi:10.1109/aike.2019.00011.

[19] K. Pugdeethosapol, M. Bishop, D. Bowen, and Q. Qiu. Automatic image labeling with click supervision on aerial images. 2020 International Joint Conference on Neural Networks (IJCNN), 2020. doi:10.1109/ijcnn48605.2020.92073.

[20] C. J. Rapson, B.-C. Seet, M. A. Naeem, J. E. Lee, M. Al-Sarayreh, and R. Klette. Reducing the pain: A novel tool for efficient ground-truth labelling in images. 2018 International Conference on Image and Vision Computing New Zealand (IVCNZ), 2018. doi:10.1109/ivcnz.2018.8634750.

[21] C. K. Singh, A. Majumder, S. Kumar, and L. Behera. Deep network based automatic annotation for warehouse automation. 2018 International Joint Conference on Neural Networks (IJCNN), 2018. doi:10.1109/ijcnn.2018.8489424.

[22] H. Song, Y. Zhou, Z. Jiang, X. Guo, and Z. Yang. ResNet with global and local image features, stacked pooling block, for semantic segmentation. 2018 IEEE/CIC International Conference on Communications in China (ICCC), 2018. doi:10.1109/iccchina.2018.8641146.

[23] Tuhin Shukla, Nishchol Mishra, and Sanjeev Sharma. Automatic image annotation using SURF features. International Journal of Computer Applications, 68(4), 0975–8887, April 2013.

[24] Venkatesh N. Murthy, Subhransu Maji, and R. Manmatha. Automatic image annotation using deep learning representations. 2015.

[25] P. Wang, P. Chen, Y. Yuan, D. Liu, Z. Huang, X. Hou, *et al.* Understanding convolution for semantic segmentation. 2017.

[26] Xiang Zhang, Wei Zhang, Jinye Peng, and Jianping. Automatic image labelling at pixel level fan. 2020.

[27] X. Yao, J. Han, G. Cheng, X. Qian, and L. Guo. Semantic annotation of high-resolution satellite images via weakly supervised learning. IEEE Transactions on Geoscience and Remote Sensing, 54(6), 3660–3671, 2016. doi:10.1109/tgrs.2016.2523563.

9

Environmental Weather Monitoring and Predictions System Using Internet of Things (IoT) Using Convolutional Neural Network

M. P. Karthikeyan[1], C. V. Banupriya[2], R. Pandiammal[2], M. Vijetha[3], and N. Karunya[3]

[1]Department of BCA, School of CS & IT, Jain (Deemed-to-be) University, India
[2]Department of Computer Application, Hindusthan College of Arts and Science, Coimbatore, Tamilnadu
[3]Department of Computer Science, Sri Ramakrishna College of Arts and Science, India
E-mail: karthi.karthis@gmail.com; Banupriya.venkat@gmail.com; panbhavya@gmail.com; Vijethachandran1983@gmail.com; karunyakarun@gmail.com

Abstract

Weather is a multifaceted, dynamic, and chaotic process that occurs in real time. Forecasting and monitoring the weather is difficult due to these characteristics. Wireless gadgets are crucial fragments not only important to the organizations for development control, yet, moreover, in step-by-step life for security of building's and movement stream assessing, common parameters estimation. In atmosphere checking, factors, for instance, temperature, soddenness, and weight, are to be assessed for this wander; thus, sensors have reliably been given the endeavor for doing all things considered. Data getting structures are, to a great degree, surely understood for purchasers and present-day users. The proposed shape has three sensors that process uncommon factors as communicated beyond and for rain fall recognizable evidence and storm bearing tempo estimation environment tool is included to stored data and compared to the previous 60 years of weather data to predict future weather using convolutional neural networks. Previously, meteorologists employed a variety of approaches for weather predicting, ranging from simple temperature, rain, air pollution, and moisture observations to complicated computerized mathematical models. Convolutional neural networks (CNNs) are a strong deep-learning-based data modeling tool for capturing and representing complex input/output interactions. The actual strength and advantage of convolutional neural networks is their ability to model both linear and nonlinear relationships straight from the data. Based on the experimental approach performed in MATLAB 2013a, the quality and performance of these algorithms are evaluated. The proposed system performances are 93.56%, 94.12%, and 94.32% in terms of accuracy, sensitivity, and precision. The application of convolutional neural networks has produced the most accurate weather prediction when compared to the existing technique such as support vector machine and decision tree. For the most part, the modeling results show that reasonable forecast accuracy was attained in the proposed system.

Keywords: Convolutional neural networks, weather prediction, temperature, moisture, rain fall detection, MATLAB 2013a.

9.1 Introduction

Weather simply refers to the state of the environment factor on the planet at a specific location and time. It is a process that is ongoing, data-intensive, complex, dynamic, and chaotic. These characteristics make weather forecasting a difficult task. Forecasting is the practice of making educated guesses about unknown conditions based on historical evidence. Weather forecasting is one of the most difficult scientific and technology challenges to solve in the previous century (Kalimuthu, S. *et al.*, 2021). Making an accurate prediction is, without doubt, one of the most difficult tasks that meteorologists face around the world. Weather prediction has been one of the most interesting and exciting fields since ancient times. Scientists have explored a variety of approaches to forecast meteorological characteristics, some of which are more accurate than others specified by Dehghanian, P. *et al.* (2018).

Scientific weather forecasting, which involves predicting the state of the atmosphere at a certain area, necessitates meteorology knowledge. Human weather forecasting is a good example of the need to make decisions in

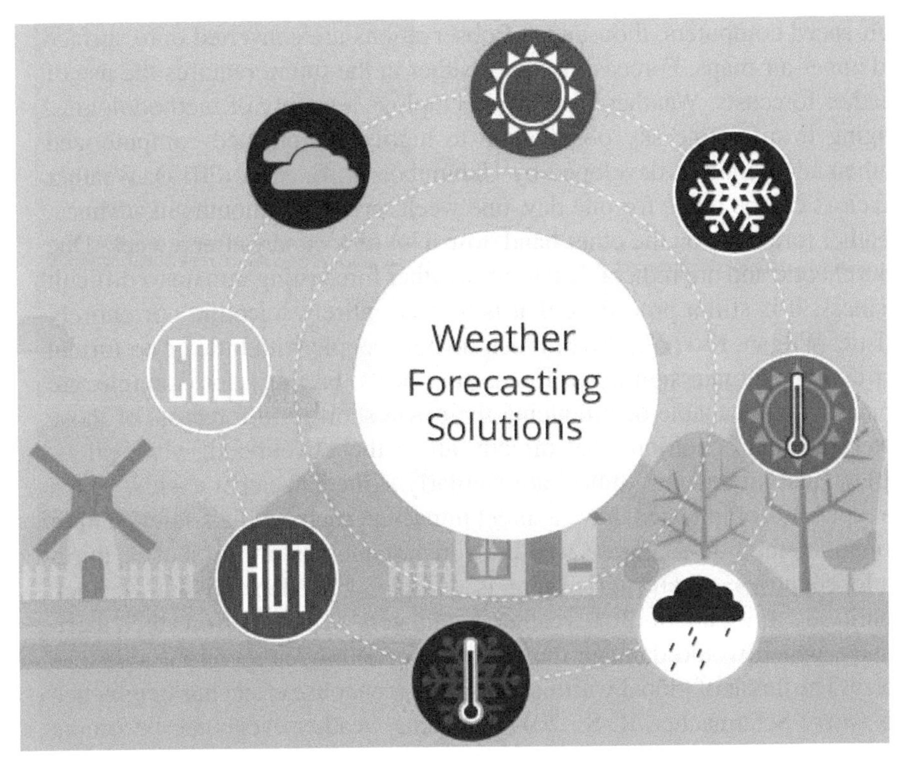

Figure 9.1 Weather forecasting features (Galanis, G., 2017).

the face of uncertainty. Weather predictions are often generated by gathering quantitative data about the current condition of the atmosphere and using scientific knowledge of atmospheric processes to estimate how the environment will evolve in the future, as proven by Galanis, G. (2017). The value of learning about the cognitive process in weather forecasting has been recognized in recent years. Even while most human forecasters employ ways based on meteorology to deal with the challenges of the job, as Gómez-Romero, J. *et al.* (2018) point out, forecasting the weather becomes a duty for which the specifics can be quite personal.

Weather forecasting entails predicting how the current area of the atmosphere will change in the future. Ground observations, observations from ships, observations from aero planes, radio noises, Doppler radar, and satellites are all used to determine current weather conditions. This information is delivered to meteorological centers, which collect, analyze, and present the information in various charts, maps, and graphs. Using contemporary high-speed computers, thousands of observations are converted onto surface and upper-air maps. Forecasting the weather in the future requires the use of weather forecasts. Weather forecasting employs a variety of methodologies, ranging from simple sky observation to highly complicated computerized mathematical models developed by Ukhurebor, K. E. *et al.* (2017). Weather forecasts can be made for one day, one week, or several months in advance. Weather forecasts, on the other hand, lose a lot of accuracy after a week. Due to its chaotic and unpredictable nature, weather forecasting remains a difficult business. It is still a procedure that is neither entirely scientific nor entirely artistic. Wilgan, K. *et al.* (2017) illustrate how people with little or no formal instruction can gain significant forecasting skills. Farmers, for example, are typically quite capable of producing their own short-term forecasts of those meteorological conditions that directly affect their livelihood, while pilots, anglers, and mountain climbers are similarly skilled. Weather events, which are typically complicated, have a direct impact on such people's safety and/or economic stability. Accurate weather forecast models are critical in third-world countries, where agriculture is entirely dependent on the weather. Identifying any patterns for weather parameters to depart from their periodicity, which would damage the country's economy, is therefore a serious worry. The threat of global warming and the greenhouse effect has heightened this worry Schumacher, R. S., 2017. Extreme weather events are becoming increasingly costly to society, inflicting infrastructure damage, injury, and even death (Singh, M. *et al.*, 2021).

Weather forecasting, as conducted by professionally educated meteorologists, is now a highly developed skill based on scientific principles and methods, utilizing advanced technical tools. Since 1950, technology advancements, basic and applied research, and the application of new information and procedures by weather forecasters have resulted in a significant improvement in forecast accuracy, as Pierce, F. J., & Lal, R. (2017) demonstrate. High-speed computers, meteorological satellites, and weather radars are examples of tools that have helped improve weather forecasting. A number of other elements have aided in the improvement of predicting accuracy. Another advantage of meteorological satellites is their enhanced observational capability. The ongoing improvement of the initial conditions prepared for the forecast models is a third main cause for the increase in accuracy (Kalimuthu, S., 2021).

Statistical approaches can anticipate a wider range of meteorological factors than models alone can, and they can adjust the less exact model forecasts to specific places. On a worldwide scale, satellites now allow for practically continuous viewing and remote sensing of the atmosphere. An increase in the quantity of observations and greater use of the observations in computational approaches has resulted in improved initial conditions.

9.1.1 Types of Weather Forecasting

A daily weather forecast is made possible by the contributions of thousands of observers and meteorologists from all around the world. Cloud photos are captured from space by weather satellites orbiting the planet, and modern computers make forecasts more precise than ever. Forecasters create their predictions based on observations from the ground and space, as well as algorithms and principles based on historical experience. Meteorologists create daily weather forecasts by combining various distinct methodologies (Maleki, H. *et al.*, 2019). They really are.

a) Computer forecasting
b) Synoptic forecasting
c) Persistence forecasting
d) Statistical forecasting

9.1.1.1 Computer Forecasting

Forecasters use their observations to enter numbers into complex calculations. These many equations are executed on several ultra-high-speed computers to create computer "models" that provide a forecast for the following several days (Poterjoy, J. *et al.*, 2019). Because different equations often produce different outcomes, meteorologists must always combine this strategy with other forecasting methods (Khandakar, A. *et al.*, 2019).

9.1.1.2 Synoptic Forecasting

The basic rules for predicting are used in this strategy. Meteorologists use their observations and the laws they have learned to generate a forecast for the next few days (Kang, G. K. *et al.*, 2018).

9.1.1.3 Persistence Forecasting

Persistence forecasting is the most basic approach of weather prediction. When the weather is stable, such as during the summer season in the tropics, this can be a good technique to forecast the weather. The occurrence of a stationary weather pattern is critical for this type of forecasting (Kumar, K. R., 2018). It can be used in both short- and long-term projections. This presupposes that the weather will continue to behave as it does presently.

Meteorologists perform weather observations to determine how the weather is behaving.

9.1.1.4 Statistical Forecasting

Meteorologists speculate on what the weather this is time of year or in future. Forecasters can gain a sense of what the weather is "supposed to be like" at a certain time of year by looking at historical data of rainfall, snowfall, and typical temperatures (Hosseini, S. M., & Mahjouri, N., 2018). The following is a list of the book chapter's remainders. Section 9.2 discusses past weather predictions for various datasets and related work, Section 9.3 discusses the proposed feature selection mechanism with a convolutional neural network, and Section 9.4 compares the experimental outcomes of CNNs and existing systems. Finally, part five contains the work's concluding thoughts and future scope.

9.2 Literature Review

For the scientific community, accurate weather forecasting is a big challenge. Computer models, observation, and knowledge of trends and patterns are all used in weather prediction modeling. Different weather forecasting methods can be used with these methods to get reasonably accurate forecasts. They are enumerated below. For 24-hour weather forecasting, Zubaidi, S. L. *et al.* (2020) used soft computing models based on radial basis function network. In comparison to the multilayer perceptron network, they found that radial basis function neural networks give the most accurate forecasts.

As-syakur, A. R. *et al.* (2019) showed that the ANN model may be utilized as a suitable forecasting tool for rainfall prediction, outperforming the autoregressive integrated moving average (ARIMA) model. Also Erickson *et al.* (2018) employed artificial intelligence algorithms to estimate regional rainfall, and they discovered that this technique has a reasonable level of accuracy for monthly and seasonal forecasts. Huntingford, C. *et al.* (2019) provided a method for classifying and forecasting future weather using a back-propagation algorithm, as well as a discussion of previous weather forecasting models. Finally, the study shows that the new wireless medium technology can be utilized in the weather forecasting process.

Waliser, D. *et al.* (2018) demonstrate a weather prediction program using support vector machines. Using ideal kernel values, the system's performance is measured over time periods ranging from 2 to 10 days. Using ten years of meteorological data (1996–2006), Verbois, H. *et al.* (2018) investigated artificial neural networks built on multilayer perceptrons. The findings suggest that the multilayer perceptron network has the lowest predicting error and can be used to construct short-term temperature forecasting systems.

Yemane, S. *et al.* (2021) describe a weather prediction application using a back-propagation neural network. The real-time dataset is used to test their proposed proposal. The results were compared to the actual work of the meteorological department, and they confirmed that real-time weather data processing indicates that back-propagation network-based weather forecasts outperform not only numerical model guidance forecasts but also official local weather service forecasts. Talavera, J. M., *et al.* (2017) developed a feature-based neural network model for weather forecasting.

To forecast maximum temperature, relative humidity, and minimum temperature, this model employs an FFANN with back propagation for supervised learning. A trained artificial neural network was used to forecast future weather conditions. A feed forward neural network was utilized by Blair, G. S. *et al.* (2019) to predict typhoon rainfall. FNN was used to estimate the residuals from the linear model to the variations between anticipated rainfalls and data from a typhoon rainfall or snowfall climatology model, and the findings were good.

One of the most popular supervised training methods is BPNN. Iterative weight updating based on minimizing the mean square error is commonly used in training. The error signal is then transmitted back to the lower layers using the steepest descent algorithm, which updates the network's weights. The algorithm adjusts the weights of the network in such a way that the error decreases in a downward direction. The activation function for the back-propagation algorithm must be continuous and differentiable (Qing, X., & Niu, Y., 2018).

The most widely used learning method for feed forward neural networks is back propagation. In terms of information flow direction, the feed forward neural network is the simplest ANN architecture. The back-propagation technique can be implemented in two different ways: batch updating and online updating. The batch updating method, like the standard gradient method, accumulates the weight adjustment across all training samples before conducting the update. The online updating strategy, on the other hand, adjusts the network weights instantly after each training sample is fed (Fu, M. *et al.*, 2015).

Back propagation is the most common learning strategy for feed forward neural networks. The feed forward neural network architecture is the most basic in terms of information flow direction. Many neural network topologies employ the feed forward neural network. The back-propagation technique can be implemented in two different ways: batch and online. The batch updating strategy, like the classic gradient method, accumulates the weight correction across all training samples before conducting the update. The network weights are updated instantly after each training sample is fed in the online updating strategy, on the other hand (Zhao, H. *et al.*, 2019).

The vast majority of artificial neural network systems were supervised during training. The artificial neural network must be trained before it can be used in supervised learning. The network is trained using input and output data. The training set is the name given to this collection of data. The actual output of a neural network is compared to the desired output in this mode. In the next iteration, or cycle, the network adjusts the weights, which are generally set randomly at initially, so that the expected and actual outputs are closer. The learning method tries to reduce all processing elements' current flaws. Adjusting the input weights until the network accuracy is adequate achieves this global error decrease over time (Kumar, Y. J. N. *et al.*, 2020).

Learning without supervision has a bright future ahead of it. It demonstrates how, in the future, computers may be able to learn on their own in a robotic sense. Self-supervised learning is the name given to this promising field of unsupervised learning. External factors have no effect on the weights of these networks. Instead, they keep track of their own performance. These networks search for patterns or trends in the input signals and adjust the network's function accordingly. Even if it is not told whether it is correct or incorrect, the network must have some knowledge of how to arrange itself. The network topology and learning rules contain this information. Unsupervised learning is still a research topic because it is not well understood (Mihai, A. *et al.*, 2019). The proposed convolution-based weather prediction approach used overcome the above-mentioned drawbacks to improve the performance within the time duration are discussed in the following section.

9.3 System Design

For capturing and displaying intricate input/output relationships, an NN is a common data modeling technique. The goal to design an artificial system capable of performing cognitive tasks similar to those done by the human brain fueled the creation of neural networks (Barton, T., & Musilek, P, 2019). A layered neural network's neurons are organized into layers. An input layer of source nodes projects onto an output layer of neurons in the simplest version of a layered network but not the other way around (da Silva Fonseca

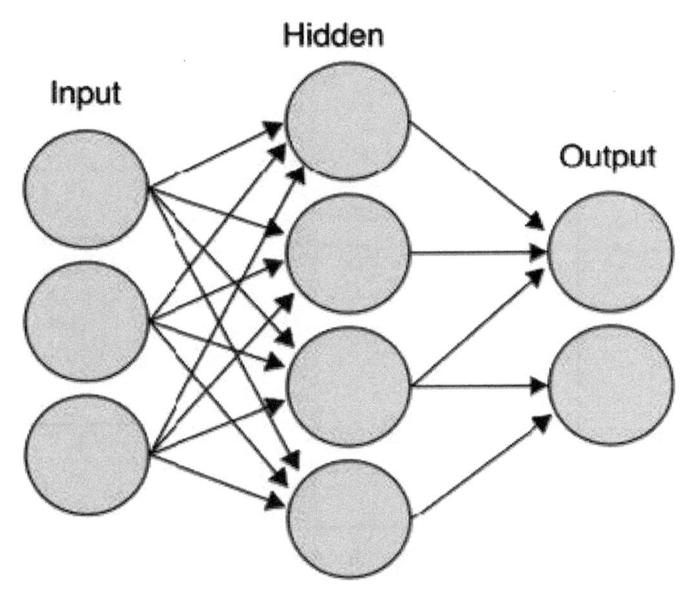

Figure 9.2 Multilayer feed forward network architecture.

Jr., J. G. *et al.*, 2012). A single-layer network is essentially a feed forward network because it only has one input and output layer. The input layer is not counted as a layer because it performs no mathematical operations. The input layer is not counted as a layer because it performs no mathematical operations.

A feed forward network, to put it another way, is one in which data can only travel from one input layer to the hidden layers, and then to the output layer. In this type of paper, there are no feedback links (Liu, Y. *et al.*, 2016). Figure 9.2 depicts the architecture of a multilayer feed forward network. The hidden neuron's job is to boost the amount of processing that happens between the input and output layers. This improves the accuracy of the network in use, allowing it to handle more difficult tasks. By adding more hidden layers, the network can analyze more weather data and extract higher order. The input signal is sent to the neurons in the second layer. The second layer's output signal is sent into the third layer, and so on (Xu, G. *et al.*, 2019). Figure 9.3 shows the input features and output predictions in the weather rain fall prediction UCI dataset using CNN hidden layer architecture. The proposed system architecture is depicted in Figure 9.4 along with a comparison for SVM and the decision tree algorithm architecture system flow.

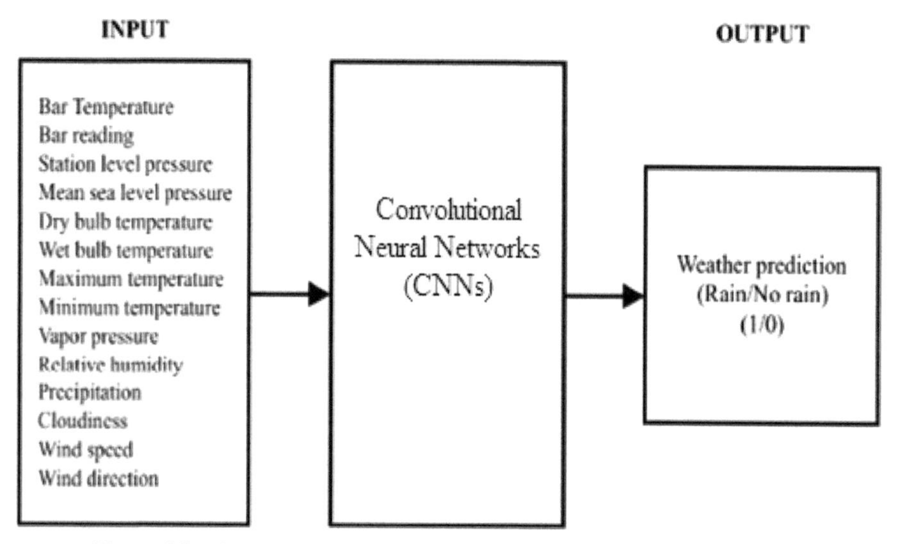

Figure 9.3 General structures for weather forecasting system using CNNs.

Back-propagation learning is the process of encoding an input–output relationship, represented by a set of x, d, with a back-propagation network that has been sufficiently trained to generalize to the future. This can be trained multiple times in the same network, with each training run yielding distinct synaptic connections. Cross-validation, a standard statistical method, serves as a guiding principle. The supplied dataset is divided into two groups at random: training and testing. After that, the training set is separated into two parts (Wen, J. *et al.*, 2020). The majority of meteorological systems are characterized by temporal and spatial variability, as well as physical process nonlinearity, spatial and temporal scale conflict, and parameter estimation uncertainty. Neural networks can extract the relationship between a process's inputs and outputs even if the physics are not explicitly stated. As a result, neural networks' features are ideally adapted to the challenge of weather forecasting at hand. There are two steps to the back-propagation learning algorithm: propagation and weight update (Yoo, C. *et al.*, 2019).

Phase 1: Propagation

The following steps are involved in each propagation:

1. In order to generate the propagation's output activations, forward propagation of a training pattern's input is delivered through the neural network.

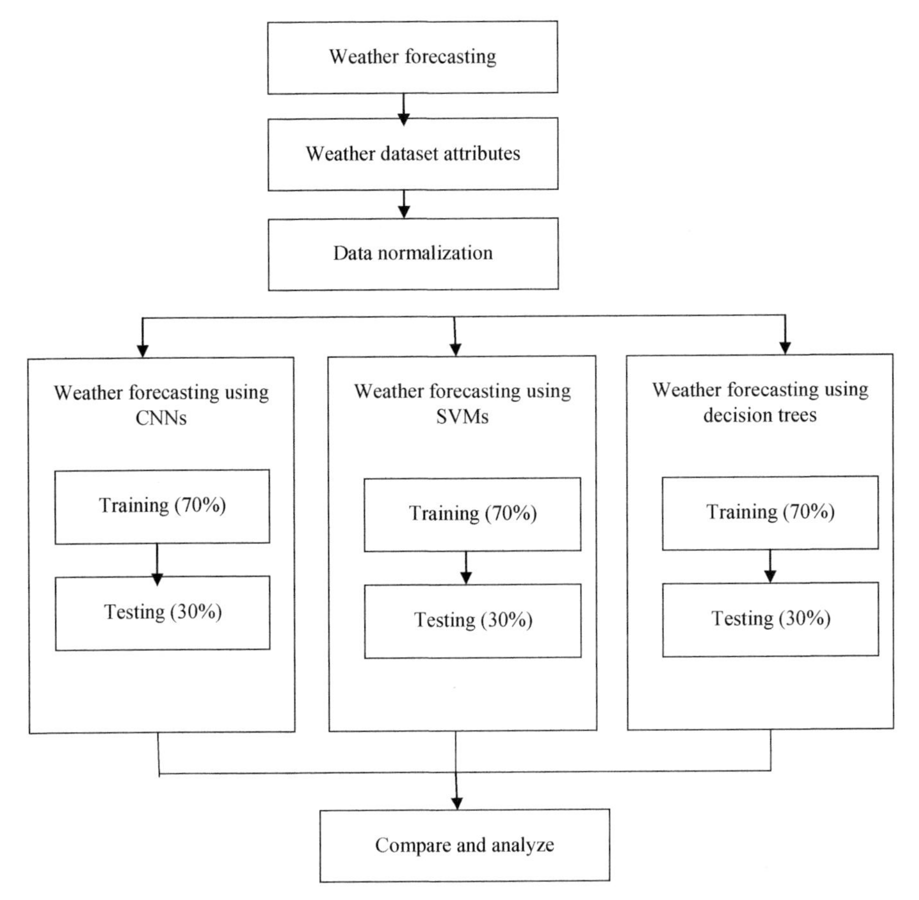

Figure 9.4 General framework of the proposed study.

2. Using the training pattern's target, back propagate the output activation through the neural network to generate the deltas of all output and hidden neurons.

Phase 2: Weight Update

For each weight-synapse:

1. To calculate the weight's gradient, multiply its input activation by its output delta.
2. Add a ratio from the weight to move the weight in the gradient's direction.

The learning rate is a fraction that improves the time duration and quality of the proposed approach. The sign of a weight's gradient must be updated in the opposite area because it indicates where the inaccuracy is rising. The first and second phases are repeated until the network's performance is acceptable. The following are the steps in back propagation. Figure 9.3 shows a completely linked feed forward back-propagation neural network, in which each layer's neuron is connected to the layer before it. The network's signal flow is in a forward direction, layer by layer, from left to right.

Two distinct calculation passes may be detected when using the back-propagation process. The forward pass and the backward pass are both referred to as such. The synaptic weights are left unchanged in the forward pass, and the network's function signals are computed neuron by neuron, as shown in Figure 9.3. As a result, the forward computation phase starts with the delivery of the input vector to the first hidden layer and ends with the computation of the error signal for each neuron in the output layer. The backward pass, on the other hand, begins at the output layer and iteratively computes the local gradient for each neuron by passing error signals layer by layer leftward through the sensitivity network. The network's synaptic weights can change in response to the delta regulations, thanks to this cyclic

Step 1. Initialize the weights in the network (often randomly)

Step 2. Do

Step 3. For each e in the training set

a. O= neural-net-output (network, e) ; forward pass

b. T = teacher output for e

Step 4. Calculate error (T - O) at the output units

Step 5. Compute Δwh for all weights from hidden layer to output layer;

Backward pass

Step 6. Compute Δwi for all weights from input layer to hidden layer;

Backward pass continued

Step 7. Update the weights in the network

Step 8. Until all e's classified correctly or stopping criterion satisfied

Step 9. Return the network

process. A neuron's gradient at the output layer is just its error signal multiplied by the first derivative of its nonlinearity. By conveying changes to all synaptic weights, the iterative computation is repeated layer by layer. Back propagation is an iterative process that begins at the top layer and works its way down until it reaches the bottom layer. It is acceptable to assume that the layer's output error is known for each layer. When the output error is known, calculating changes to the weights to reduce the error is simple. The problem is that only the inaccuracy in the final layer's output can be seen. Back propagation detects an error in a previous layer's output by using the output of the current layer as feedback. As a result, the procedure is iterative, beginning with the last layer and computing weight changes.

9.4 Result and Discussion

Based on the training set provided to NN, a BPNN is utilized to predict weather. It has been demonstrated that an intelligent system may be effectively integrated with an NN weather data prediction to predict rain and no-rain categorization through the use of this technology. This method improves convergence. This method is a simple conjugate gradient method. The back-propagation neural network approach to weather forecasting can produce good results and can be used instead of established meteorological approaches. This method can figure out the non-linear relationship between the historical data fed into the system during the training phase and create a prediction about what the forecast will be in the future.

9.4.1 Dataset

The Indian Meteorological Department of Tamil Nadu provided weather data for ten years (2001–2020). The weather data is divided into two groups: the training group, which accounts for 70% of the data, and the test group, which accounts for 30% of the data. Today's weather forecasts rely on gathering and interpreting data and measurements from all across the globe. Weather.com and AccuWeather.com provided some of the misclassified data. Rather than giving common users with the opportunity to modify and interactively discover prospective concerns linked with imminent weather hazards, it assisted meteorologists in studying and projecting personalized weather forecasts for a city or metropolitan area. There are 14 attributes in the data collection (Sheikh, F. *et al.*, 2016). They are as follows:

- Bar reading

- Wind direction
- Mean sea level pressure
- Maximum temperature
- Dry bulb temperature
- Wind speed
- Minimum temperature
- Vapor pressure
- Cloudiness
- Bar temperature
- Relative humidity
- Precipitation
- Wet bulb temperature
- Station level pressure

A confusion matrix is a table that lists the actual and predicted categories in a classification system. The matrix data is commonly used to assess the performance of such systems. The accuracy is the percentage of correct guesses in the total number of forecasts. The true positive rate (TP), also known as the recall rate, is the proportion of positive events that are accurately identified. The percentage of accurately diagnosed negative instances is known as the true negative rate (TN). The false negative rate (FN) is the percentage of positive instances that were categorized incorrectly as negative. Finally, precision (P) refers to the percentage of positive cases that are correctly predicted. Four classic performance criteria, namely sensitivity, accuracy, specificity, and precision, are used to validate all trials. In the performance metrics equations (Rajesh, P., & Karthikeyan, M., 2017), events that are defined are as follows:

$$Precision = \frac{TP}{TP + FP}$$

$$Sensitivity = \frac{TP}{TP + FN}$$

$$Specificity = \frac{TN}{FP + TN}$$

$$Accuracy = \frac{TP + TN}{TP + TN + FP + TN}$$

The back-propagation network for the given sort of weather pattern can be utilized to calculate classification statistics using a confusion matrix in Table 9.1. Figure 9.5 shows a performance analysis of weather forecasting.

Operation of the receiver: Aside from confusion matrices, another way for evaluating classifier effectiveness is to use characteristic (ROC) graphs.

Table 9.1 Confusion matrix results of CNNs in weather prediction

	Rain	No-rain	Total
Rain	192	68	260
No-rain	42	218	260
Total	260	260	

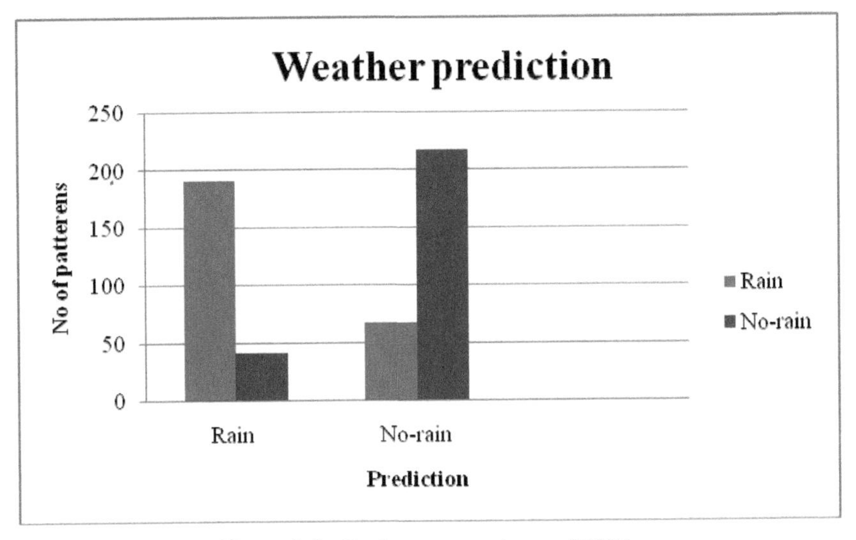

Figure 9.5 Performance analyses of CNNs.

The false positive rate is plotted on the *X*-axis, while the true positive rate is plotted on the *Y*-axis, in an ROC graph. The point (0,1) is the best classifier because it correctly categorizes both positive and negative cases. Since the false positive rate is 0 (zero) and the true positive rate is 1, the answer is affirmative (0, 1) (all). A classifier with point (0, 0) expects all cases to be negative, while a classifier with point (1, 1) expects all cases to be positive (1, 1). The classifier at point (1, 0) is wrong in all classifications. A classifier may have a parameter that can be modified to increase TP at the expense of FP or decrease FP at the expense of TP in many cases. An (FP, TP) pair exists for each parameter setting, and an ROC curve can be produced using a sequence of these pairings. A single ROC point represents the (FP, TP) pair of a non-parametric classifier.

For classification, all available rain and no-rain features are used. In table, the validation parameters, precision, sensitivity, and accuracy are 93.56, 94.12, and 94.32, respectively. The 42 misclassified datasets here are related to rain features, whereas the 46 misclassified datasets are related to no-rain

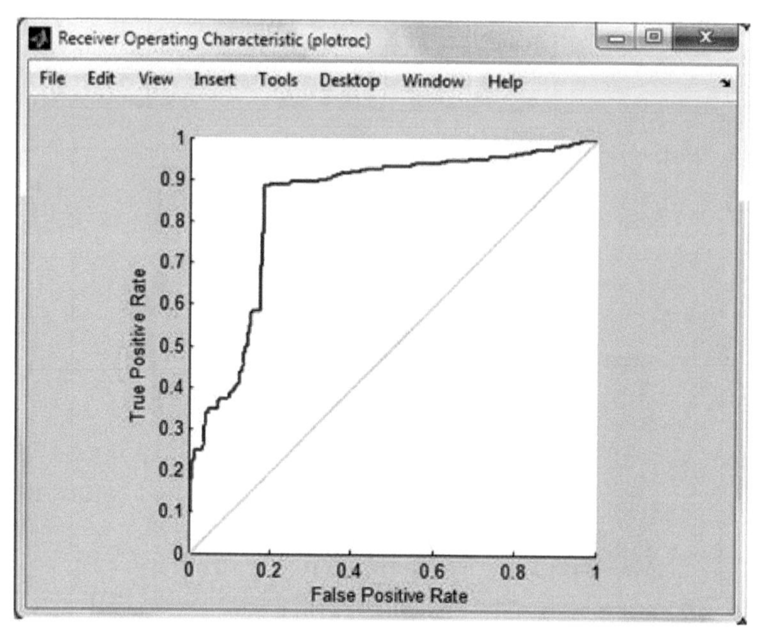

Figure 9.6 ROC curve for CNNs.

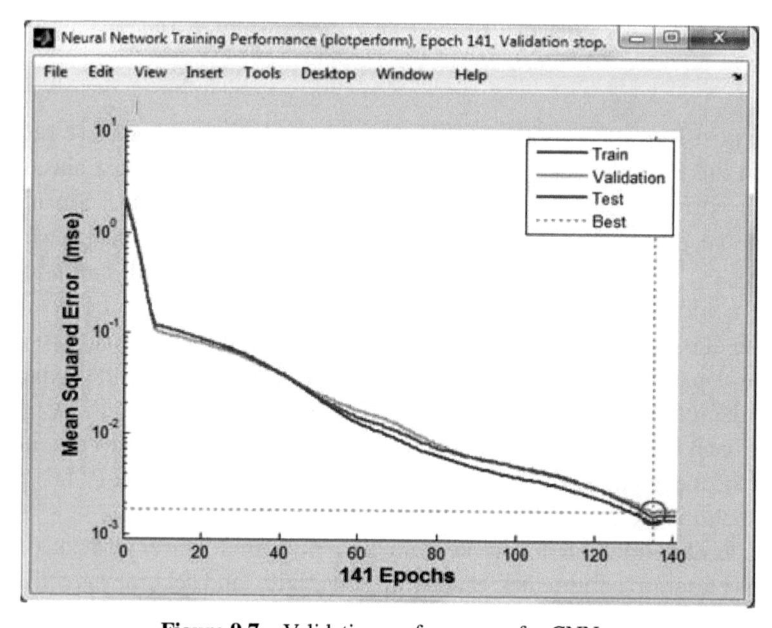

Figure 9.7 Validation performances for CNNs.

Table 9.2 Performance comparison results

	Accuracy	Sensitivity	Precision
CNNs	93.56	94.12	94.32
SVM	84.89	87.65	88.17
Decision tree	81.23	82.22	81.98

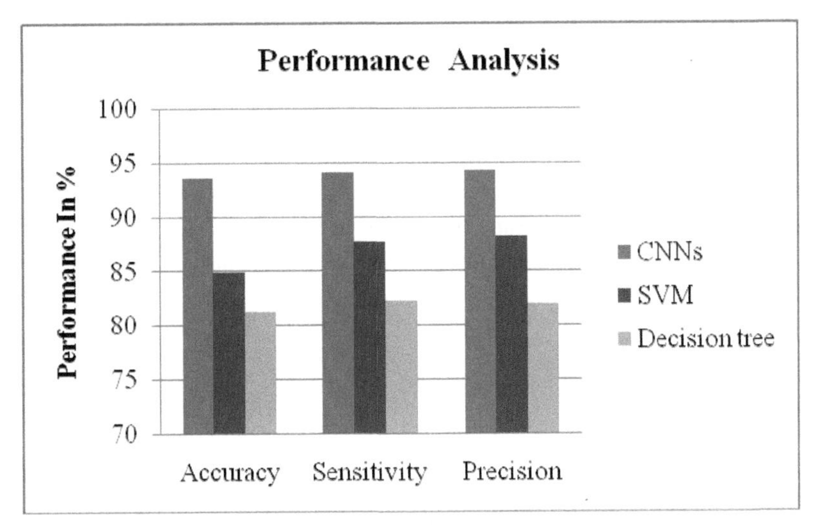

Figure 9.8 Performance comparison results.

features. The training time for the network is 18.46 seconds. Figures 9.6 and 9.7 illustrate the receiver operating characteristic (ROC) curves for this investigation, as well as the best validation performance of 0.001 at 141 epochs.

Table 9.2 and Figure 9.8 show the performance evaluation results; the CNNs achieve the better results in terms of accuracy, sensitivity, and precision in 93.56, 94.12, and 94.32. The other conventional approaches such as support vector machine achieve 84.89, 87.65, and 88.17, and decision tree algorithm achieves 81.23, 82.22, and 81.98. Comparatively CNNs produced better performance results.

9.5 Conclusion

Accurate weather forecasting is essential for day-to-day activity planning. Many real-time applications, such as weather forecasting, have used neural networks. Based on numerous factors acquired from the meteorological department, a neural network model for weather forecasting has been constructed. Because of their simplicity and robustness, neural networks have become very popular in weather prediction. In this study, an application of neural networks model was employed to forecast weather for Tamil Nadu, India. The advantage of neural networks lies in its computational speed and its capability of adapting to changing information. Neural network finds its application in weather forecasting domain. This research has focused on the application of artificial neural network in weather forecast classification which helped in identifying weather prediction for future. Online training, fault analysis system design, error prediction and removal, probability analysis, and nonlinear equalization are some of the future uses of the proposed weather prediction. Long-term data may be used for around 20 or more years. The other classifier techniques, namely statistics and genetic algorithms, can be applied in future study.

References

As-syakur, A. R., Imaoka, K., Ogawara, K., Yamanaka, M. D., Tanaka, T., Kashino, Y., & Osawa, T. (2019). Analysis of spatial and seasonal differences in the diurnal rainfall cycle over Sumatera revealed by 17-year TRMM 3B42 dataset. SOLA.

Barton, T., & Musilek, P. (2019, May). Day-ahead dynamic thermal line rating using numerical weather prediction. In 2019 IEEE Canadian Conference of Electrical and Computer Engineering (CCECE) (pp. 1-7). IEEE.

Blair, G. S., Henrys, P., Leeson, A., Watkins, J., Eastoe, E., Jarvis, S., & Young, P. J. (2019). Data science of the natural environment: A research roadmap. Frontiers in Environmental Science, 7, 121.

Dehghanian, P., Zhang, B., Dokic, T., & Kezunovic, M. (2018). Predictive risk analytics for weather-resilient operation of electric power systems. IEEE Transactions on Sustainable Energy, 10(1), 3-15.

Erickson, M. J., Colle, B. A., & Charney, J. J. (2018). Evaluation and post processing of ensemble fire weather predictions over the Northeast United States. Journal of Applied Meteorology and Climatology, 57(5), 1135-1153.

Fu, M., Wang, W., Le, Z., & Khorram, M. S. (2015). Prediction of particular matter concentrations by developed feed-forward neural network with rolling mechanism and gray model. Neural Computing and Applications, 26(8), 1789-1797.

Galanis, G., Papageorgiou, E., & Liakatas, A. (2017). A hybrid Bayesian Kalman filter and applications to numerical wind speed modeling. Journal of Wind Engineering and Industrial Aerodynamics, 167, 1-22.

Gómez-Romero, J., Molina-Solana, M., Ros, M., Ruiz, M. D., & Martin-Bautista, M. J. (2018). Comfort as a service: A new paradigm for residential environmental quality control. Sustainability, 10(9), 3053.

Hosseini, S. M., & Mahjouri, N. (2018). Sensitivity and fuzzy uncertainty analyses in the determination of SCS-CN parameters from rainfall–runoff data. Hydrological Sciences Journal, 63(3), 457-473.

Huntingford, C., Jeffers, E. S., Bonsall, M. B., Christensen, H. M., Lees, T., & Yang, H. (2019). Machine learning and artificial intelligence to aid climate change research and preparedness. Environmental Research Letters, 14(12), 124007.

da Silva Fonseca Jr., J. G., Oozeki, T., Takashima, T., Koshimizu, G., Uchida, Y., and Ogimoto, K. (Nov. 2012). Use of support vector regression and numerically predicted cloudiness to forecast power output of a photovoltaic power plant in Kitakyushu, Japan. Progress in Photovoltaics, 20(7), 874-882.

Kalimuthu, S. (2021). Sentiment analysis on social media for emotional prediction during COVID-19 pandemic using efficient machine learning approach. Computational Intelligence and Healthcare Informatics, 215.

Kalimuthu, S., Naït-Abdesselam, F., & Jaishankar, B. (2021). Multimedia data protection using hybridized crystal payload algorithm with chicken swarm optimization. In Multidisciplinary Approach to Modern Digital Steganography (pp. 235-257). IGI Global.

Khandakar, A., EH Chowdhury, M., Khoda Kazi, M., Benhmed, K., Touati, F., Al-Hitmi, M., & Gonzales, J. S. (2019). Machine learning based photovoltaics (PV) power prediction using different environmental parameters of Qatar. Energies, 12(14), 2782.

Kang, G. K., Gao, J. Z., Chiao, S., Lu, S., & Xie, G. (2018). Air quality prediction: Big data and machine learning approaches. International Journal of Environmental Science and Development, 9(1), 8-16.

Kumar, K. R., & Kalavathi, M. S. (2018). Artificial intelligence based forecast models for predicting solar power generation. Materials Today: Proceedings, 5(1), 796-802.

Kumar, Y. J. N., Spandana, V., Vaishnavi, V. S., Neha, K., & Devi, V. G. R. R. (2020, June). Supervised machine learning approach for crop yield prediction in agriculture sector. In 2020 5th International Conference on Communication and Electronics Systems (ICCES) (pp. 736-741). IEEE.

Maleki, H., Sorooshian, A., Goudarzi, G., Baboli, Z., Birgani, Y. T., & Rahmati, M. (2019). Air pollution prediction by using an artificial neural network model. Clean Technologies and Environmental Policy, 21(6), 1341-1352.

Mihai, A., Czibula, G., & Mihulet, E. (2019, September). Analyzing meteorological data using unsupervised learning techniques. In 2019 IEEE 15th International Conference on Intelligent Computer Communication and Processing (ICCP) (pp. 529-536). IEEE.

Poterjoy, J., Wicker, L., & Buehner, M. (2019). Progress toward the application of a localized particle filter for numerical weather prediction. Monthly Weather Review, 147(4), 1107-1126.

Pierce, F. J., & Lal, R. (2017). Monitoring the impact of soil erosion on crop productivity. In Soil Erosion Research Methods (pp. 235-263). Routledge.

Qing, X., & Niu, Y. (2018). Hourly day-ahead solar irradiance prediction using weather forecasts by LSTM. Energy, 148, 461-468.

Rajesh, P., & Karthikeyan, M. (2017). A comparative study of data mining algorithms for decision tree approaches using weka tool. Advances in Natural and Applied Sciences, 11(9), 230-243.

Schumacher, R. S. (2017). Heavy rainfall and flash flooding. In Oxford Research Encyclopedia of Natural Hazard Science.

Sheikh, F., Karthick, S., Malathi, D., Sudarsan, J. S., & Arun, C. (2016). Analysis of data mining techniques for weather prediction. Indian Journal of Science and Technology, 9(38), 1-9.

Singh, M., Kumar, B., Niyogi, D., Rao, S., Gill, S. S., Chattopadhyay, R., & Nanjundiah, R. S. (2021). Deep learning for improved global precipitation in numerical weather prediction systems. arXiv preprint arXiv:2106.12045.

Talavera, J. M., Tobón, L. E., Gómez, J. A., Culman, M. A., Aranda, J. M., Parra, D. T., ... & Garreta, L. E. (2017). Review of IoT applications in agro-industrial and environmental fields. Computers and Electronics in Agriculture, 142, 283-297.

Liu, Y., Racah, E., Correa, J., Khosrowshahi, A., Lavers, D., Kunkel, K., ... & Collins, W. (2016). Application of deep convolutional neural networks for detecting extreme weather in climate datasets. arXiv preprint arXiv:1605.01156.

Ukhurebor, K. E., Abiodun, I. C., Azi, S. O., Otete, I., & Obogai, L. E. (2017). A cost effective weather monitoring device. Archives of Current Research International, 1-9.

Verbois, H., Huva, R., Rusydi, A., & Walsh, W. (2018). Solar irradiance forecasting in the tropics using numerical weather prediction and statistical learning. Solar Energy, 162, 265-277.

Waliser, D. (2018). The Weather Environmental Prediction Enterprise: A System (and Social) Engineering and Data Science Challenge.

Wen, J., Yang, J., Jiang, B., Song, H., & Wang, H. (2020). Big data driven marine environment information forecasting: A time series prediction network. IEEE Transactions on Fuzzy Systems, 29(1), 4-18.

Wilgan, K., Hadas, T., Hordyniec, P., & Bosy, J. (2017). Real-time precise point positioning augmented with high-resolution numerical weather prediction model. GPS Solutions, 21(3), 1341-1353.

Xu, G., Zhu, X., Tapper, N., & Bechtel, B. (2019). Urban climate zone classification using convolutional neural network and ground-level images. Progress in Physical Geography: Earth and Environment, 43(3), 410-424.

Yemane, S. (2021). Deep Forecasting of Renewable Energy Production with Numerical Weather Predictions.

Yoo, C., Han, D., Im, J., & Bechtel, B. (2019). Comparison between convolutional neural networks and random forest for local climate zone classification in mega urban areas using Landsat images. ISPRS Journal of Photogrammetry and Remote Sensing, 157, 155-170.

Zhao, H., Wang, Y., Song, J., & Gao, G. (2019). The pollutant concentration prediction model of NNP-BPNN based on the INI algorithm, AW method and neighbor-PCA. Journal of Ambient Intelligence and Humanized Computing, 10(8), 3059-3065.

Zubaidi, S. L., Ortega-Martorell, S., Kot, P., Alkhaddar, R. M., Abdellatif, M., Gharghan, S. K., ... & Hashim, K. (2020). A method for predicting long-term municipal water demands under climate change. Water Resources Management, 34(3), 1265-1279.

10

E-Learning Modeling Technique and Convolution Neural Networks in Online Education

Fahad Alahmari, Arshi Naim, and Hamed Alqahtani

Department of Information Systems, College of Computer Science, King Khalid University, KSA
E-mail: fahad@kku.edu.sa; arshi@kku.edu.sa; hsqahtani@kku.edu.sa

Abstract

Higher educational systems are accountable for ensuring effective E-learning (EL) environments for online learners. An effective learning environment engages learners in educational activities. The chapter has three parts; in the first part, we discussed the convolutional neural network (CNN). CNN has many models, but for the purpose of this study, we have applied three models and found them to be most appropriate to measure students' engagement (SEt) in EL assignments. We have applied all convolutional networks (All-CNN), network-in-network (NiN-CNN), and very deep convolutional network (VD-CNN) because they have simple network architectures and show efficiency in conditions and categories. These categories are based on the conditions of learners for their facial expressions in an online environment. In the second part of the chapter, we cover the methods of application and benefits of machine learning (ML) and artificial intelligence (AI) in King Khalid University's (KKU) E-learning. The third part of the chapter covers the role of Internet of Things (IoT) in the education sector and defines the advantages, types of security concerns, and challenges of deployment of IoT. This research is descriptive in nature; results for the application of three models of CNN are referred for their advantages and challenges for online learners, and results for ML and AI are based on qualitative analysis done through tools and techniques of EL applied in KKU's learning management services (LMS) and blackboard (BB). Results for IoT show the benefits for both students and educators.

Keywords: Convolutional neural network, artificial intelligence, Internet of Things, E-learning, machine learning.

10.1 Introduction

The past decade has witnessed a drastic development in EL and identified applications of several web-based learning analytical tools in its practices [1]. EL has facilitated individual learning and group learning in an easy manner and also enhanced the cost advantages [2]. EL has included all types of target group in its learning areas and defined the advantages for them in higher education and professional environment and added to their expertise in their knowledge areas [3]. Disadvantages are also described for EL in comparison to face-to-face (F2F) learning. In F2F environment, the instructors have a great opportunity to understand students' behavior, their interests, physical gesture, and facial expressions. These symbols help instructors to know the students' participation in learning, but in the EL environment, this important part is found to be a limitation [4]. This limitation concludes in confusion in evaluating SEt in learning, completing assignments, and other related educational endeavors. In this situation, the CNN has brought a great advantage. A machine- or technology-based solution is also suggested for identifying the students' involvement through various ways such as following their mannerism, time they devoted on online learning, their questions, and answering them during the online learning session. Monitoring makes students more careful and they are motivated to involve in the process of learning. This is one good option to reduce stress level and anxiety amongst the students and eventually can reduce the number of withdrawn students. Learning which is based on competence and critical thinking requires more involvement at an internal level as well as an external level. These internal and external levels measure the SEt by the help of CNN. These internal and external features include perception, observable facial description, physical gesture, verbal communication, and conduct [5]. It is difficult to measure involvement internally. But the external apparent factors can be assessed using new sensing and affective computing techniques using video [6], audio [7], and physiological signals [9]. The sensitivity of students can be contingent from these measures via affective computing that is increasingly being used in learning technologies [10]; however, their applications in online learning have not been widely applied yet. Facial image analysis is the widely accepted external factor for facial gestures and sensitivity identification. Facial features represent poignant conditions such as awkwardness, antipathy, and contentment, which show imperative function in students' expressions of frustration and involvement during education. Good improvement has been applied, but there are still a lot of limitations that exist related to the appropriate mapping of facial expressions with the students' emotion and involvement identification. Deep learning techniques (DPT) have given great progress in computer vision, but, to some extent, this has not been used to measure the SEt in EL. In this chapter, we examine the applicability of the (CNN), DPT for SEt detection in EL. Specifically, three popular models of CNN and a proposed model are tested for SEt recognition using facial expressions. The three popular models include All-CNN [11], NiN-CNN [12], and VD-CNN [13]. We observed that each of these architectures has some unique characteristics that can help to increase the accuracy of a typical CNN model. The application of multilayer perceptron can increase the depth of the network by using small convolutional filters and, finally, replace some max-pooling layers by convolutional layer with an increased stride. We identified that the accuracy of the above three models and the proposed model in estimating students' perceived SEt (external observable factors, such as facial expressions) in online learning. Since teachers rely on external observable factors to judge perceived SEt to adapt their teaching behavior, the automation of perceived SEt identification is likely to be useful in EL to offer special help to the students in need.

EL is witnessing its growth as never before, from the educational sector to the commercial environment, but without ML and AI, this growth would have never been possible. The introduction of ML and AI has filled the distances of communication [14]. ML and AI are sketching the future growth of EL by using the techniques of prediction, and algorithms to develop more individual-centric EL practices. This paper attempts to answer the methods and processes to achieve this objective. As an example, KKU EL deanship is used to observe the role of ML and AI in EL growth. This paper has tried to answer the possibilities of changes in EL with more advancement in ML and AI. The paper is segmented into three parts. The first part presents the literature review where a description and introduction of EL in KKU is given specifically. Also, the general historic vision of EL is given along with the growth of ML and AI from past to present. The second part is the discussion on various areas of ML and AI in general and in context to EL

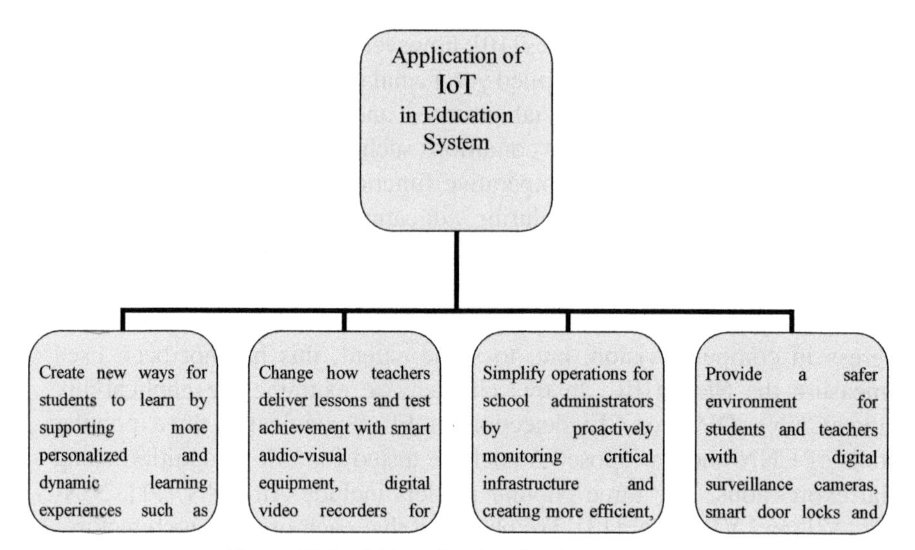

Figure 10.1 IoT applications in education [20].

in KKU specifically as of educational sector that covers the definition of ML and AI, ML classification, benefits of ML and AI in EL, EL transformation due to ML, and AI and ML application. All these areas are discussed with reference to KKU EL deanship. The last part provides the qualitative results on expected changes in EL in KKU after the application and determination of ML and AI [15, 16]. Besides the above-mentioned three parts, a short note on predictive limitations is also given which is completely based on researchers' experiences with ML and AI in EL environment.

The Internet of Things (IoT) has the capabilities to transform education by intensely altering how educational academies collect data, interface with users, and automate processes [17]. The applications of IoT are related to the networking of physical objects through the use of embedded sensors, actuators, and other devices. This process collects and transmits information about real-time campus activity in educational sector [18]. New learning environment is developed by the implementation of IoT because it is able to integrate user's mobility and data analytics. Figure 10.1 shows the tasks that can be performed by IoT [19].

This research paper contributes in explaining the role of emerging technologies in education sectors and illustrates the guidelines as well as advantages of their applications which can be followed by similar types of academies [21]. Figure 10.2 gives the list of the main components of IoT that can be used in educational sector.

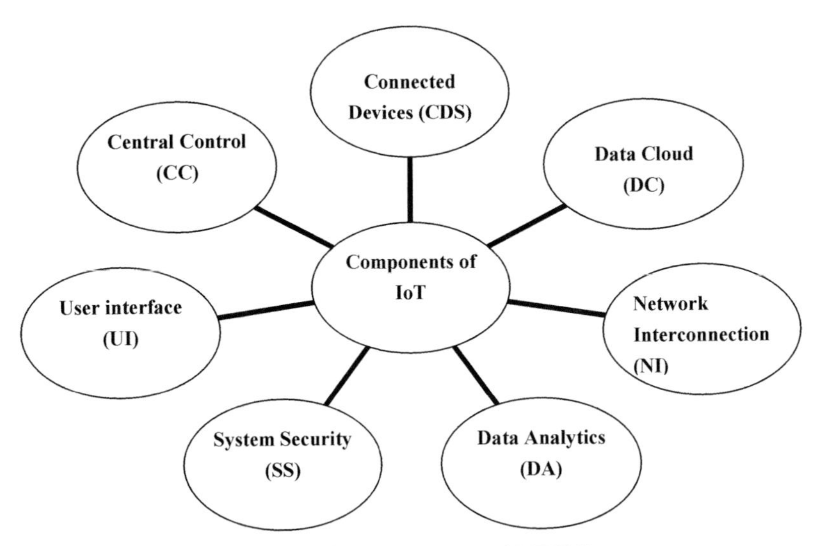

Figure 10.2 Components of IoT [22].

CDS are the primary physical devices which are connected to the educational learning system. These devices establish the connectivity of all the equipment necessary for the learning operations. CC is also a device, but their job is to manage the data and ensure the security, follow protocol, and manage traffic on network. CC also scales the hardware as per the requests of clients [23]. These clients are students, faculty, IT professionals, or other university staff. DC stores the learning material and other useful data of university students and employees. IoT provides UI for the education clients for two-way communication. A major issue of using IoT is security problems, but SS helps in solving this problem and assures academicians to use the system with our apprehensions. Finally, DA provides analysis of data in a systematic manner. DA is an example of computational analysis which can be applied for various benefits in education system such as data interpretation and other decision-making processes for learning and teaching [24].

10.2 Literature Review

At a high level, SEt identification methods can be divided into three main categories: manual, semi-automatic, and automatic [25]. The manual category refers to the methods where learners' direct involvement is needed in SEt detection. This category includes observational checklist and self-reporting techniques. The observational checklist relies on questions completed by external observers (e.g., teachers) instead of the learners. The self-reporting poses a set of questions in which students report their own level of attention, distraction, excitement, or boredom [26]. Self-reporting is of great interest to many researchers because it is easy to administer and can provide some useful insight into students' SEt [27]. However, their validity depends on a number of factors like learners' honesty, their willingness to report their emotions, and the accuracy of learners' perception of their emotions [28]. The semi-automatic category includes the methods of SEt tracing. The SEt tracing utilizes the timing and accuracy of learner responses to practice problems and test questions [30]. Although these methods have been used in classroom-based learning and intelligent tutoring systems, not many applications of these methods can be found in online learning [31]. The automatic category includes computer-vision-based methods, sensor data analysis, and log-file analysis. Among these methods, the computer-vision-based methods are found to be more suitable to use in online learning as these are unobtrusive to the users and the hardware and software to capture and analyze the data are wide-spread available at low cost. The typical computer-vision-based methods use facial expressions, gestures and posture, and eye movement. Some research studies combine more than one modality to achieve better accuracy. A good deal of information used by humans to make SEt judgment is based on human faces, and it has been hypothesized that facial expressions are directly linked to the perceived SEt [32]. Cameras provide a continuous and non-intrusive way of capturing face images when the learner uses a mobile device or a personal computer for learning activities. The captured facial information is used to understand certain facets of the learner's current state of mind. Many different methods have been proposed to automate this detection process by analyzing the face images [33]. Based on how the information from a face appearance is used, these methods are divided into two groups: part-based and appearance-based [34]. Part-based methods deal with different parts of a face (e.g., eyes, mouth, nose, forehead, chin, and so on) for the SEt detection. A comprehensive way to analyze the parts of a face is the Facial Action Coding System (FACS) [35]. In appearance-based methods, features from whole-face regions are extracted and used to generate patterns for SEt classification. Among different feature extraction techniques, local binary patterns (LBP) and histogram of oriented gradients (HOG) are found to be popular in appearance-based methods [36]. SEt detection using CNN includes a generic architecture of CNN which is a combination of deep learning technology with ANNs. A generic CNN architecture typically

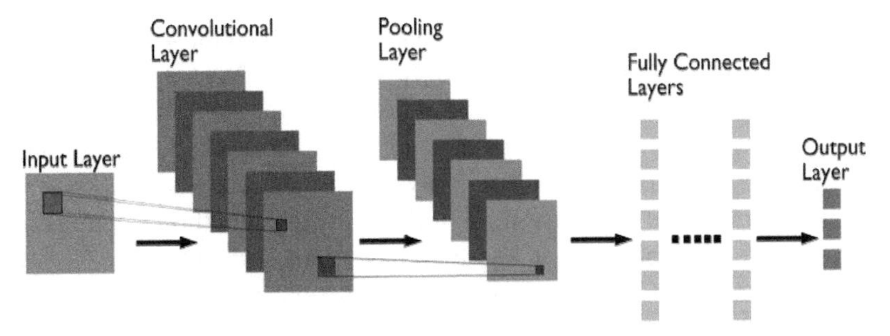

Figure 10.3 Typical architecture of CNN [35].

contains an input layer, multiple hidden layers, and an output layer. In the hidden layers, there can be a different combination of convolutional layers, activation layers, pooling layers, normalization layers, and fully connected layers. Figure 10.3 shows a typical architecture of CNN as presented in [35], where basically the input, convolutional, pooling, fully connected, and the output layers are illustrated. Each feature map can detect the presence of single feature at all possible locations. Each output of the convolutional layer is then passed through an activation layer which uses an activation function to decide the final value of a neuron. The activation function transforms the linear combination of features into non-linear so that the neural network can learn faster with high accuracy.

Most CNN architectures use one or more fully connected layers before the output layer, which is a typical MLP. All neurons in this layer are fully connected to all activations in the previous layer. A typical CNN model works with minimizing a loss function which is computationally feasible and represents the price paid for inaccurate predictions. The main reason for the model's popularity of ALL-CNN is that the model achieves high accuracy in different benchmarking datasets with its simple network architecture in classification. This model is different from a standard CNN model mainly in two key aspects. First, this model replaces the pooling layers by using standard convolutional layers with an increased stride. Second, this model makes use of small convolutional layers with kernel size which greatly reduces the number of parameters in a network and thus serves as a form of regularization. NiN-CNN Typical architecture of CNN [37]. uses a "micro-network" structure to approximate the nonlinear functions in CNN and install "micro-network" architecture (MLP). This deep model leads to severance between latent features, which help to achieve better abstraction and accuracy. In a traditional CNN, the feature maps of the last convolutional layer are flattened and passed on to one or more fully connected layers, which are then passed on to softmax for the classification. In VD-CNN [37], the authors address the aspect of depth in CNN. In this architecture, authors increase the depth of the network by adding more convolutional layers with stride 1. The authors argue that a stack of two 3×3 layers without any pooling layer in between has an effective receptive field of 5×5, where the three such layers have a 7×7 effective receptive field which has various advantages.

The EL deanship (ELD) at King Khalid University was established in the year 1426 H (as per Arabic Calendar) as part of the continuous online learning in KKU. EL has tried to use the best of techniques in developing and improving its educational services. In a general context, preeminent examples of using ML and AI are in the development process of LMS for any educational system. From the introduction time, EL canters have been performing various researches and using trained IT staff to enhance education systems, develop online learning skills, and apply the best of online expertise in imparting knowledge [38]. In the current scenario, EL deanship in KKU has launched many new online services to achieve a new level of success in the online education sector. KKU's vision for EL is to take "KKU's human resource at the highest skills and empower them to fulfil their changing needs and aspirations through using embedded EL" [39].

To meet these objectives EL in KKU has recently introduced Advanced Google Classroom, Mediasite for online streaming of lectures, etc. EL has also focused on advancing LMS for effective communication and feedback beyond the traditional learning environment [40]. Many external applications have also been developed by focusing on education sharing through Google Classroom and other tools such as YouTube, Google Docs, Google Mail, Task Manager, and Google Analytics [40]. These are the focal tools of the development programs of EL in the extensive references. These applications have enabled all online users to share information at all levels for learning and teaching (L&T) purposes [41].

History of EL: The expression "EL" is not very old; it was introduced in 1999, initially employed at a CBT [42] systems seminar. As a synonym, many other expressions too are used like "online learning" and "virtual learning" for EL [42]. EL has taken extensive development over a period of time and has achieved a much deeper meaning in the 19th century [42]. Arthur Samuel was an American initiator in the field of PC gaming and AI, and, in 1959, while working at IBM, he introduced the concept of ML [43]. In the 1960s, Nilsson authored a book on learning machines, explaining the management systems of ML for design classification [43].

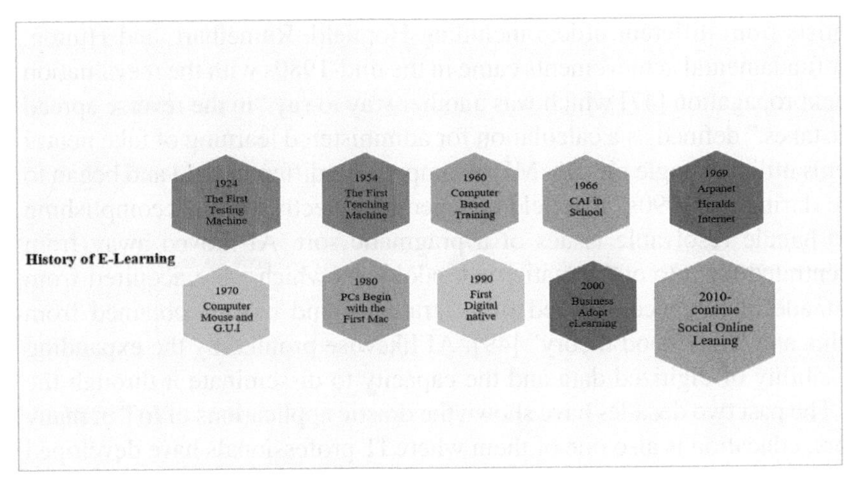

Figure 10.4 History of EL [42].

The enthusiasm for the design and application of ML was identified in 1970 and 1973; Duda and Hart portrayed the same in their book "Pattern Classification" [44]. As a logical undertaking, ML and AI have become very popular applications from past to present. Scholars, scientists, professionals, and academicians all use this for machining the information. They use ML and AI for the application of representative strategies, named as "neural systems"; these are for the most important part of perceptrons [45]. Another application of ML and AI is probabilistic thinking [45] which is mostly utilized and mechanized in the clinical analysis by the medical practitioners [45].

ML and AI provide accentuation on a methodology that helps in expanding the sensible processing of information. In 1980, probabilistic frameworks were beset by hypothetical and viable issues of information procurement and representation [46], and in the same year, master frameworks were developed to command and measure man-made intelligence [46]. Work on emblematic/information-based learning continued inside man-made intelligence, prompting inductive rationale programming. However, the more factual line of researches is outside the field and domain of AI appropriateness in design acknowledgment and data retrieval [46]. Neural systems exploration had been relinquished by simulated intelligence and software engineering around a similar time during the 1980s. This line, as well, was preceded outside the man-made intelligence/CS field, called "connectionism," by the scientists from different orders including Hopfield, Rumelhart, and Hinton. Their fundamental achievements came in the mid-1980s with the reevaluation of backpropagation [47] which was another way to say "in the reverse spread of mistakes," defined as a calculation for administered learning of fake neural systems utilizing angle plunge. ML revamped as a different field and began to thrive during the 1990s. The field changed its objective from accomplishing AI to handle resolvable issues of a pragmatic sort. AI moved away from concentrated form to emblematic methodologies which were acquired from man-made intelligence, derived from strategies and models obtained from insights and "likelihood theory" [49]. AI likewise profited by the expanding accessibility of digitized data and the capacity to disseminate it through the web. The past two decades have shown the drastic applications of IoT in many sectors; education is also one of them where IT professionals have developed a new platform of learning and teaching by the use of IoT. It is expected in 2022 that IoT will take a great lead in secured online assessments for higher educational system [49]. Research works in 2017 have discussed that 67% of primary and secondary education systems in developed nations such as US, UK, Germany, and Australia have already applied IoT in their teaching system and this percentage is increasing sharply every year [49]. Research conducted in 2018 showed the concerns of educational systems about security and privacy for the IoT in education. IoT has caused cyber security threats and aggravated the network attacks. Also, studies conducted in 2019 show that major attacks were planned by the students for altering their grades and attendance. IT professionals have developed more advanced security systems in 2019 to prevent such cyber-attacks in the academies and revolutionize the application of IoT in education. In the year 2017, IoT has facilitated in preventing distributed denial of service (DDoS) attacks that were designed to intermittently bring down the University of Michigan's computing network. The year 2020 explained the relevance of IoT in education sectors for all countries because offline learning was suspended due to the spread of COVID 19 [49]. Past research works have focused on single application of technologies in learning system, whereas this paper presents the example from the higher education system presenting how these technologies were successfully implemented.

10.3 Discussion

This chapter proposes a new CNN architecture where we incorporate different advantageous features from these three base models. Unlike using

Table 10.1 Specific architectures for the combined model and the three base models

All-CNN	VD-CNN	NIN-CNN	Combined models
Input 32×32 Grayscale Image			
3×3 conv. 96 ReLU 3×3 conv. 96 ReLU	3×3 conv. 96 ReLU 3×3 conv. 96 ReLU 3×3 conv. 96 ReLU	5×5 conv. 192 ReLU 1×1 conv. 192 ReLU 1×1 conv. 192 ReLU	3×3 conv. 192 BatchNorm ReLU 1×1 conv. 192 ReLU
3×3 conv. 96 ReLU with stride 2	3×3 max-pooling with stride 2	3×3 max-pooling dropout (0.5)	3×3 conv. 192 BatchNorm ReLU with stride 2 dropout (0.5)
3×3 conv. 192 ReLU 3×3 conv. 192 ReLU	3×3 conv. 192 ReLU 3×3 conv. 192 ReLU 3×3 conv. 192 ReLU	3×3 conv. 96 ReLU 1×1 conv. 96 ReLU 1×1 conv. 96 ReLU	3×3 conv. 96 BatchNorm ReLU 1×1 conv. 96 ReLU 1×1 conv. 96 ReLU
3×3 conv. 192 ReLU with stride 2	3×3 max-pooling with stride 2	3×3 max-pooling dropout (0.5)	3×3 max-pooling with stride 2 dropout (0.5)
3×3 conv. 192 ReLU 1×1 conv. 192 ReLU	3×3 conv. 192 ReLU 1×1 conv. 192 ReLU	3×3 conv. 32 ReLU 1×1 conv. 32 ReLU 1×1 conv. 32 ReLU	3×3 conv. 32 BatchNorm ReLU 3×3 conv. 32 BatchNorm ReLU 3×3 conv. 32 BatchNorm ReLU
1×1 conv. 3 global average pooling 3-way softmax	1×1 conv. 3 global average pooling 3-way softmax	1×1 conv. 3 global average pooling 3-way softmax	1×1 conv. 3 global average pooling 3-way softmax

homogeneous blocks as in the base models [50], we use heterogeneous blocks where we keep the network depth limited for achieving computational efficiency. This chapter suggests applying the combination of three CNN models and the specific architectures for the combination of model, and the three base models are described in Table 10.1.

For the models training and testing, Figure 10.5 shows the relationship between the loss function and epoch times for the training and validation sets for All-CNN, VD-CNN, NiN-CNN, and the combined model.

The relationship among the loss function of the training set, validation set, and epoch times, and the relationship among the accuracy of the training set, validation set, and epoch times are shown in Figure 10.6.

10.3.1 Definition of ML and AI

ML is not a separate branch; it has emerged from AI that includes algorithms for performing predictions and giving outcomes. The entire process depicts a type of pattern and then concludes learning extracted through data. Also, a parallel process runs where all novice information received is first analyzed and used to predict the user's behavior. It is relevant to note here that LMS [51] is benefitted from this process as it provides a personalization outlook for the users.

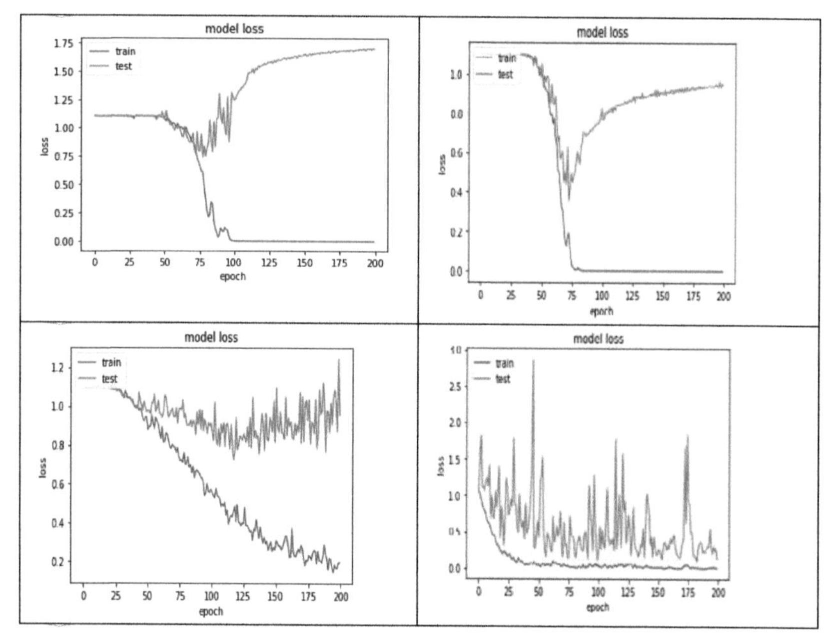

Figure 10.5 Relationship between the loss function and epoch times for the training and validation sets for (a) All-CNN, (b) VD-CNN, (c) NiN-CNN, and (d) the combined model.

10.3.2 Definition of ML and AI in KKU EL

EL in KKU has also utilized the significance of ML and AI in LMS, where student's online access of data is evaluated and aids in calculating the duration and determining which tool students have used while working on LMS in EL.

In defining ML, two types of ML frameworks are identified: *proprietary and open-source* [52].

Proprietary and open-source are two types of examples of deep learning [52]. EL in KKU is using proprietary deep learning software more than open-source. EL in KKU is using different tools developed by Google for its performance such as tensor process units [53] and other academies have a choice to implement proprietary or open-source as per their educational requirements.

10.3.3 ML Classifications for KKU EL

ML is respected to identify data of a user and give results in form of patterns. These patterns are given by ML using algorithms to forecast the effect. EL in KKU has successfully implanted ML's algorithm classification [53] in LMS.

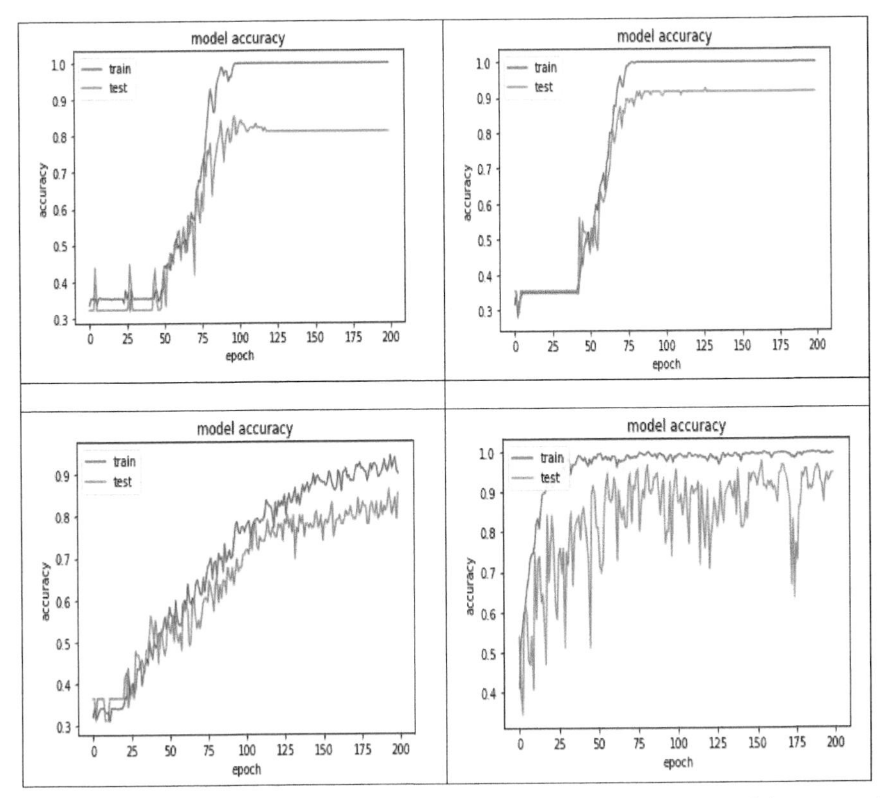

Figure 10.6 Relationship between the accuracy and epoch times for the training sets and validation sets for All-CNN, VD-CNN, NiN-CNN, and the combined model.

Table 10.2 explains the classification of ML [54] into three categories: supervised, unsupervised, and reinforcement [54]. EL in KKU is using all three applications in the development of various online tools and techniques in LMS services like category one is used by information technology (IT) specialists in developing a new interface for BB on LMS services where the IT human resources (HR) are predicting new datasets based on past datasets. For the second category, EL in KKU is using a sub-set category known as "semi-supervised" [14], where IT HR tries to offer the system with accurate input and output relationship in building any new LMS platform. To meet this objective, EL deanship conducts extensive IT HR training. IT HR follows optimum methods of evaluating output sets in the third category of ML in EL in KKU. This is an example of the process of learning from reward and errors.

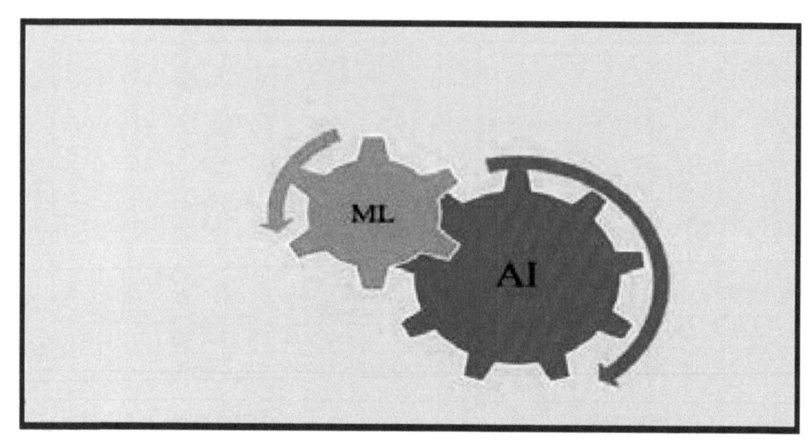

Figure 10.7 ML as a subpart of AI [51].

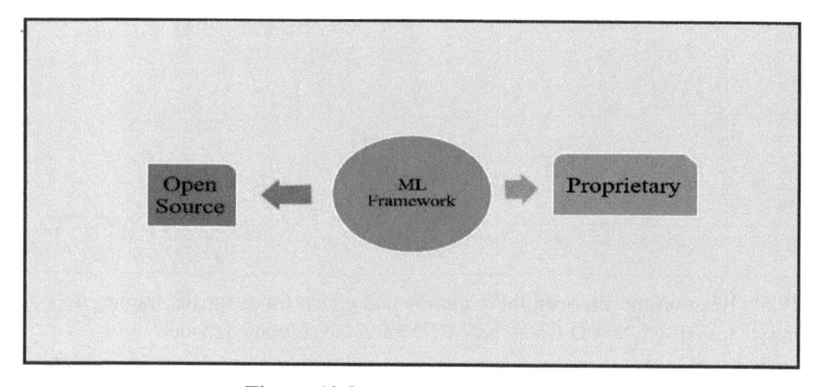

Figure 10.8 ML framework [51].

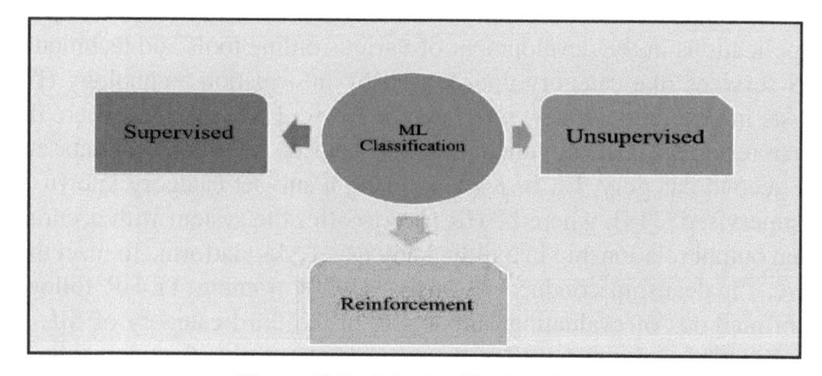

Figure 10.9 ML classification [54].

Table 10.2 Definition of ML classification

Classification of ML [2, 11]		
Supervised [2, 11]	Unsupervised [2, 11]	Reinforcement [2, 11]
This classification is based on trend analysis where the system takes instances from previous extracted and novice to give expected data. Organizational programmer has to provide both input and output to the systems for the perfection of the software. Repeated practice makes the autonomous process for constructing targets and new datasets.	This explains no fixed rules on classifications. The system analyzes the datasets to conclude the patterns and give some conclusions. This classification is very helpful in case of ambiguous data fashion; however, it is not concerned with the alignment of input and output.	As from the meaning, it focuses on fixed objectives that a system has to reach. This classification is based on reviews and feedback to achieve the desired milestone. It is like learning by mistakes and effectiveness.

10.3.4 The Benefits of ML and AI in KKU EL [55]

EL in KKU is using the best of benefits provided by ML and AI for its online learners and instructors. The focus is on the present and future of LMS practices. This is mostly for new LMS [55] where the idea is to confer predictive algorithms and robotic delivery of EL contents. Figure 10.10 gives the benefit of ML and AI in KKU EL in general.

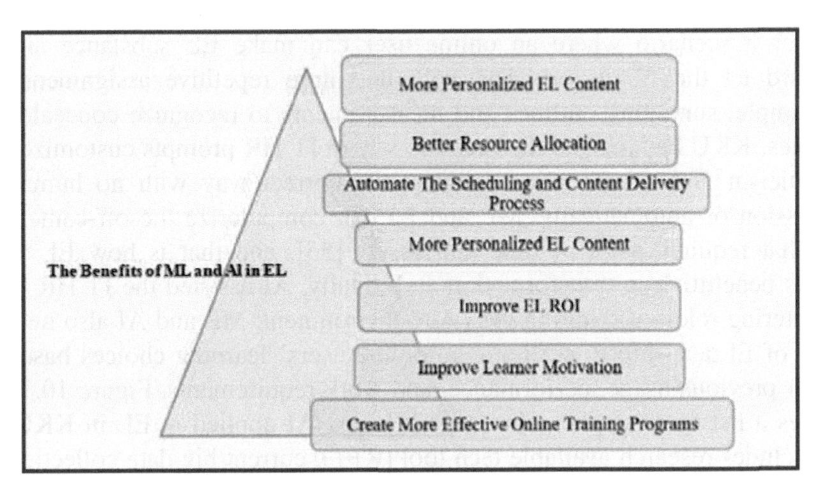

Figure 10.10 Benefits of ML and AL in EL [55].

Table 10.3 Benefits of ML and AL in EL [55]

Benefit of ML and AI in EL KKU		
Customized EL Content [4] EL in KKU is using ML algorithms to predict results, which allow to provide specific EL content based on past performance and individual learning goals. If the online user is more active and frequent user of any tool on LMS and BB, the systems robotically provides the recommendation to the users[5] also EL is using ML and AI to show the gap and excellence in learning skills presented in its users. ML and AI benefit the KKU online users how to give more customized learning materials. The system recommends more integrated, basic or complex courses to the online learners based on their online behavior.	**Resource Allocation[4]** ML and AI offer two aids in resource allocation. One for educational sector and one for corporate[6]. EL in KKU is benefitted by educational aspect where online learners on LMS gets the accurate online resources. This builds the bridge for their skills and facilitate in meeting learning outcomes. they require to fill gaps and achieve their learning goals[6].The second benefit of ML and AI in resource allocation ,EL in KKU identified is for HR team. Now they are taking minimum time in analysis and developing powerful contents for LMS [7].	**Automate Content Delivery and Scheduling Process [4]** Developing tools on LMS is not an easy practices and task but with ML and AI such complex, ambiguous and time taking tasks become easier. EL deanship in KKU schedule the coursework for online learners and deliver online resources. This technique depends on their EL assessment results or simulation presentation. EL in KKU expects that AI in coming days with make this process robotic and create a niche in EL modules [8] for all online learner who are participants on LMS.
Improve KKU EL Return On Investment 4] EL in KKU witnessed better return on investment with the ML and AI.; resultant in creating more customized approach and better revenue earnings. With the help of predictive analysis online learners takes online training promptly that cause in time effectiveness for other endeavors. Apart from this EL in KKU use AI-equipped software that could monitor and predict online user's behavior on LMS.	**Improve Learner Motivation[4]** EL deanship believes that the ML systems and AI in future can be likened to a private virtual teaching. Many online earning programs have been developed following the concepts of ML and AI by the IT HR. For example KKU X under EL deanship , used ML and AI in establishing distinguished contents for young Saudi students to develop skills to prepare the graduate for potential employment[9] .	**Online Training Programs[4]** ML and AI is implemented by EL in KKU in making peer-to-peer communication productive. Also AI created a mapping between Electronic experts and learning on LMS platform in KKU.

Researchers have discussed how EL in KKU is taking the benefits from ML and AI in its LMS practices. Table 10.3 gives a brief description of receiving benefits from ML and AI in LMS and other online platforms.

10.3.5 ML and AI are Transforming the EL Scenario in KKU

Consider a scenario where an online user can make EL substance and afterward let the framework deal with the more repetitive assignments, for example, surveying outlines and measurements to recognize concealed examples. KKU has imaged the scenario where IT HR prompts customized EL criticism and steers online students the correct way with no human intercession or automatically. ML and AI can computerize the off-camera work that requires a lot of time and assets [55], and that is how EL in KKU is benefitting in transformation and, finally, AI-assisted the IT HR in encountering relevant errors in the LMS environment. ML and AI also help IT HR of El deanship in customizing online users' learning choices based on their previous usage, performance, and work requirements. Figure 10.11 provides a list of an application under ML and AI applied in EL in KKU. This includes research available tech tool (RTT), current big data collection (BDC), ML's role in online training strategy (MLOTS), and future game plan for online learning (FGOL).

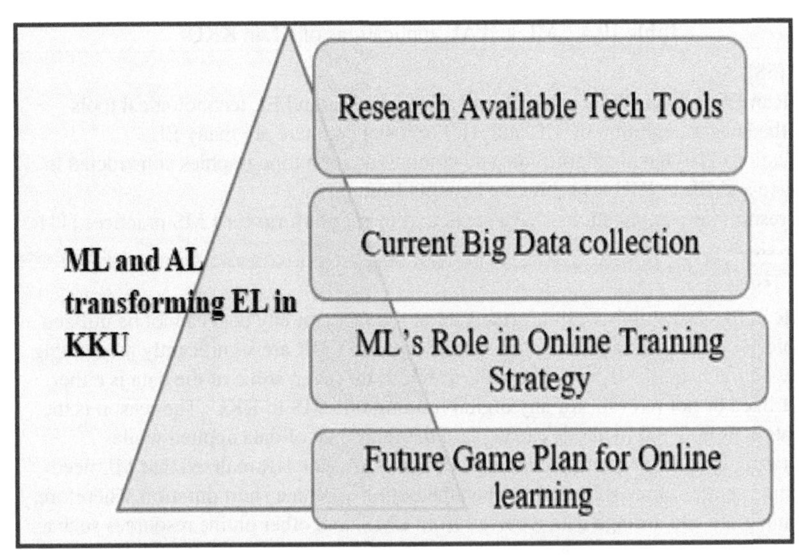

Figure 10.11 ML and AI application of EL in KKU [56].

All the applications of ML and AI are explained in the context of EL in KKU and similar plans that other institutions can work too. Table 10.4 gives a brief account of the application.

IT HR in EL in KKU did a research and selected LMS and EL technological tools with the latest integration of ML and AI. For instance, there are many EL applications [56] having algorithms and computerization topographies constructed in advance and EL in KKU is achieving benefits from them. As a result, these applications have created an efficient platform for LMS practices [57] in KKU.

IT HR in El deanship is well informed about the fact that any data cannot be utilized completely unless ML and AI give a factual report. IT HR is significantly assembling data with the help of ML and AI for EL in KKU; however, some of the data is either not utilized or not relevant for any digital training on LMS in KKU. The reason is the limitation as no good methods can explain the relevance of data applied while integrating in algorithm and predictive analytics. Also, IT HR realized that ML needs to achieve an entire impression and not just the outline of some short duration. Therefore, they integrate and arrange data received from LMS, and other online resources such as social network and web sources to elaborate the pattern.

EL in KKU has understood the advantages of ML in describing the online learning and teaching (L&T) strategies and, therefore, applied in the LMS

Table 10.4 ML and AL applications of EL in KKU

RTT [58] IT HR in EL in KKU did a research and selected LMS and EL technological tools with the latest integration of ML and AI. For instance, there are many EL applications [10] having algorithms and computerization topographies constructed in advance and EL in KKU is achieving benefits from them. As a result, these applications shave created efficient platform for LMS practices [11] in KKU.
BDC [58] IT HR in EL deanship is well informed about the fact that any data cannot be utilized completely unless ML and AI give a factual report. IT HR are significantly assembling data with the help of ML and AI for EL in KKU; however, some of the data is either not utilized or not relevant for any digital training on LMS in KKU. The reason is the limitation as no good methods can explain the relevance of data applied while integrating in algorithm and predictive analytics. Also, IT HR realized that ML needs to achieve entire impression and not just the outline of some short duration. Therefore, they integrate and arrange data received from LMS, and other online resources such as social network and web sources to elaborate the pattern. This concluded IT HR in EL deanship to provide trend for online training for LMS users in KKU.
MLOTS [58] EL in KKU has understood the advantages of ML in describing the online learning and teaching (L&T) strategies and, therefore, applied in the LMS platform. However, limitations have not been over looked for the complete dependence on ML and AI by the EL in KKU. Therefore, EL deanship developed and applied a combination of big data and in-person contribution to develop online L&T strategies for LMS. EL in KKU has a vision for ML and AI; all the IT HR are working to result the effective of online L&T strategies and how it will be met by the use of ML and AI. These strategies are being checked on trend analysis. Unless EL in KKU does not become confident, it decides not to overrule the in-person contribution and make the entire system automatic.
FGOL [58] Based on the working of MLOTS, EL in KKU cannot elaborate the fixed date for complete application of ML and AI for its LMS and other online services like online training for L&T. Rather, IT HR is focusing on tentative plan and applying ML and AI for the effective outcome for each activity on LMS and other services.

platform; however, limitations have not been overlooked for the complete dependence on ML and AI by the EL in KKU. Therefore, El deanship developed and applied a combination of big data and in-person contribution to develop online L&T strategies for LMS. EL in KKU has a vision for ML and AI, and all the IT HR are working on results for effective online L&T strategies and how it will be met by the use of ML and AI.

10.3.6 Customized EL Content [57]

EL in KKU is using ML algorithms to predict results, which allow providing specific EL content based on past performance and individual learning goals. If the online user is a more active and frequent user of any tool on LMS and BB, the systems robotically provide the recommendation to the users [57] and EL is using ML and AI to show the gap and excellence in learning skills presented in its users. ML and AI benefit the KKU online users by teaching how to give more customized learning materials. The system recommends more integrated, basic, or complex courses to the online learners based on their online behavior.

10.3.7 Resource Allocation [57]

ML and AI offer two aids in resource allocation: one for the educational sector and one for corporate [6]. EL in KKU is benefitted by educational aspect where online learners on LMS get the accurate online resources. This builds the bridge for their skills and facilitates in meeting learning outcomes. They require filling gaps and achieving their learning goals [57]. The second benefit of ML and AI is resource allocation; HR teams in EL in KKU apply ML and AI in resource allocation. Now, they are taking minimum time in analysis and developing powerful contents for LMS [57].

10.3.8 Automate Content Delivery and Scheduling Process [57]

Developing tools on LMS is not an easy practice and task, but with ML and AI, such complex, ambiguous, and time-consuming tasks become easier. EL deanship in KKU schedules the coursework for online learners and delivers online resources. This technique depends on their EL assessment results or simulation presentation. EL in KKU expects that AI in coming days will make this process robotic and create a niche in EL modules [58] and for all online learners who are participants on LMS.

10.3.9 Improve KKU EL Return on Investment [58]

EL in KKU witnessed better return on investment with the ML and AI. This has resulted in creating more customized approach and better revenue earnings. With the help of predictive analysis, online learners take online training promptly that cause in time effectiveness for other endeavors. Apart from this EL in KKU, using AI-equipped software can monitor and predict online user's behavior on LMS.

10.3.10 Improve Learner Motivation [58]

EL deanship believes that the ML systems and AI in future can be likened to a private virtual teaching. Many online learning programs have been developed by following the concepts of ML and AI by the IT HR. For example, KKU X under EL deanship used ML and AI in establishing distinguished contents for young Saudi students to develop skills to prepare the graduate for potential employment [58].

10.3.11 Online Training Programs [58]

ML and AI are implemented by EL in KKU in making peer-to-peer communication productive. Also, AI has created a mapping between electronic experts and learning on LMS platform in KKU.

EL in KKU has assumed that ML and AI will certainly make a noticeable development in the future growth of EL. These applications have particularly given various advantages such as connecting individual online students to a group of other online users and finally to the wider remote associations. Researchers have identified the limitation in the scope of ML and AI. Results have not adequately proved the range of advantages and capacity of the assistance of using ML and AI.

KKU has applied the IoT solutions for enhancing the security and preventing cyber-attacks. The academy has a strong cybersecurity unit that applies IoT to prevent all types of network attacks and threats to the system. This university is using IoT for effective learning and redefining the roles of all users such as students, teachers, and administrators. Now, these users can interact and connect to technology and devices in classroom environments, aiding learning experiences, improve educational outcomes, and reduce costs. Some of the examples of IoT solutions used by KKU are given in Figure 10.12.

There are many benefits for which IoT can be used, and KKU is using some of them very effectively. Other benefits of IoT in education are given in Figure 10.13 which are used by other academies in developed nations.

There are many benefits of IoT that educational sectors have identified; however, they have to face major challenges also in its applications. IoT is the reason for huge data flow that has helped in increasing performance at operational as well as management level but, at the same time, has raised the security issues too. To solve these issues, universities and schools have to increase the infrastructure to manage the security at the network level [59]. Many academies including KKU have adopted a strategy of using traditional

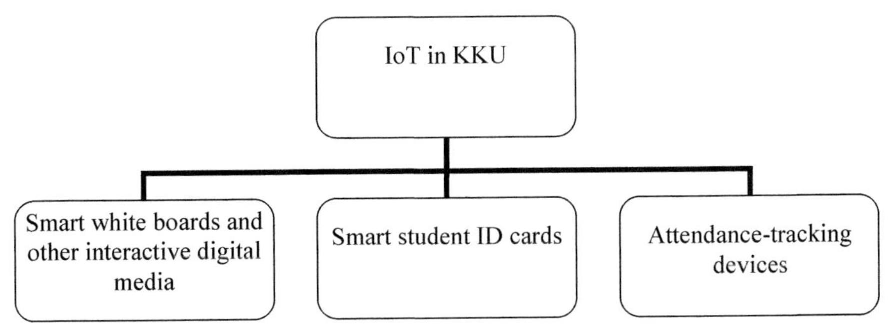

Figure 10.12 IoT in learning systems in KKU.

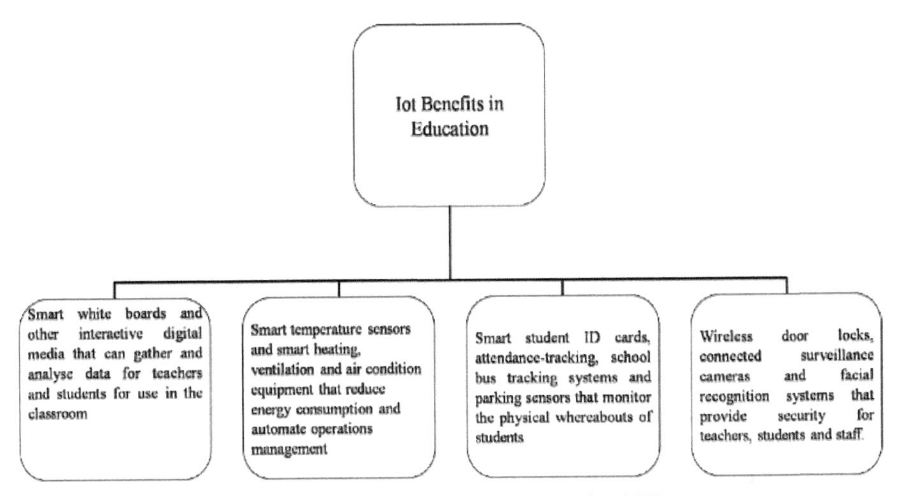

Figure 10.13 IoT benefits in education [59].

network designs to offer innovative methods of solving security problems and introducing new hierarchy of network intelligence, automation, and security. With this objective, the university has developed a simple, automated process for IoT device on boarding and avoided applying large IoT (LIoT) systems [59]. This option helped the academy because LIoT systems can have a big number of sensors or mechanics that can cause issues in management and cause more complex errors. It can provide a secure environment against cyber-attacks and data loss. KKU has implemented the security at multiple levels, including control of the IoT networks. For protecting IoT traffic and devices in the academy, KKU has its own strategic approach which is taking advantage of multi-security safeguards instead of single security technology. KKU introduced high quality of network connectivity for its users which in return assures better working and secured data delivery [59]. Therefore, KKU applied IoT system in its branches and colleges to implement smart planning, smart learning systems, and smart design in the online as well as face-to-face learning.

10.4 Results

The results for the first part shows the averages of the recognition accuracy in test datasets for the three-level SEt and the combined CNN model achieved higher accuracy than All-CNN, VD-CNN, and NiN-CNN models. Reliable models that can recognize learners' SEt during a learning session, particularly in contexts where there is no instructor present, play a key role in allowing learning systems to intelligently adapt to facilitate the learner in online learning.

This research chapter is based on a complete qualitative approach where IT HR and other online users such as online instructors, online students, and EL admin people were given a close-ended questionnaire. Researchers concluded this report on the responses filled by the respondents. Close-ended questions cover four domains namely ExL, EER, OnT, and AC [61] which are given in Figure 10.14. Questions on AI and ML determining EL in KKU are referred to in Table 10.5.

The time Internet got commercialized, L&T took great advantage of this. EL in KKU has acknowledged that the activities for EL and online education today will become the standards of academic instruction tomorrow, [62] and the ML and AI will decide its pedagogy.

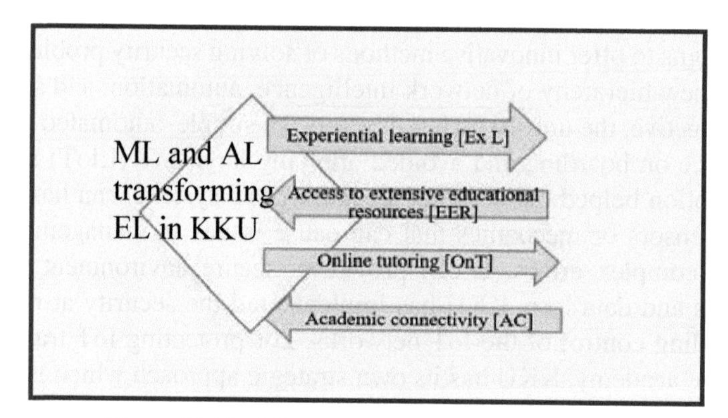

Figure 10.14 ML and AI transforming EL in KKU.

Table 10.5 Questionnaire; ML and AI transforming EL in KKU

Domain	Questions	Strongly Agree	Agree	Neutral	Disagree	Strongly Disagree
ExL	EL provides LMS as first development in online L&T	45	–	5	–	–
	EL in KKU used various applications in student's learning on LMS and BB	44		6	–	–
	LMS BB is easy to use for all online users	2	46	2	–	
	LMS BB helped online instructors to achieve CLOs and build curriculum as focusing learners	4	46	–	–	–
EER	EL in KKU provides online learning database	4	46	–	–	–
	EL gives access to Saudi digital library for good learning resources	4	46	–	–	–
	EL has developed full online course to support massive needs of the KKU learners' massively online open courses (MOOCs) [21]	4	46	–	–	–
	EL has developed full online course to train KKU online experts under massively online open courses (MOOCs) [21]	4	46	–	–	–
	EL has introduced more online learning environment such as Google class room KKUX [29]	4	46	–	–	–
OnT	EL has used some video conferencing options such as BB collaborate [21]	2	48	–	–	–
	EL online communication works effectively for L&T purpose for all online users	1	45	4	–	–

(Contiuued)

Table 10.5 Continued.

Domain	Questions	Strongly Agree	Agree	Neutral	Disagree	Strongly Disagree
	EL has provided a variety of L&T and assessment tools on BB, which are effective for L&T purpose		46	4	–	–
AC	EL provides platform for online research in education sectors by providing access to learning resources and software	2	5	43	–	–
Determinants	Questions	Strongly Agree	Agree	Neutral	Disagree	Strongly Disagree
	EL on LMS BB is using automated grading methods for some online assessments through grade center [21]	2	48	–	–	–
	Online grade center tool on LMS facilitates in analyzing the students' performance	2	48	–	–	–
AG&M	Online instructor's user LMS BB and other tools of EL to check student's participation and monitor their educational growth	2	48	–	–	–
CC	LMS BB and other tools of EL are used to develop online learning materials, curriculum, and design	5	10	35	–	–
CSL	EL in KKU provides crowd-sourced knowledge allocation and association on its LMS BB	5	6	39	–	–
	EL provides tools to monitor the methods; students use and transform LMS BB services	5	10	35	–	–
SLS	LMS BB and other tools of EL provide communication between online users and build a liaison between different participants in L&T	3	40	7	–	–

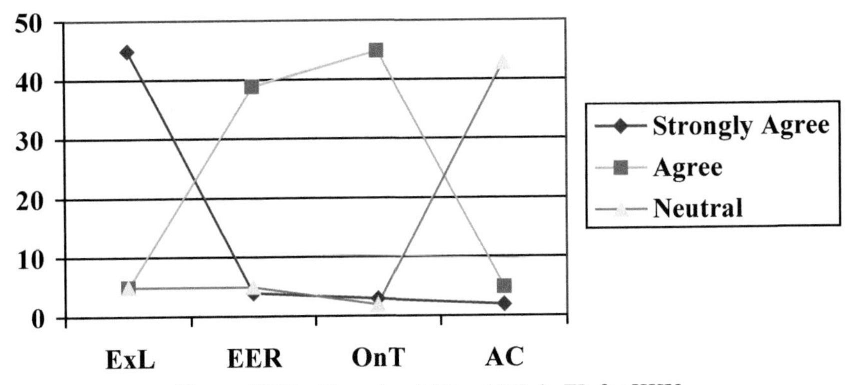

Figure 10.15 Domain of AI and ML in EL for KKU.

Respondents were asked to rate the questions from strongly agree to strongly disagree on 5 scales given in four domains, as mentioned in Figure 10.15. Questions given in Table 10.5 provide close-ended inquiries on how ML and AI are transforming EL in KKU. In total, 50 respondents (online users) were given this questionnaire.

Researchers did not inform respondents about the objective of the research because they may not know the technical aspect of the concepts ML and AI, and they were just requested to rate their experiences on various tools of EL in KKU. However, many abbreviations were used in the questionnaires that were described in the notes of the questionnaire for better understanding.

10.4.1 ExL

For analyzing ExL, four questions were asked pertaining to EL tools and techniques, and most of the respondents agreed on effectiveness and believed that EL LMS BB provides the pioneer online services through LMS BB in KKU. IT HR added that ML and AI played a vital role in developing the platform in EL deanship.

10.4.2 EER

More than 45% of the respondents agreed that EL tools and techniques have provided an excellent online interface for the fully online learning environment. ML and AI aided in developing massively online open courses (MOOCs) and other online learning programs for EL in KKU such as BB ultra.

10.4.3 OnT

EL in KKU has witnessed that without real-time communication, EL cannot work effectively. More than 46% of the respondents agreed on the effectiveness of LMS blackboard collaboration. ML and AI are used by IT HR to develop video conferencing apps for LMS services.

10.4.4 AC

It is assumed that respondents are not having an absolute notion about the online research advantages available on EL tools and techniques; therefore, results show a neutral conclusion. However, IT HR in EL explained various methods and applications developed in EL KKU for online research options such as providing access to an online digital library, strategic alliance to other universities' learning resources, and having a liaison with research and development site on KKU. Also, IT HR extended their expression using algorithms and natural language processing for showing technological advancement in ten years in EL in KKU with the help of ML and AI [62]. Figure 10.16 identifies the determinants of ML and AI in EL in KKU.

Based on these determinants, close-ended report was collected that defined more appropriately the role of ML and AI in EL deanship in KKU. Figure 10.17 shows the determinants of ML and AI for EL in KKU.

10.4.5 AG and M

Respondents are highly satisfied with this determinant as results show an absolute percentage for this. IT HR is utilizing AI language developers using

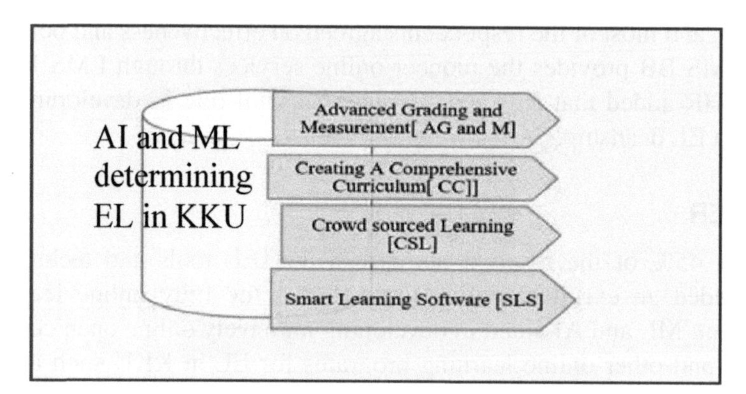

Figure 10.16 AI and ML determining EL in KKU.

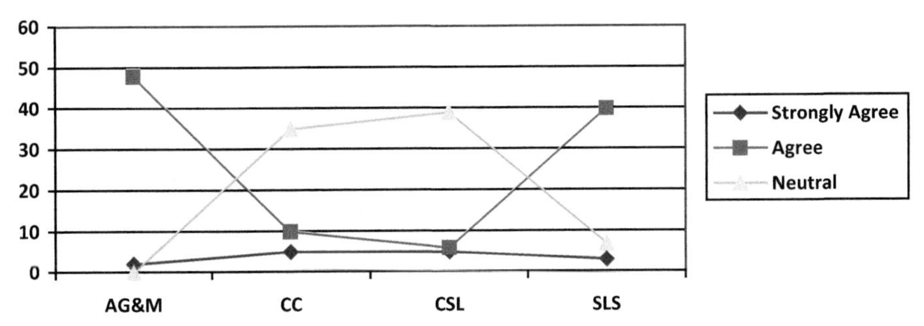

Figure 10.17 Determinants of AI and ML in EL in KKU.

algorithms and ML builds up to check assessments' tools [62]. EL in KKU applied optical recognition technique for grade center tool on LMS BB. ML and AI facilitated the online instructors in two ways [62]: develop a congenial learning environment in general and develop customized learning packages to enhance course learning outcomes (CLOs).

10.4.6 CC

EL in KKU has realized the importance of deep learning [63] (DL) which is a more extensive type of ML; the idea of DL is to develop a smart application for service industries like finance, legal, education, etc. [64]. These ideas are to apply in developing learning material, creating attractive designs for learning modules, and also developing inclusive and integrated curriculum [64]. The domino effect shows neutral results, but IT HR elaborates the implication in developing exceptional online modules.

10.4.7 CSL

Respondents show pretty similar results such as those of CC where they could not identify the EL in KKU dealing with the presence of information reservoirs such as information available on social networking or Wikipedia options [64]. IT HR in EL in KKU does realize the significance of information gathering and creating a pool of information for online users in KKU. With ML and AI, they have developed an FAQ information pool dealing with online routine queries [64].

10.4.8 SLS

In this section, results show a bit of variation, and respondents were neutral for methods used to transform in LMS BB services, whereas when it is about online communication between all users and sharing other online services, respondents fairly agreed. EL in KKU explained that ML and AI technologies have facilitated all sections of users in EL deanship along with the strategic participants. ML and AI contributed to building the online educational infrastructure for EL in KKU.

IoT has aided KKU in building its learning strategies; it has made the students get more involved in learning procedures and to retail for future applications in their professional and educational advancements. KKU has been applying IoT in many ways like in collaborative studies, review process, smart surveys, and their analysis, carrying out smart quality and accreditation processes for national and international societies.

10.5 Conclusion

CNN and a combined model for the students' SEt classification show high accuracy. In the experiments, three-level (not-engaged, normally engaged, and highly engaged) decisions on SEt detection have been made, where the combined model shows high accuracy in SEt classification. The SEt detection method can help improve learners' learning experiences in different online learning activities, such as reading, writing, watching video tutorials, and participating in online meetings.

ML and AI have facilitated the development of tools, techniques, and online services of EL in KKU. These services include LMS BB, blackboard collaborates, KKUX, etc. IT HR of EL in KKU comprehended that ML and AI have helped in developing, designing, and implementing tools and techniques helpful to educators and learners. Also, EL in KKU looks forward to applying ML algorithms and AI in the prediction of potential advantages in its services' development to conclude in meeting CLOs. Also, in future, ITHR plans to develop interoperability with other KKU digital services such as students' registration site and faculty self-services to automate some features such as attendance, grades, faculty profile, etc.

This short study concludes the growth and benefits of IoT in education for good communication and smart learning by the implementation of various IoT applications. KKU has also realized the importance of IoT and therefore applied many applications of IoT for meeting course learning outcomes and monitoring students' success and growth in academics.

References

[1] Agrusti, F., Mezzini, M., & Bonavolontà, G. Deep learning approach for predicting university dropout: A case study at Roma Tre University. Journal of e-Learning and Knowledge Society, 16(1), 44–54. 2020

[2] Dolianiti, F. S., Iakovakis, D., Dias, S. B., Hadjileontiadou, S., Diniz, J. A., & Hadjileontiadis, L. Sentiment analysis techniques and applications in education: A survey. In International Conference on Technology and Innovation in Learning, Teaching and Education (pp. 412–427). Springer, Cham. June 2018.

[3] Iqbal, R., Doctor, F., More, B., Mahmud, S., & Yousuf, U. Big data analytics: Computational intelligence techniques and application areas. Technological Forecasting and Social Change, 153, 119253. 2020

[4] Sirshar, M., Baig, H. S., & Ali, S. H. A Systematic Literature Review of Research Methodologies Used for Evaluation of Augmented Reality Based Learning Applications. 2019.

[5] Khan, I. U., Aslam, N., Anwar, T., Aljameel, S. S., Ullah, M., Khan, R., & Akhtar, N. Remote diagnosis and triaging model for skin cancer using EfficientNet and extreme gradient boosting. Complexity, 2021.

[6] Chen, X., Zou, D., Xie, H., & Cheng, G. Twenty years of personalized language learning. Educational Technology & Society, 24(1), 205–222. 2021.

[7] Jianqiang, Z., Xiaolin, G., & Xuejun, Z. Deep convolution neural networks for twitter sentiment analysis. IEEE Access, 6, 23253–23260. 2018.

[8] Yoo, H. J. Deep convolution neural networks in computer vision: A review. IEIE Transactions on Smart Processing and Computing, 4(1), 35–43. 2015.

[9] Li, X., Ding, Q., & Sun, J. Q. Remaining useful life estimation in prognostics using deep convolution neural networks. Reliability Engineering & System Safety, 172, 1–11. 2018.

[10] Rahmouni, N., Nozick, V., Yamagishi, J., & Echizen, I. Distinguishing computer graphics from natural images using convolution neural networks. In 2017 IEEE Workshop on Information Forensics and Security (WIFS) (pp. 1–6). IEEE. December 2017.

[11] Koushik, J. Understanding convolutional neural networks. 2016. arXiv preprint arXiv:1605.09081.

[12] Ketkar, N. Convolutional neural networks. In Deep Learning with Python (pp. 63–78). Apress, Berkeley, CA. 2017.

[13] Jmour, N., Zayen, S., & Abdelkrim, A. Convolutional neural networks for image classification. In 2018 International Conference on Advanced Systems and Electric Technologies (IC_ASET) (pp. 397–402). IEEE. March 2018.

[14] Cengil, E., Çinar, A., & Güler, Z. A GPU-based convolutional neural network approach for image classification. In 2017 International Artificial Intelligence and Data Processing Symposium (IDAP) (pp. 1–6). IEEE. September 2017.

[15] Guo, T., Dong, J., Li, H., & Gao, Y. Simple convolutional neural network on image classification. In 2017 IEEE 2nd International Conference on Big Data Analysis (ICBDA) (pp. 721–724). IEEE. March 2017.

[16] Liu, S., & Deng, W. Very deep convolutional neural network based image classification using small training sample size. In 2015 3rd IAPR Asian Conference on Pattern Recognition (ACPR) (pp. 730–734). IEEE. November 2015.

[17] Arsenov, A., Ruban, I., Smelyakov, K., & Chupryna, A. Evolution of convolutional neural network architecture in image classification problems. In Selected Papers of the XVIII International Scientific and Practical Conference on IT and Security (ITS 2018).–CEUR Workshop Processing (pp. 35–45). November 2018.

[18] Tarasenko, A. O., & Yakimov, Y. V. Convolutional neural networks for image classification. 2020.

[19] Karimi, H., Derr, T., Huang, J., & Tang, J. *Online Academic Course Performance Prediction Using Relational Graph Convolutional Neural Network*. International Educational Data Mining Society. 2020.

[20] Yang, R. Vocational education reform based on improved convolutional neural network and speech recognition. Personal and Ubiquitous Computing, 1–12. 2021.

[21] Shen, Y., Heng, R., & Qian, D. Smart classroom learning atmosphere monitoring based on FPGA and convolutional neural network. Microprocessors and Microsystems, 103488. 2020.

[22] Xie, X., & Tang, C. Simulation of art design course development based on FPGA and convolutional neural network. Microprocessors and Microsystems, 103385.2020.

[23] Wan, Q., Sharbati, M. T., Erickson, J. R., Du, Y., & Xiong, F. Emerging artificial synaptic devices for neuromorphic computing. Advanced Materials Technologies, 4(4), 1900037. 2019.

[24] Han, D., & Yoo, H. J. Direct feedback alignment based convolutional neural network training for low-power online learning processor. In IEEE/CVF International Conference on Computer Vision Workshops (pp. 0–0). 2019.

[25] Dubey, S. R., Chakraborty, S., Roy, S. K., Mukherjee, S., Singh, S. K., & Chaudhuri, B. B. Diffgrad: An optimization method for convolutional neural networks. IEEE Transactions on Neural Networks and Learning Systems, 31(11), 4500–4511. 2019.

[26] Garg, K., Verma, K., Patidar, K., & Tejra, N. Convolutional neural network based virtual exam controller. In 2020 4th International Conference on Intelligent Computing and Control Systems (ICICCS) (pp. 895–899). IEEE. May 2020.

[27] Wang, X., Wu, P., Liu, G., Huang, Q., Hu, X., & Xu, H. Learning performance prediction via convolutional GRU and explainable neural networks in e-learning environments. Computing, 101(6), 587–604. 2019.

[28] Ariza, J., Jimeno, M., Villanueva-Polanco, R., & Capacho, J. Provisioning computational resources for cloud-based e-learning platforms using deep learning techniques. IEEE Access, 9, 89798–89811. 2021.

[29] Zakka, B. E., & Vadapalli, H. Detecting learning affect in E-learning platform using facial emotion expression. In International Conference on Soft Computing and Pattern Recognition (pp. 217–225). Springer, Cham. December 2019.

[30] Hesham, H., Nabawy, M., Safwat, O., Khalifa, Y., Metawie, H., & Mohammed, A. Detecting education level using facial expressions in e-learning systems. In 2020 International Conference on Electrical, Communication, and Computer Engineering (ICECCE) (pp. 1–6). IEEE. June 2020.

[31] Cioffi,R, Travaglioni, M., Piscitelli, G., Petrillo, A., & De Felice, F. Artificial intelligence and machine learning applications in smart production: Progress, trends, and directions, MDPI. Sustainability, 12(2), 492. 2020. https://doi.org/10.3390/su12020492.

[32] Castro F., Vellido A., Nebot À, & Mugica F. Applying Data Mining Techniques to e-Learning Problems. In: Jain L. C., Tedman R. A., & Tedman D. K. (eds). Evolution of Teaching and Learning Paradigms in Intelligent Environment. Studies in Computational Intelligence, vol. 62. 2007. Springer, Berlin, Heidelberg. https://doi.org/10.1007/978-3-540-71974-8_8.

[33] Naim, A., Sattar, R. A., AL Ahmary, Nalah., & Razwi, M. T. Implementation of quality matters standards on blended courses: A case study. Finance India Indian Institute of Finance, XXXV(3), 873–890. September 2021.

[34] Koh, J. H. L., & Kan, R. Y. P. Students' use of learning management systems and desired e-learning experiences: Are they ready for next generation digital learning environments? Higher Education Research & Development 0:0, Interactive Learning Environments, 1–16. Published online: 03 December 2012. ISSN: 1049-4820 (Print) 1744-5191 (Online). Journal homepage: https://www.tandfonline.com/loi/nile20 https://doi.org/10.1080/10494820.2012.745433. December 2012

[35] Moore, J. L. Dickson-Deane, C., & Galyen, K. e-Learning, online learning, and distance learning environments: Are they the same? The Internet and Higher Education, 14(2), 129–135. March 2011. Available online: 3 November 2010. https://doi.org/10.1016/j.iheduc.2010.10.001.

[36] Gikandi, J. W., Morrow, D., & Davis, N. E. Online formative assessment in higher education: A review of the literature. Computers & Education, 57(4), 2333–2351. 2011. https://doi.org/10.1016/j.compedu.2011.06.004.

[37] Cheah, S.-M., Chan, Y.-J., Khor, A. C., and Say, E. M. P. Artificial intelligence-enhanced decision support for informing global sustainable development: A human-centric AI-thinking approach. Information, 11, 39. 2020. MDPI Publishers.

[38] Horvitz, E. J., Breese, J. S., and Henrion, M. Decision theory in expert systems and artificial intelligence. International Journal of Approximate Reasoning, 2(3), 247–302. 1988. https://doi.org/10.1016/0888-613X(88)90120-X.

[39] Jing, Y., Bian, Y., Hu, Z. *et al.* Deep learning for drug design: An artificial intelligence paradigm for drug discovery in the big data era. The AAPS Journal, 20, 58. 2018. Springer. https://doi.org/10.1208/s12248-018-0210-0.

[40] Hamed, M. A., & Samy S. A. N. An intelligent tutoring system for teaching the 7 characteristics for living things. International Journal of Advanced Research and Development. 2(1), 31–35. 2017. Phil Archive copy v2: https://philarchive.org/archive/HAMAIT-7v2.

[41] Popenici, S. A. D., & Kerr, S. Exploring the impact of artificial intelligence on teaching and learning in higher education. Research and Practice in Technology Enhanced Learning, 12, 22. 2017. Springer. https://doi.org/10.1186/s41039-017-0062-8.

[42] Moubayed, A., Injadat, M., Nassif, A. B., Lutfiyya, H., & Shami, A. E-Learning: Challenges and research opportunities using machine learning & data analytics. IEEE Access, 6, 39117–39138. 2018. doi: 10.1109/ACCESS.2018.2851790.

[43] Russell, R. *et al.*, Automated vulnerability detection in source code using deep representation learning. In 17th IEEE International Conference on Machine Learning and Applications (ICMLA) (pp. 757–762). Orlando, FL. 2018. doi: 10.1109/ICMLA.2018.00120.

[44] Maccagnola, D., Messina, E., Gao, Q., & Gilbert, D. A machine learning approach for generating temporal logic classifications of complex model behaviours. In Winter Simulation Conference (WSC) (pp. 1–12). Berlin. 2012. doi: 10.1109/WSC.2012.6465202.

[45] Kukar, M., & Kononenko, I. Reliable Classifications with Machine Learning. In: Elomaa T., Mannila, H., & Toivonen, H. (eds). Machine Learning: ECML 2002. Lecture Notes in Computer Science, vol. 2430. 2020. Springer, Berlin, Heidelberg. https://doi.org/10.1007/3-540-36755-1_19.

[46] Zawacki-Richter, O., Marín, V. I., Bond, M. *et al.* Systematic review of research on artificial intelligence applications in higher education – where are the educators?. International Journal of Educational Technology in Higher Education, 16, 39. 2019. Springer. https://doi.org/10.1186/s41239-019-0171-0.

[47] Naim, A., Alahmari, F., & Rahim, A. Role of artificial intelligence in market development and vehicular communication. Smart Antennas: Recent Trends in Design and Applications, 2, 28. 2021.

[48] Prasad, P. W. C., Alsadoon, A., Khan, S. S., & Maag, A. A systematic review: Machine learning based recommendation systems for e-learning. Received: 15 April 2019/Accepted: 12 November 2019/Published online: 14 December 2019 # Springer Science+ Business Media, LLC, part of Springer Nature. December 2019.

[49] Cui, Q. *et al.*, Stochastic online learning for mobile edge computing: Learning from changes. IEEE Communications Magazine, 57(3), 63–69. March 2019. doi: 10.1109/MCOM.2019.1800644.

[50] Goksel, N. (Anadolu University, Turkey) and Bozkurt, A. (Anadolu University, Turkey). Artificial Intelligence in Education: Current Insights and Future Perspectives, Source Title: Handbook of Research on Learning in the Age of Transhumanism, ch. 14. 2019. Copyright: ©2019|Pages: 13, doi: 10.4018/978-1-5225-8431-5.

[51] Kolb, A. Y., & Kolb, D. A. Learning styles and learning spaces: Enhancing experiential learning in higher education academy of management learning & education. 4(2). Articles, https://doi.org/10.5465/amle.2005.17268566. Published Online: 30 November 2017.

[52] Naim, A., & Alahmari, F. Reference model of e-learning and quality to establish interoperability in higher education systems. International Journal of Emerging Technologies in Learning (iJET), 15(02), 15–28. 2020. https://onlinejour.journals.publicknowledgeproject.org/index.php/i-jet/article/view/11605.

[53] Naim, A., Hussain, M. R., Naveed, Q. N., Qamar, S., Khan, N., & Hweij, T. A. Ensuring interoperability of e-learning and quality development in education. In IEEE Jordan International Joint Conference on Electrical Engineering and Information Technology (JEEIT). 2019. https://ieeexplore.ieee.org/abstract/document/8717431.

[54] Govindasamy, T. Successful implementation of e-Learning: Pedagogical considerations. The Internet and Higher Education, 4(3–4), 287–299. 2001. https://doi.org/10.1016/S1096-7516(01)00071-9.

[55] Derouin, R. E., Fritzsche, B. A., & Salas, E. E-learning in organizations. Journal of Management. Sage Publications, Research Article https://doi.org/10.1177/0149206305279815. First Published: 01 December 2005.

[56] Naim, A. Application of quality matters in digital learning in higher education. Texas Journal of Multidisciplinary Studies, 1(1), 3–12. 2021. https://zienjournals.com/index.php/tjm/article/view/11.

[57] Abed, S., Alyahya, N., & Altameem, A. IoT in education: Its impacts and its future in Saudi universities and educational environments. In 1st International Conference on Sustainable Technologies for Computational Intelligence (pp. 47–62). Springer, Singapore. 2020.

[58] Mathews, S. P., & Gondkar, D. R. Solution integration approach using IoT in education system. International Journal of Computer Trends and Technology (IJCTT), 45(1). 2017.

[59] Naim, A. Realization of diverse electronic tools in learning and teaching for students with diverse skills. Global Journal of Enterprise Information System, 12(1), 72–78. 2020. https://gjeis.com/index.php/GJEIS/article/view/451.

[60] Marquez, J., Villanueva, J., Solarte, Z., & Garcia, A. IoT in education: Integration of objects with virtual academic communities. In New Advances in Information Systems and Technologies (pp. 201–212). Springer, Cham. 2016.

[61] Mohanty, D. Smart learning using IoT. International Research Journal of Engineering and Technology, 6(6), 1032–1037.
[62] Moubayed, A., Injadat, M., Nassif, A. B., Lutfiyya, H., & Shami, A. E-learning: Challenges and research opportunities using machine learning & data analytics. IEEE Access, 6, 39117–39138. 2018.
[63] Rasheed, F., & Wahid, A. Learning style detection in E-learning systems using machine learning techniques. Expert Systems with Applications, 174, 114774. 2021.
[64] Ayodele, T., Shoniregun, C. A., & Akmayeva, G. Towards e-learning security: A machine learning approach. In International Conference on Information Society (i-Society 2011) (pp. 490–492). IEEE. June 2011.

11

Quantitative Texture Analysis with Convolutional Neural Networks

C. Aldrich and X. Liu

Western Australian School of Mines: Minerals, Energy and Chemical Engineering, Curtin University, Australia
E-mail: chris.aldrich@curtin.edu.au

Abstract

Texture analysis plays an important role in computer vision in that it is critical to both the characterization and segmentation of regions in images. Its application is wide-ranging in different technical disciplines, such as the food industry, materials characterization, remote sensing, and medical image analysis. Over the last decade, deep learning has redefined research frontiers in image recognition, including texture analysis. Most of the current advances are driven by transfer learning with convolutional neural networks, which do not require large volumes of data to develop and deploy models.

In this chapter, comparative analyses of textures based on the use of transfer learning with different neural network architectures and traditional approaches are presented via three case studies. In the first, Voronoi simulation of material textures is discussed and the ability of convolutional neural networks to discriminate between different textures is considered. Textures like these play a critical role in the geometallurgical and quantitative structure property relationship modeling in material science. In the second case study, it is shown that transfer learning can significantly improve froth imaging sensors to expedite advanced real-time control of mineral flotation plants.

In the final case study, textures associated with historic gold price data are considered and it is shown that these methods can be used to visualize changes in the stock price data or any other signals more generically.

Keywords: Convolutional neural networks, texture analysis, transfer learning, froth images, Voronoi texture, gold price.

11.1 Introduction to Transfer Learning with Convolutional Neural Networks

Artificial neural networks have a long history rooted in the early work of pioneers such as McCulloch and Pitts in the 1940s. As a family of machine learning models, they form the backbone of what is referred to as connectionist artificial intelligence, as opposed to symbolic artificial intelligence that is associated with expert systems and logic formalisms.

While experiencing successive periods of growth and stagnation, artificial neural networks have diversified from perceptrons to kernel-based architectures, self-organizing architectures, autoencoders, as well as deep learning architectures, including convolutional neural networks (CNNs).

Although CNNs trace their history back to the 1980s [1–3], it is only recently that their application in industry has seen explosive growth. *Ab initio* development of CNNs requires a) extensive computational resources, and b) large labeled data sets. Of these, the second requirement can be a critical hindrance; as such data may more often than not be available or very costly to acquire. As a consequence, the wide acceptance of CNNs in industry has been and continues to be driven by what is known as transfer learning.

Transfer learning [4, 5] is achieved when learning or knowledge acquired in one domain, referred to as the source, is transferred to another, referred to as the target. For example, a CNN trained on a large labeled data set is applied to a different data set that is related to the training data. This is particularly important in image analysis, where CNNs that have been trained on the large ImageNet data base as source can be applied directly to other tasks in image analysis in a target domain or can be retrained with comparatively small data sets and little retraining from this domain.

As discussed by Pan and Yang [4], different transfer learning approaches can be followed, based on the characteristics of the source and target data and domains, as summarized in Figure 11.1. The most popular approach is inductive transfer learning, where labeled data are available in both the source and target domains, as indicated by the solid red line in Figure 11.1. This is also the approach that was followed in the case studies presented in this chapter.

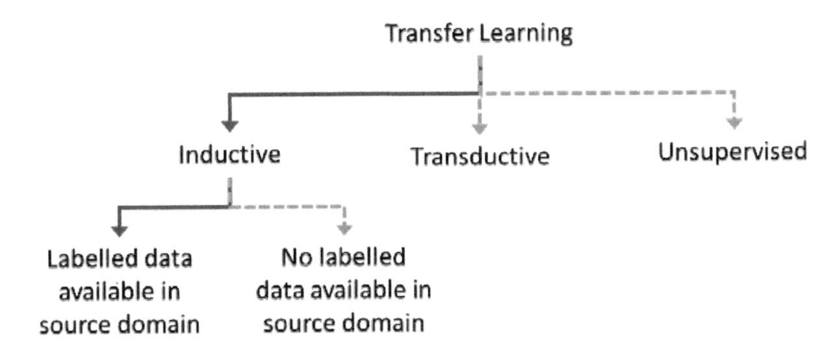

Figure 11.1 Transfer learning approaches.

11.1.1 The ImageNet Large-Scale Visual Recognition Challenge (ILSVRC)

ILSVRC is a challenge designed for the recognition of large-scale objects, based on the ImageNet data set and it has been running since 2010 [6]. It contains approximately 15,000,000 labeled images of 1000 common objects, and competing algorithms are required to identify these objects. ImageNet is the most common data base used in transfer learning and the winners of the competition are among other all publically available for transfer learning applications.

11.1.2 Transfer Learning Strategies

Different transfer learning strategies can be employed, as determined by the availability of computational resources and data from the application domain. For example, if very few or no data are available, transfer learning can be used directly, without any further training. Examples of this would be the extraction of features from single images. Under these circumstances, the images would simply be passed through the CNN and the features would be extracted from any of the various feature layers in the network (usually the last). If a small labeled data set is available, it may be possible to use these data to further train some of the last feature layers in the network, after adapting the dense layer of the network to the application at hand, whether that is classification or regression. If a larger data set is available, extensive retraining or further training of the network may be possible. This is illustrated in the left panel in Figure 11.2.

As indicated in the right panel in Figure 11.2, deep neural networks trained by means of transfer learning can outperform the same networks

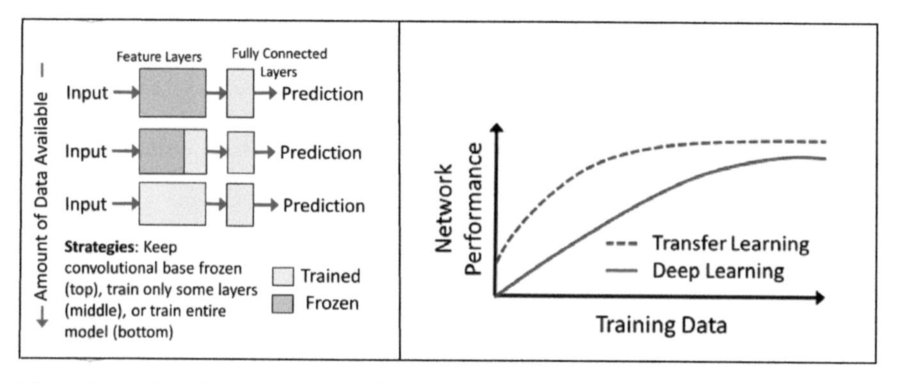

Figure 11.2 Transfer learning strategies for CNNs (left) and comparative performance with *ab initio* deep learning of the same models (right)

trained from an initial position with randomized weights, particularly if relatively few data are available for training.

11.2 Texture Analysis

Texture in images is not formally defined, but, essentially, it can be seen as repeating patterns of local variation in the intensities of the image pixels, and texture analysis is essentially aimed at quantifying these variations in intensities and patterns.

11.2.1 Textures in Nature and the Built Environment

Textural patterns occur very widely in nature and the built environment, as which just a few examples are shown in Figure 11.3. Broadly speaking, the primary goals of image texture analysis are four-fold, focusing on texture shapes, categorization, segmentation, and synthesis. The classification of texture is used to find regions in images that have different textures; texture segmentation focuses on the identification of boundaries between such regions. Texture synthesis is a relatively recent development associated with algorithms that can generate realistic textures similar to real ones. Finally, identification of texture can also facilitate the recovery of three-dimensional shapes of objects in images.

For example, the derivation of quantitative descriptors for ore textures has grown rapidly in tandem with development in analytical instruments, such as QEMSCAN, MLA [7, 8], and X-ray computer tomography [9, 10].

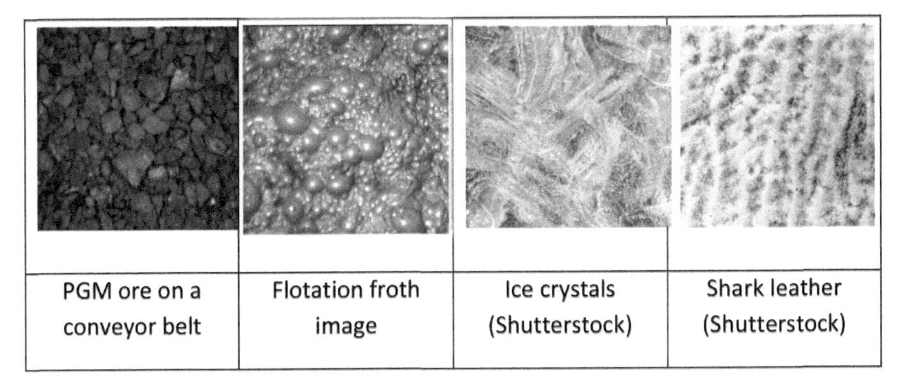

Figure 11.3 Examples of natural textures.

Similarly, quantification of texture can be used in failure diagnostic models of metallic structure [11–14].

Shark skin textures serve as a biomimetic template for low-friction materials [15, 16] used among other athletic swimwear, non-toxic biofouling control [17, 18], and marine vessels [19]. In the meat industry, quantification of meat texture has attracted considerable interest recently [20–22]. These are just a few examples of the many diverse applications that have been reported to date. Textures in the built environment include textures on different scales. On a microscopic scale, this could be associated with the crystallographic orientations in scanning electron micrographs. On a macroscale, these could be the surface characteristics of materials captured photographically, such as the surface finish of a steel plate in a cold rolled state during production.

11.2.2 Traditional Approaches to Texture Analysis

Traditionally, four main classes of approaches are recognized in texture analysis, viz. statistical, spectral, structural, and model-based approaches, as indicated in Table 11.1. These and other methods are considered in detail, among other by Ghalati *et al.* [23] and are only briefly discussed.

11.2.2.1 Statistical methods

Statistical methods focus on analysis of the spatial distributions of the values of gray levels in the neighborhoods of pixels. First-order (pixel value frequencies), second-order (pair-wise relationships of pixel intensities), or higher order (relationships between pixel intensities beyond pixel pairs) methods can be defined.

Generally, second-order statistical methods, such as GLCMs, are more powerful discriminators than first-order methods. LBPs are examples of the latter, combining the occurrence of pixel intensity values with local spatial structures [23].

Table 11.1 A taxonomy of texture analytical methods

Class	Approach	Method	References
Classical	Statistical	First-order statistics	Bonnin *et al.* [24]
		GLCM, 3D-GLCM	Haralick *et al.* [25]; Kim *et al.* [26]
		LBP	Ojala *et al.* [27]; Ojala *et al.* [28]; Ahonen *et al.* [29]
		LDP	Shabat and Tapamo [30]
		LBP-TOP	Nanni *et al.* [31]; Fu and Aldrich [32]
		LDP-TOP	Bonomi *et al.* [33]; Arita *et al.* [34]
		LTP	Nanni *et al.* [31]
		LTP-TOP	Nanni *et al.* [31]
		NGLDM	Arita *et al.* [34]
		SGLDM	Moolman *et al.* [35, 64, 65]
	Structural	Morphological operations	Soille [36]
		Primitive measurement	Jing and Shan [37]
		Skeleton representation	Wang and Na [38]
	Spectral	Wavelets	Unser [39]
		Laws filters	Laws [40, 41]
		Gabor transforms	Jain and Farokhnia [42]; Kim *et al.* [26]
	Model-Based	Random fields	Cross and Jain [43]; Yang and Liu [44]
		Fractals	Pentland [45]
		Autoregressive	Kashyap and Khotanzed [46]
		Texems	Xie and Mirmehdi [47]
Recent	Graph-based		Li *et al.* [48]; Bashier *et al.* [49]; Gaetano *et al.* [50]
	Entropy		Jernigan and D'Astous [51]; Silva *et al.* [52]
Learning	Vocabulary		Varma and Zisserman [53]
	Deep learning (CNNs)		Bastidas-Rodriguez *et al.* [14, 54]

11.2.2.2 Structural methods

Structural methods are based on the premise that texture consists of sets of texture elements or primitives that can have a regular or irregular arrangement. Structural elements are designed to identify these primitives (e.g., blobs, line segments, or regions characterized by uniform gray level values), as well as inferring their arrangement in the image.

11.2.2.3 Spectral methods

Spectral methods are also referred to as transform-based or filter-based methods. These methods can be used to analyze the frequency components of texture in the spatial domain, frequency domain, or both. Respective examples of these are laws filters [40, 41], Fourier transforms [55], and Gabor filters [42]. With these approaches, the frequency content is represented by a filter response set that is obtained through convolution of the image with the filters in question, the statistics of which comprise the textural features of the images.

11.2.2.4 Modeling approaches

With model-based methods, textural features are represented by the parameters of models. Identification of the correct models and mapping of textural features onto these models are critical to the success of these approaches, examples of which are indicated in Table 11.1.

11.2.3 More Recent Approaches to Texture Analysis

More recent approaches in the analysis of textures include graph-based and entropy methods. With graph-based methods, graphs consisting of vertices (nodes) and edges (connections) are defined over image textures, after which features associated with the graphs can be extracted.

Entropy-based methods are derived from information theory and are established approaches in time series analysis, where they provide a means to quantify the complexity of the time series. They are less well established in image analysis.

11.2.4 Learning Approaches to Texture Analysis

11.2.4.1 Vocabulary-based approaches

Vocabulary learning is a bag of words method [56], designed to learn a dictionary containing textural elements computed by local descriptors. Common local descriptors are Leung–Malik filters [56], maximum response filters [53], scale-invariant feature transforms [57], rotation-invariant feature transforms [58], and local binary patterns [27].

Extracted local descriptors are clustered to construct the dictionary, followed by the encoding of features and pooling into a global descriptor. Feature encoding is a key element of vocabulary-based learning methods. In the generation of texton features for example, a histogram of the features is obtained by counting the numbers of local features assigned to code words in the dictionary.

11.2.4.2 Deep learning approaches

CNNs are particularly suitable for texture analytical applications, given that they are generally designed for image analysis and that their trainable filter banks are finely attuned to the detection and repetitive textural patterns. Since AlexNet's success in the ImageNet competition, pretrained CNNs have been applied highly successfully in textural image analysis. In particular, image texture is represented by the features learnt by these networks when trained to identify labels associated with images, whether discrete or numeric. In the majority of cases, this is accomplished by transfer learning, as discussed in Section 11.1.

More recently, CNNs have been customized for texture analysis, e.g., T-CNN [59, 60], B-CNN [61], FASON [62], Deep-TEN [63], and deep adaptive wavelet network (DAWN) [54]. Although some studies have indicated the advantages of these architectures over more general ones [54], the performance of these customized networks compared to classical CNNs has not been widely investigated as yet.

11.3 Methodology of Texture Analysis with Convolutional Neural Networks

The texture analytical methodology considered in the rest of the chapter is described in more detail in this section. This includes brief summaries of GLCMs, LBPs, and textons that served as a baseline for comparative assessment of the performance of the deep learning methods that were investigated.

11.3.1 Overall Analytical Methodology

The general analytical methodology for analysis of texture based on the use of machine learning and transfer learning is illustrated in Figure 11.4.

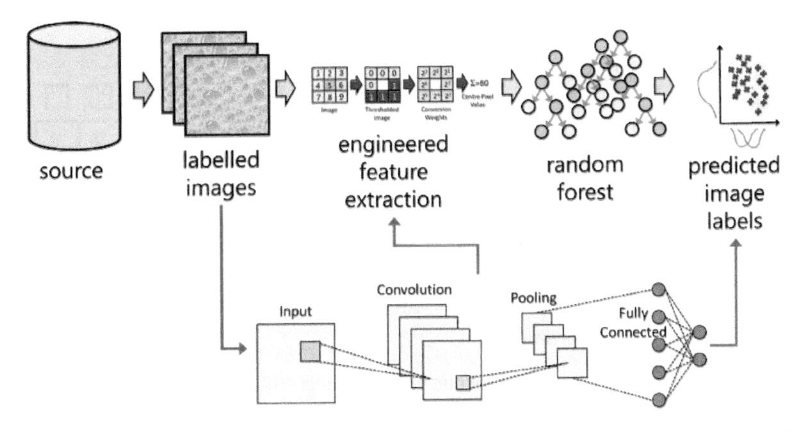

Figure 11.4 Analytical approach to texture analysis with machine and transfer learning.

a) *Image acquisition*: Images representing some system are first acquired. These images could be videographic, optical, scanning electron micrographs, or images generated from other data sources, such as signals or hyperspectra.
b) *Feature extraction*: Textural features are subsequently extracted from these images. This can be accomplished by use of traditional algorithms, such as GLCMs, LBPs, etc., as summarized in Table 11.1. It can also be accomplished by using CNNs and other deep learning algorithms.
c) *Modeling*: The extracted features serve as predictors or input variables to a model, the design of which depends on the image labels to be predicted. The labels can be categorical, for example, to designate a common object, such as in the ImageNet data base. The labels could also be continuous, as would be the case with image-based soft sensors.

11.3.2 Traditional Algorithms

Three traditional algorithms were considered in the case studies in this chapter, namely algorithms based on gray level co-occurrence matrices (GLCM), local binary patterns (LBP), and textons. These algorithms are briefly summarized below.

11.3.2.1 Gray level co-occurrence matrices (GLCM)

Consider the GLCM, $A_{I(D,G)}$, of an image I parameterized by D and G. More specifically, D is some measure of the distance between pairs of pixels

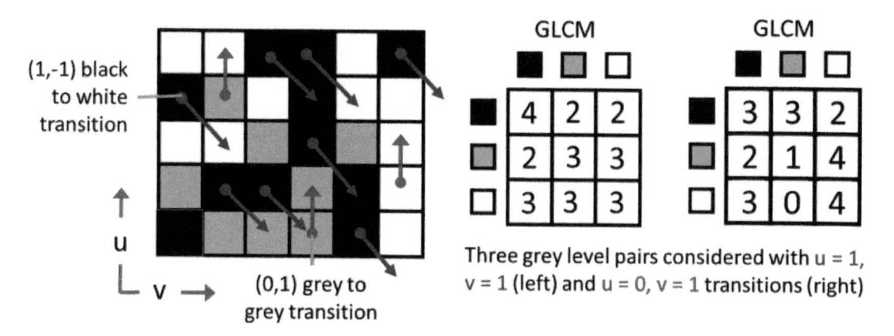

Figure 11.5 An image (left) with co-occurrence matrices (middle and right) illustrating the frequency of pair-wise combinations of three gray levels, represented by Cartesian coordinates (u,v).

in the image and G is the number of gray levels in the image. Each entry a_{ij} in the $G \times G$ matrix is the number of times that a gray level is associated with a pair of pixels displaced in the image by a distance D, as indicated in Figure 11.5.

The Haralick descriptor set [25] derived from GLCM images are widely used in image analysis. Four of these features were used in this investigation.

$$ENE = \sum_{ij} \breve{a}_{ij}^2 \tag{11.1}$$

$$CON = \sum_{ij} |i-j|^2 \breve{a}_{ij} \tag{11.2}$$

$$COR = \sum_{ij} \frac{(i-m_i)(j-m_j)\breve{a}_{ij}}{s_i s_j} \tag{11.3}$$

$$HOM = \sum_{ij} \frac{\breve{a}_{ij}}{1+|i-j|}. \tag{11.4}$$

In Equations (11.1)–(11.4), $\breve{a}_{ij}$ is the (i,j)th element of the normalized GLCM and m_i, m_j, s_i, and s_j are the means and standard deviations of the matrix rows (i) and columns (j).

Methods based on GLCM have been considered extensively in a range of applications in various technical domains.

11.3.2.2 Local binary patterns (LBPs)

LBPs are derived from images based on the differences between intensities of neighboring pixels in images [27]. The difference in the intensity of a given pixel (g_c) and that of one of its neighboring pixels (g_p) is set to either 0 or 1, after thresholding with a function *s*, as in the following equation:

$$\left.\begin{matrix} s\left(g_p - g_c\right) = 0 \\ s\left(g_p - g_c\right) = 1 \end{matrix}\right\}, \text{ if } \begin{matrix} g_c < g_p \\ g_c = g_p \end{matrix} \tag{11.5}$$

The next step after thresholding is to computer a local binary pattern (LBP) in accordance with eqn (11.6), for all $p = 1, 2 \ldots P$.

$$\text{LBP} = \sum_{p=1}^{P} 2^p \mathrm{s}\left(g_p - g_c\right) \tag{11.6}$$

The LBP operator in eqn (11.6) is applied to each pixel in each image with *G* gray levels, as indicated in Figure 11.6, with the results that images can be represented by LBPs that range from 0 to *G*, while the images themselves can be referred to as LBP images.

Feature extraction from images based on LBP is used widely in a range of disciplines, including mineral processing [32, 66, 67], some aspects of which are discussed in the case study in Section 11.2.

11.3.2.3 Textons

Textons are cluster centers located in a space defined by the responses of localized filters. The filter response space is generated by convolving a set of training images with spatially configured basis functions that are contained in a bank of filters [56], as shown in Figure 11.7.

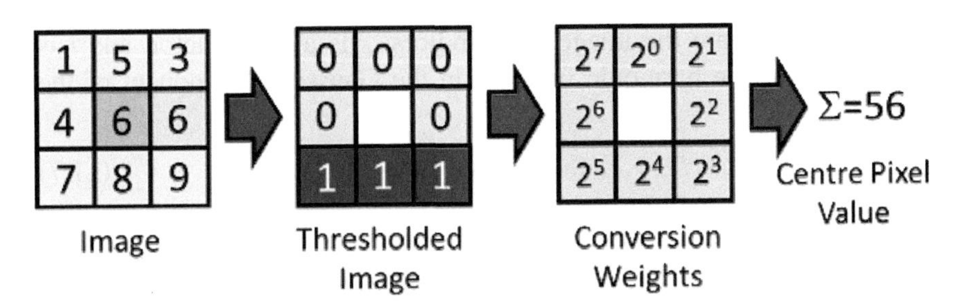

Figure 11.6 Center image pixel (shaded), surrounded by eight neighboring pixels (left), binary thresholded pixel values (middle). Thresholded values are multiplied with the conversion weights shown to generate the decimal LBP value replacing the center pixel (right).

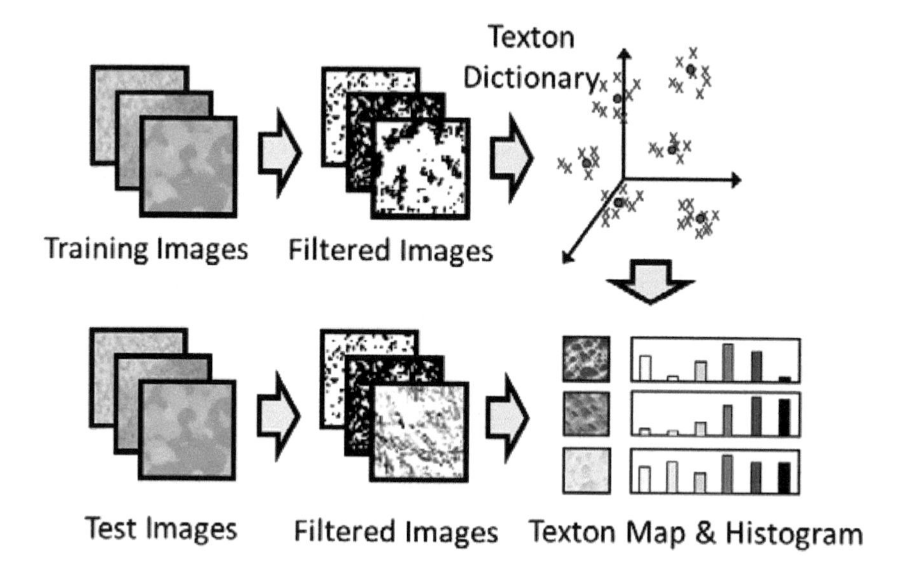

Figure 11.7 Texton image features obtained from image filtering, extraction, and clustering of localized filter vectors. These vectors comprise a texton dictionary to which images can be mapped and features extracted based on the resulting histogram counts.

The cluster centers in the filter response space are referred to as a texton dictionary and are typically determined by use of *k*-means cluster analysis. Image pixels are mapped to this feature space and texton features extracted from an image consisting of counts of the numbers of pixels assigned to specific texton channels [66].

In the case studies considered in this chapter, the Schmid filter bank [68] with 13 rotationally invariant filters was used, as given by eqn (11.7). In this equation, *r* are the image pixel indices, *s* is a scale factor, and *t* is the frequency of the harmonic function in the Gaussian component of the filter [53].

$$(\mathbf{r}, s, t) = F_0(s,t) + \cos\left(\frac{\pi \mathbf{r} t}{s}\right) e^{\mathbf{r}^2/2s^2}. \tag{11.7}$$

Studies in mineral processing have indicated that textons are better able to capture textures associated with mineral processing systems, such as flotation froths, bulk particulates, or slurry flows, compared with other widely used feature extraction algorithms, including GLCM and LBP methods [66, 69].

Table 11.2 Convolutional neural networks used in this chapter

Neural network	Depth (weights)	Reference
AlexNet	8 (61 million)	Krizhevsky *et al.* [70]
VGG19	19 (144 million)	Simonyan and Zisserman [71]
GoogLeNet	22 (7 million)	Szegedy *et al.* [72]
ResNet50	50 (25.6 million)	Tian and Chen [73]

11.3.3 Deep Learning Algorithms

Four deep learning algorithms were considered in this chapter. These were AlexNet, VGG19, GoogLeNet, and ResNet50. These network architectures are described in detail elsewhere, as indicated in Table 11.2.

11.4 Case Study 1: Voronoi Simulated Material Textures

11.4.1 Voronoi Simulation of Material Textures

In this first case study, the textural structures of natural material were simulated with the graphic representations of Voronoi tessellation due to the similarity between the textures of natural materials and Voronoi.

Image textures grouped into two equisized Classes A and B of 1000 images each were both generated by the same bivariate uniform distribution. The only difference was that Class A and B images were obtained from tessellation of realizations of the distribution consisting of 100 and 120 data points, respectively. As a result, the average size of the grains in Class A was larger than that in Class B, as indicated in Figure 11.8.

11.4.2 Comparative Analysis of Convolutional Neural Networks and Traditional Algorithms

A PyTorch backend was used to build the CNNs used in the case studies in this chapter. Experiments were run on the Google Colab platform and Voronoi images were identified using features based on GLCMs, LBPs, and textons, as well as using features from AlexNet, VGG19, GoogLeNet, and ResNet50 that had been pretrained on the ImageNet data base. This latter approach using the CNNs can be referred to as deep feature extraction.

Three variants of these deep features were extracted as previously indicated in Figure 11.2. That is, features were extracted by training of the fully connected layers of the networks only or, in addition to this, also training some or all of the feature layers of the networks.

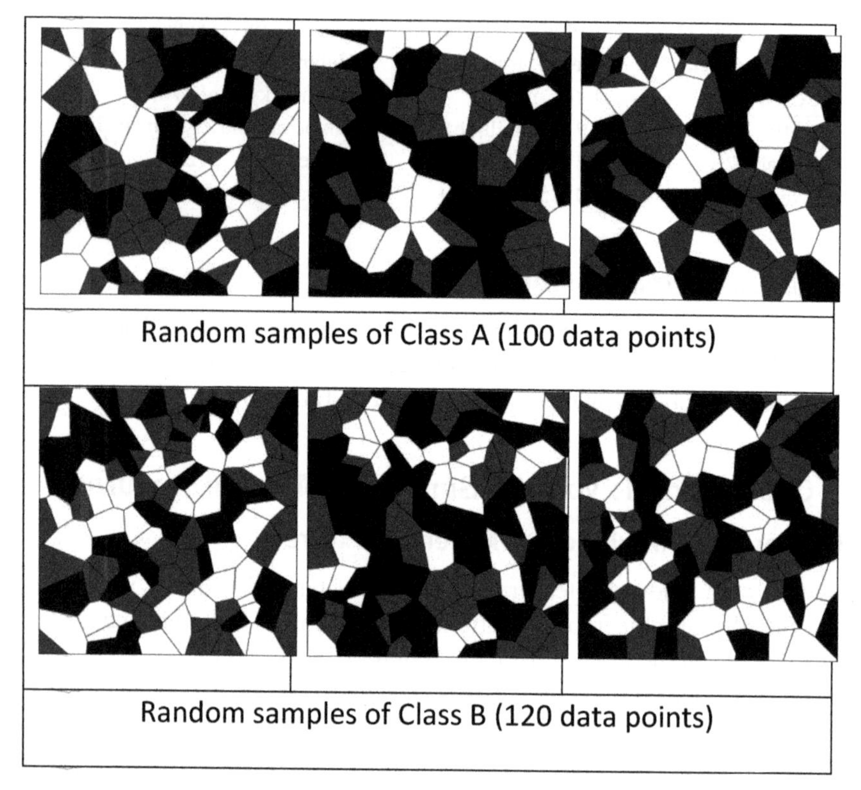

Figure 11.8 Exemplars of simulated Voronoi textures in Classes A and B (top and bottom rows, respectively) in Case Study 1.

To make the comparison as exact as possible, feature sets were subsequently used as predictors in the random forest (RF) models trained to discriminate between images from Class A and B, based on the average out-of-bag (OOB) errors recorded over 30 runs. Table 11.3 gives a summary of the hyperparameters used to develop the random forest models.

The exception to this was where features were extracted from CNNs with partially or fully retrained feature layers. In this case, the ability of the network to correctly classify the images from Classes A and B was taken as an indicator of the quality of the features. For this purpose, the image data sets were split into training (70%) and test (30%) data sets. The test sets were used to assess the performance of the CNNs that were optimized with the ADAM (adaptive momentum estimation) algorithm [74] during training.

Table 11.3 Hyperparameters of random forest models used in Case Studies 1 and 2

Hyperparameter	Description	Value
m_{try}	Number of candidate variables evaluated at each node split	$\sqrt{m}$
n_{tree}	Proportion of total number of samples evaluated by each tree	70%
K	Number of trees in the forest	500
Node size	Smallest number of samples supporting a terminal node	1
Replacement	Drawing samples with or without replacement	With
Splitting rule	Criterion used for splitting nodes	Gini

It was assumed that the pretrained neural networks already had near-optimal weights and, therefore, a small learning rate of 0.0001 was used so as not to unnecessarily disrupt the weight settings of the networks. As an additional measure to prevent overfitting, image augmentation was used during training. This included horizontally flipping, randomly rotating, shearing, and shifting the original images.

Table 11.4 gives a summary of the performance of the different models. The RF models using the traditional feature sets (GLCM, LBP, and textons) as predictors performed satisfactorily, reaching an accuracy of 87%–89%. Use of the features generated by the partially retrained networks improved the accuracy to at least 93%, while features obtained with the fully retrained networks further improved the accuracy to at least 95%.

The top performer was ResNet50, shown in the first row of Table 11.4. Interestingly, this network consistently outperformed the other deep learning architectures within each of the feature variant groups, i.e., when compared with its counterparts with fixed, partially retrained, or fully retrained feature layers.

Further insight into the performance of the algorithms can be gained by mapping the features to bivariate score plots with a *t*-distributed stochastic neighbor embedding (t-SNE) algorithm. The algorithm embeds features in a low-dimensional space so as to optimally preserve the similarities of the points in the original high-dimensional space. This is achieved by minimizing the Kullback–Leibler divergence of the distributions of the points in the high- and low-dimensional feature spaces.

The t-SNE scores of the image features are shown in Figure 11.9 together with their associated classification accuracies. Although all the traditional

Table 11.4 Assessment of texture features as predictors in Case Study 1

Model	Number of features	Accuracy (%)
ResNet50**	2048	99.0
VGG19**	4096	98.5
GoogLeNet**	1024	98.3
ResNet50*	2048	98.3
AlexNet**	4096	95.8
GoogLeNet*	1024	93.8
VGG19*	4096	93.8
AlexNet*	4096	93.3
ResNet50	2048	91.4
LBP	59	89.2
AlexNet	4096	88.3
Textons	20	87.1
GLCM	4	87.0
VGG19	4096	81.6
GoogLeNet	1024	78.3

Note: In models indicated by one or more "*," accuracy represents the performance of the end-to-end classifier. "*" and "**" indicate networks in which some or all the feature layers were retrained, respectively. All other models were random forests, where the accuracies indicated are out-of-bag values derived from the training data set.

methods and the direct deep feature extraction methods can produce reasonably good separation between two classes, it is more breath-taking to see that two separate clusters form when using deep features extracted from the partially retrained and fully retrained networks.

These score plots shown in Figure 11.9 confirm the advantage of retraining the feature layers of the CNNs. Retraining of some of the final feature layers ("*") led to an improvement in the features over CNNs where no such retraining was done. In turn, retraining of all the feature layers ("**") led to further markedly improvement in the performance of the classifiers, as clearly indicated by the separation of the clusters in the score plots.

11.5 Case Study 2: Textures in Flotation Froths

11.5.1 Froth Image Analysis in Flotation

Froth flotation is an important operation in mineral processing, where valuable material is separated from waste or gangue in a slurry in an agitated tank or flotation cell. Chemical reagents added to the slurry support the generation

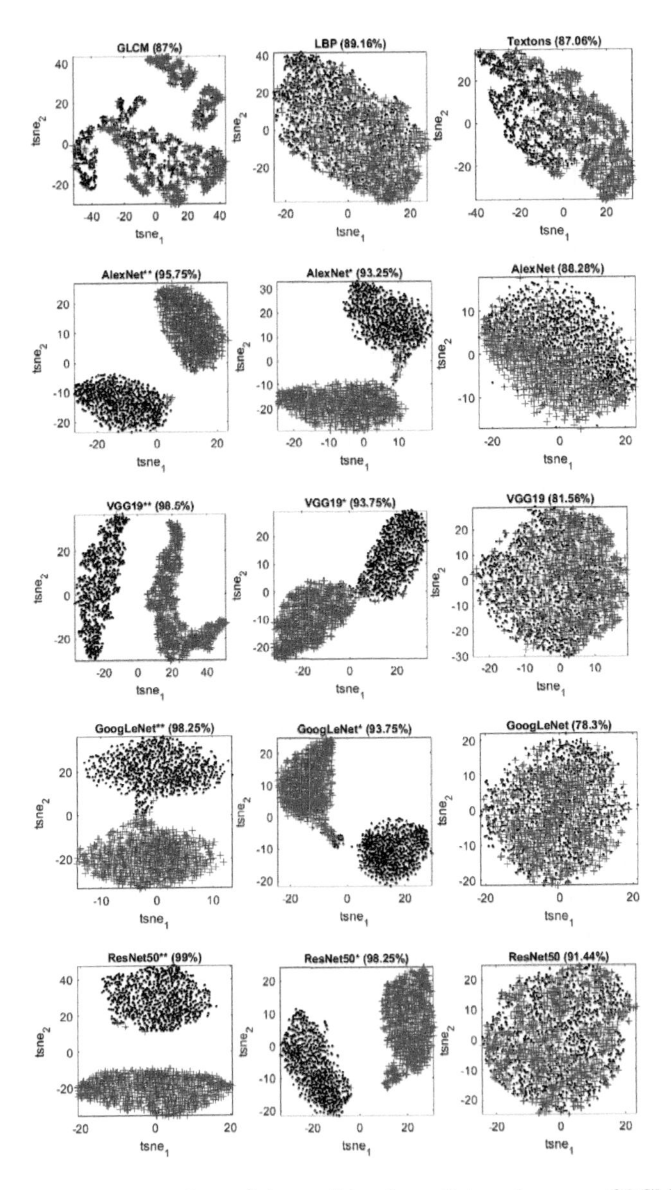

Figure 11.9 t-SNE score plots of the traditional predictors (top row, GLCM, LBP, and textons), AlexNet (second row), VGG19 (third row), GoogLeNet (third row), and ResNet50 (bottom row) in Case Study 1. Superscripts "*" and "**" indicate CNNs where feature layers were partially or fully retrained. Black dots and red "+" markers indicate Classes A and B, respectively.

of a froth layer on top of the slurry, as well as enhancing the hydrophobicity of the valuable material. This facilitates the concentration of the valuable species in the froth, from where it can be easily recovered.

It is a complex process and the appearance of the froth is a useful indicator of the performance of the flotation cell. However, operators may find it very challenging to discriminate between different froth structures and computer vision has long been investigated as a more reliable approach toward decision support on flotation plants. In this case study, the application of deep learning to froth images obtained from a platinum metal group flotation plant in South Africa is considered. The collection of the images and previous analyses are discussed in more detail by Marais and Aldrich [75] and Horn *et al*. [76].

6856 images of the froth in a primary cleaner flotation cell of size 256 × 256 pixels were collected over a 4-hour period. During this time, the air flow to the cell was periodically varied. In addition, the platinum content of samples of the froth was analyzed in a laboratory, and these results could be associated with the froth images that were collected.

Froths associated with normal operating conditions bearing high concentrations of platinum were labeled as Class A. Three other operating regimes were also identified, as indicated by froth structures labeled as Class B, Class C, and Class D. Progressively lower platinum values were associated with these three classes. The four operating regimes are represented by 1960, 1260, 1722, and 1940 images, respectively.Figure 11.10 shows an example of an image of each of these ordinal classes, as well as the relative concentration of platinum in parentheses.

Figure 11.10 also shows that the coarser froth typical of Class A was comparatively distinct from the structures of the other three classes that exhibited progressively finer bubble size distributions. These structures were linked to the different platinum grades in the images from left to right in Figure 11.10. As a consequence, identification of a particular froth class would also give an estimate of the grade associated with the image.

11.5.2 Recognition of Operational States with Convolutional Neural Networks

The same framework that was used in Case Study 1 was used for classification of Classes A–D in this case study. The same hyperparameters summarized in Table 11.3 were used in this case for optimizing the random forest models. However, in this case study, only GoogLeNet is compared with the three

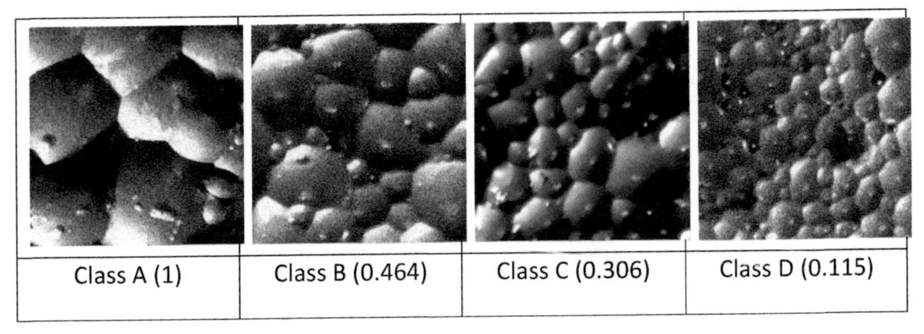

Figure 11.10 Typical images associated with Class A, Class B, Class C, and Class D operational regimes. Relative platinum concentrations are shown in parentheses.

traditional models. This is again done based on different levels of training of the feature layers of the network, as was done in Case Study 1.

Table 11.5 gives a summary of the performance of the different models, as well as the corresponding number of features used as predictors in each model. The classification accuracies with GoogLeNet with feature layers retrained in part were 94.4% and 99.5% when feature layers were retrained in full.

This was markedly higher than the accuracies (70%–82%) obtained with any of the other feature sets, including the traditional feature sets and the untrained GoogLeNet features. This observation confirms the effectiveness of retraining CNNs.

Principal component analysis (PCA) was used to visualize the features extracted from the froth images, by projecting the features to a two-dimensional space, as shown in Figure 11.11. More specifically, the principal component model was constructed from the features of froth Class

Table 11.5 Assessment of texture features as predictors in Case Study 2

Model	No. of features	Accuracy (%)
GoogLeNet**	1024	99.5
GoogLeNet*	1024	94.4
Textons	20	81.8
GoogLeNet	1024	79.9
GLCM	4	71.8
LBP	59	69.8

Note: In models indicated by one or more "*," accuracy represents the performance of the end-to-end classifier. "*" and "**" indicate networks in which some or all the feature layers were retrained, respectively. All other models were random forests, where the accuracies indicated are out-of-bag values derived from the training data set.

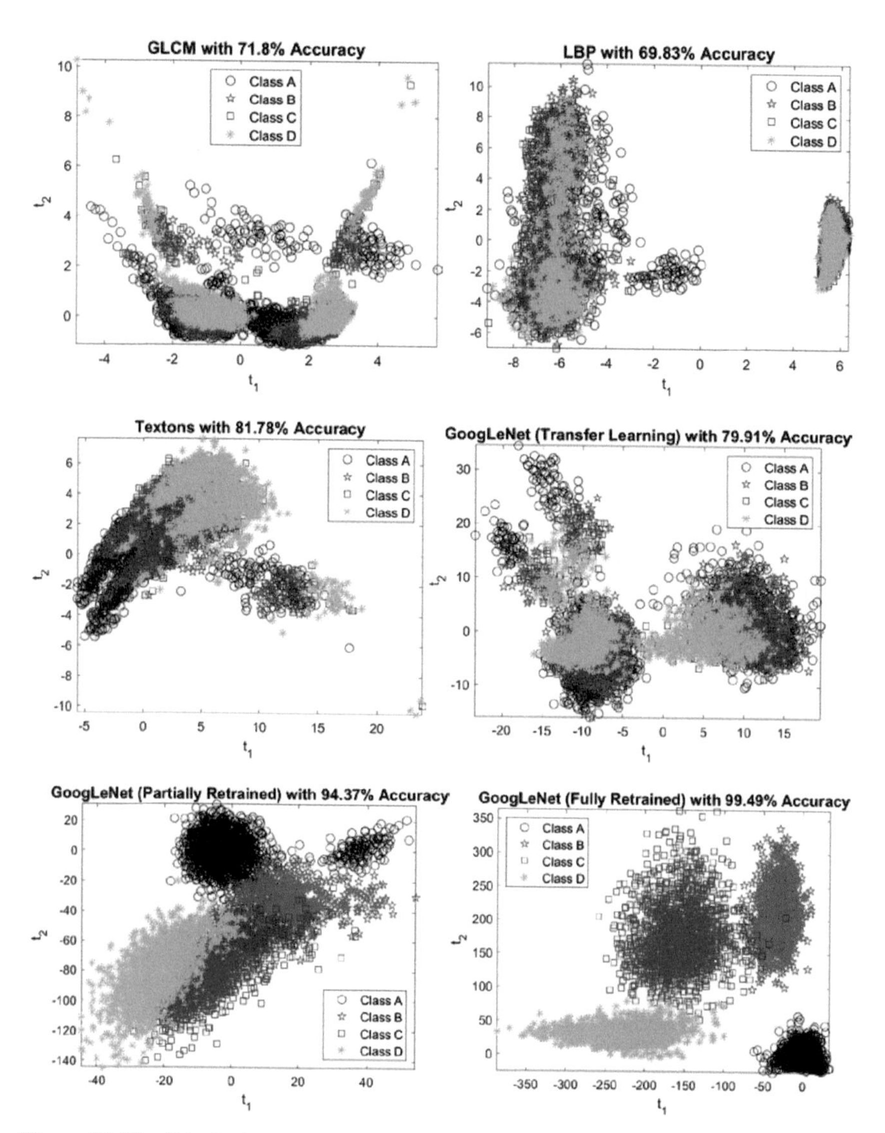

Figure 11.11 Principal component score plots of the GLCM (top, left), LBP (top, right), textons (middle, left), GoogLeNet with transfer learning (middle, right), GoogLeNet with partially retraining (bottom, left), and GoogLeNet with fully retraining (bottom, right) in Case Study 2. Classes A–D are denoted by black circles, red stars, blue squares, and green "*" markers, respectively.

A only, followed by projection of the features from the images of the other classes onto the model.

11.6 Case Study 3: Imaged Signal Textures

11.6.1 Treating Signals as Images

Transfer learning can also play an important role in signal processing with deep neural networks. The ImageNet architectures can be used directly by converting the signals to images, as illustrated in Figure 11.12. As indicated in Figure 11.12, a sliding window of a user specified length (b) is moved across a multivariate time series with step size s.

The data in each segment is consequently used as a basis for generating an image that captures the information in the window. Distance plots of size $b \times b$ are easy to construct, as their (i,j)th elements are the distances between the point i and point j captured by the window. These images will display a texture commensurate with the behavior of the time series.

Following this, features can be extracted from the image by any of a number of algorithms designed for this purpose. These features, together with labels derived from the time series, are collected in a data matrix that can then serve as the basis for analysis of the time series or development of time series models.

Examples of image representations of time series or signals are shown in Figure 11.13 for Euclidean distance plots, Gramian angular fields, Markov transition fields, and continuous wavelet transforms, all of which have recently been used in time series analysis.

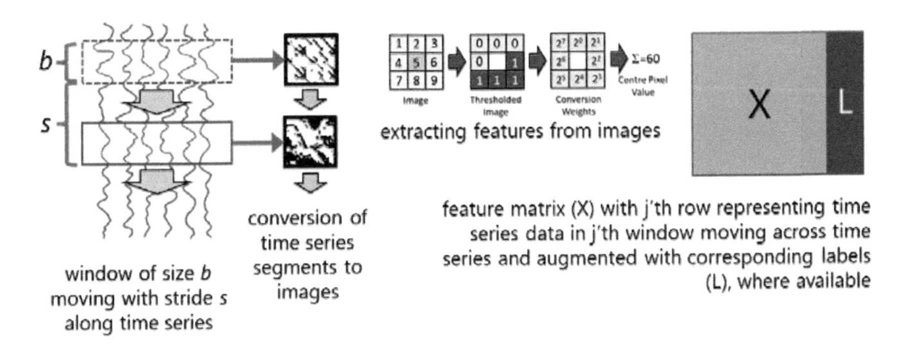

Figure 11.12 Multivariate image analysis of signals.

11.6.2 Monitoring of Stock Prices by the Use of Convolutional Neural Networks

In this subsection, a historic daily gold price data set is considered over the period from 1979 to early 2021. The data set includes over 11,000 values of the daily closing price of gold in USD.

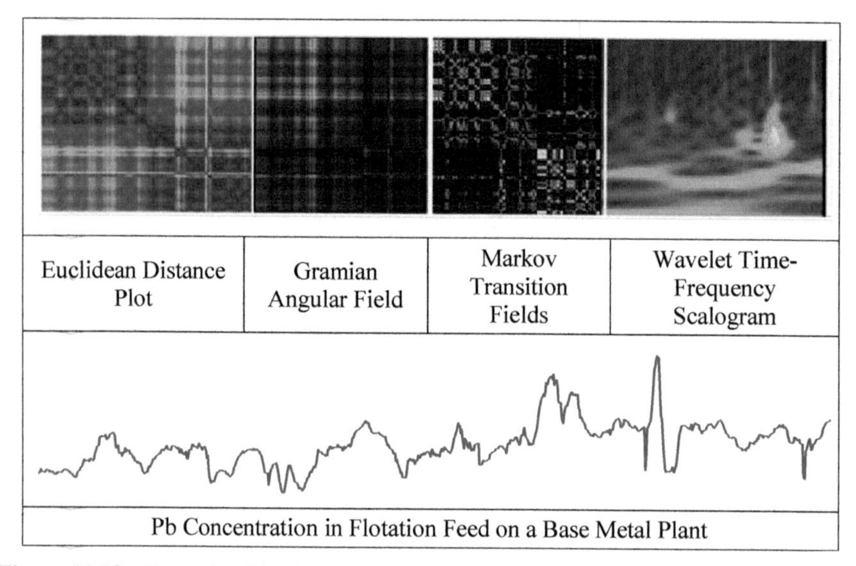

Figure 11.13 Example of image representations of time series measurements on a South African base metal flotation plant.

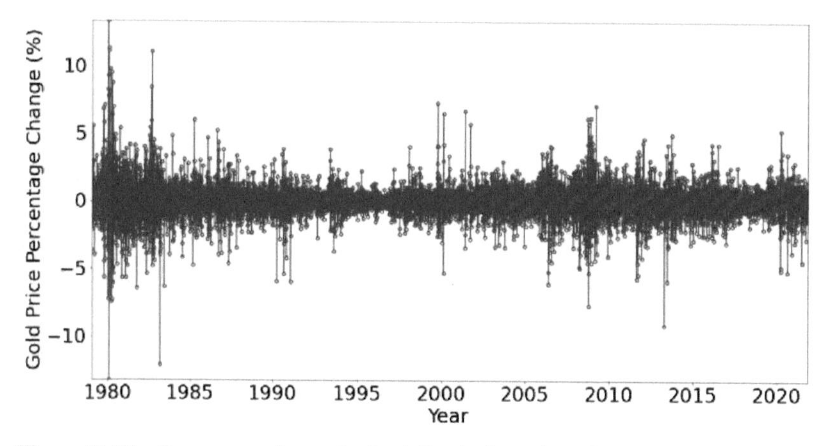

Figure 11.14 Percentage change in the daily closing price of gold from 1979 to 2021.

Table 11.6 Basic statistics for the % change of gold price data set for Case Study 3

Count	Mean	St. Dev.	Min	Max	25%	50%	75%
11170	0.026	1.183	−13.24	13.32	−0.465	0.000	0.517

Table 11.7 A Basic statistics for the absolute % change of gold price data for Case Study 3

Count	Mean	St. Dev.	Min	Max	25%	50%	75%
11170	0.755	0.910	0.000	13.315	0.190	0.489	0.995

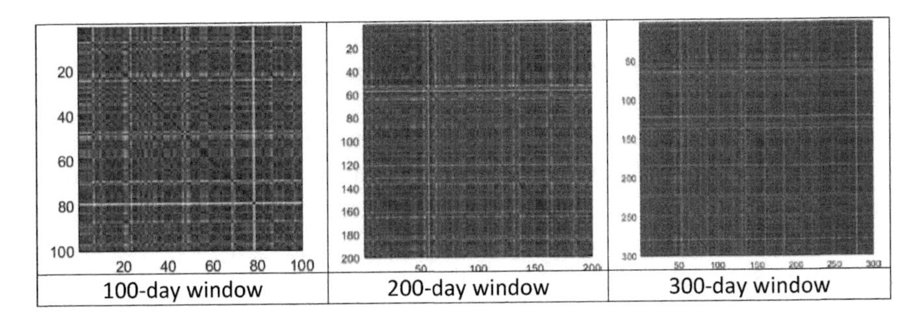

Figure 11.15 100-day (left), 200-day (middle), and 300-day (right) Euclidean distance plots of the daily percentage change in the historic gold price.

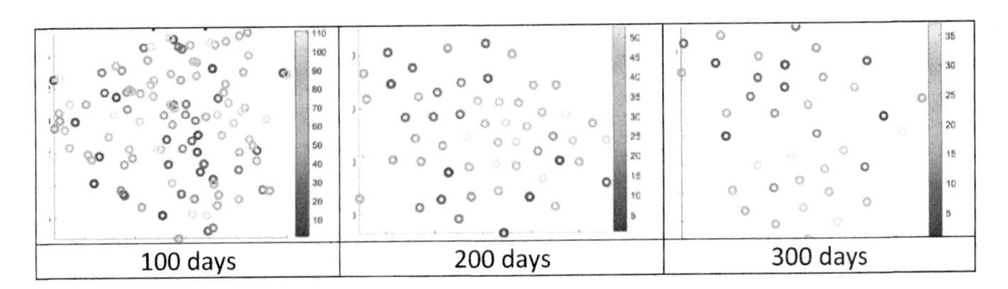

Figure 11.16 t-SNE plots of the 100-day (top, left), 200-day (top right), and 300-day (bottom) Euclidean distance plots of the daily percentage change in the historic gold price.

The percentage change of the gold price with time is shown in Figure 11.14. The basic statistics about the change and the absolute value of the change are shown in Tables 11.6 and 11.7. The gold price changes considerably on some days (greater than 10%), while, during most of the time (more than 75% of the days), the change is within the absolute amplitude of 1%. The mean absolute change over the whole period is 0.755%. The positive change frequency is almost the same as the negative change frequency.

In this subsection, the effects of the window sizes of 100, 200, and 300 are studied, which indicate an interval of 100, 200, and 300 days. The time series of gold price percentage change is segmented into 111 $\times$ 100, 55 $\times$ 200, and 37 $\times$ 300 matrices, respectively, after which each small time series segment is converted to a normalized distance matrix of dimension of 100 $\times$ 100, 200 $\times$ 200, and 300 $\times$ 300. The distance matrix is then converted to a real image matrix by color mapping. The respective last image from each image set is shown in Figure 11.15.

The following numbers of textural images were extracted from the time series with the different window sizes: 111 images (100 $\times$ 100 $\times$ 3), 55 images (200 $\times$ 200 $\times$ 3), and 37 images (300 $\times$ 300 $\times$ 3). These were passed through GoogLeNet, untrained on the target data to get the respective direct deep features. These deep features are visualized using the t-SNE algorithm, as shown in Figure 11.16. The color bars in these figures indicate time, with yellow indicating a more recent period than blue.

As can be seen from Figure 11.16, the clearest patterns are associated with the 300-day moving window, where more recent patterns (yellow and orange) appear to be segregated to some degree from earlier patterns (blue hues).

11.7 Discussion

With their deep architectures, CNNs continue to outperform classical methods in image recognition in an increasing number of domains [77]. Since *ab initio* design and training of large CNNs is costly and the availability or acquisition of sufficient data for this purpose may be unfeasible, the use of pretrained networks may be the only viable approach.

All the case studies demonstrate advantage of features extracted with CNNs over those extracted by traditional methods, specifically GLCM, LBP, and textons algorithms. This is particularly the case when the feature layers of the CNNs are retrained through end-to-end classification.

Even when it is not possible to retrain the feature layers of the CNNs, for example, when training data are available, these studies suggest that CNNs pretrained on ImageNet data are able to achieve results equivalent to those achievable with traditional methods.

The reliability of the CNNs may have been affected by the differences between the ImageNet images used as source data and the textural images used as target data. The relatively small sizes of the target data sets used in Case Studies 1 and 2 would likely also have had an inhibitory effect on the models. Retraining of the feature layers of the networks would generally compensate for the dissimilarity of the source and target data sets.

For the retraining of the pretrained networks, backpropagation is used to fine-tune the weights of the pretrained networks. Earlier layers of the pre-trained networks tend to capture generic features that are useful for different tasks across different domains, while layers further down in the processing paths of the networks tend to capture more complex target domain-specific features. It, therefore, makes sense to retrain the weights of later layers only, to force the CNNs to learn high-level features that are specific to the target data set. Furthermore, full retraining of all the layers led to near-perfect accuracy when using ResNet50 and GoogLeNet in Case Studies 1 and 2, respectively.

Finally, it should be noted that one of the drawbacks of the deep learning approaches compared to the traditional ones is the cost of model development. Large-scale development of these models is best done in high performance computing environments. While these are becoming more accessible via cloud computing, for example, it is likely to remain a challenge, as large-scale today may pale in comparison to what would be large-scale in future. To some extent, this can be addressed by a future focus on texture analysis with more compact CNNs (e.g., MobileNet or Squeezenet).

11.8 Conclusion

In this chapter, texture analysis with CNNs is explored, from which the following conclusions can be made.

a) Transfer learning with CNNs, such as ResNet50, GoogLeNet, VGG19, and AlexNet pretrained on the ImageNet data base of common objects, can be used without any further training to generate highly discriminative features of image textures.
b) Further improvement was achieved by partial or full retraining of these networks. This led to significantly higher or even near perfect classification of the different Voronoi textures in Case Study 1 and the recognition of the froth states on a platinum metal group flotation plant in Case Study 2.
c) Of the four CNNs that were compared in Case Studies 1 and 2, i.e., AlexNet, VGG19, GoogLeNet, and ResNet50, the latter yielded the most reliable features. All the networks performed as well as or better than the traditional methods, i.e., LBP, GLCM, and textons. These results are in line with those of other emerging investigations.

d) Finally, the impact of transfer learning extends beyond textural feature analysis of optical images but is also having a major impact on (non)linear signal processing.

References

[1] Fukushima, K., *Neocognitron: A self-organizing neural network model for a mechanism of pattern recognition unaffected by shift in position.* Biological Cybernetics, 1980. 36(4): p. 193-202.

[2] Fukushima, K., Miyake, S., and Ito, T., *Neocognitron: A neural network model for a mechanism of visual pattern recognition.* IEEE Transactions on Systems, Man, and Cybernetics, 1983. SMC-13 (5): p. 826-834.

[3] LeCun, Y., *et al.*, *Backpropagation applied to handwritten zip code recognition.* Neural Computation, 1989. 1(4): p. 541-551.

[4] Pan, S. J. and Yang, Q., *A survey on transfer learning.* IEEE Transactions on Knowledge and Data Engineering, 2010. 22(10): p. 1345-1359.

[5] Weiss, K., Khoshgoftaar, T. M., and Wang, D., *A survey of transfer learning.* Journal of Big Data, 2016. 3(1).

[6] Russakovsky, O., *et al.*, *ImageNet large scale visual recognition challenge.* International Journal of Computer Vision, 2015. 115(3): p. 211-252.

[7] Pérez-Barnuevo, L., Lévesque, S., and Bazin, C., *Automated recognition of drill core textures: A geometallurgical tool for mineral processing prediction.* Minerals Engineering, 2018. 118: p. 87-96.

[8] Lund, C., Lamberg, P., and Lindberg, T. *A new method to quantify mineral textures for geometallurgy. in Cape Town: Paper Presented at Process Mineralogy.* 2014.

[9] Jouini, M. S. and Keskes, N., *Numerical estimation of rock properties and textural facies classification of core samples using X-Ray Computed Tomography images.* Applied Mathematical Modelling, 2017. 41: p. 562-581.

[10] Jardine, M. A., Miller, J. A., and Becker, M., *Coupled X-ray computed tomography and grey level co-occurrence matrices as a method for quantification of mineralogy and texture in 3D.* Computers & Geosciences, 2018. 111: p. 105-117.

[11] Das, P., *et al.*, *Characterization of impact fracture surfaces under different processing conditions of 7075 Al alloy using image texture analysis.* International Journal of Technology And Engineering System(IJTES), 2011. 2(2): p. 143-147.

[12] Dutta, S., *et al.*, *Automatic characterization of fracture surfaces of AISI 304LN stainless steel using image texture analysis*. Measurement, 2012. 45(5): p. 1140-1150.

[13] Naik, D. L. and Kiran, R., *Identification and characterization of fracture in metals using machine learning based texture recognition algorithms*. Engineering Fracture Mechanics, 2019. 219: p. 106618.

[14] Bastidas-Rodriguez, M. X., Prieto-Ortiz, F.A., and Espejo, E., *Fractographic classification in metallic materials by using computer vision*. Engineering Failure Analysis, 2016. 59: p. 237-252.

[15] Li, X., *et al.*, *Tribological behavior of ZrO2/WS2 coating surfaces with biomimetic shark-skin structure*. Ceramics International, 2019. 45(17): p. 21759-21767.

[16] Huang, Q., *et al.*, *Optimization of bionic textured parameter to improve the tribological performance of AISI 4140 self-lubricating composite through response surface methodology*. Tribology International, 2021. 161: p. 107104.

[17] Kesel, A. and Liedert, R., *Learning from nature: Non-toxic biofouling control by shark skin effect*. Comparative Biochemistry and Physiology Part A: Molecular & Integrative Physiology, 2007. 146(4).

[18] Chien, H. W., *et al.*, *Inhibition of biofilm formation by rough shark skin-patterned surfaces*. Colloids Surf B Biointerfaces, 2020. 186: p. 110738.

[19] Natarajan, E., *et al.*, *The hydrodynamic behaviour of biologically inspired bristled shark skin vortex generator in submarine*. Materials Today: Proceedings, 2021. 46: p. 3945-3950.

[20] Kaleda, A., *et al.*, *Physicochemical, textural, and sensorial properties of fibrous meat analogs from oat-pea protein blends extruded at different moistures, temperatures, and screw speeds*. Future Foods, 2021. 4: p. 100092.

[21] Rigdon, M., *et al.*, *Texture and quality of chicken sausage formulated with woody breast meat*. Poultry Science, 2021. 100(3): p. 100915.

[22] Schreuders, F. K. G., *et al.*, *Texture methods for evaluating meat and meat analogue structures: A review*. Food Control, 2021. 127: p. 108103.

[23] Ghalati, M. K., *et al.*, *Texture analysis and its applications in biomedical imaging: A survey*. IEEE Reviews in Biomedical Engineering, 2021: p. 1.

[24] Bonnin, A., *et al.*, *CT texture analysis as a predictor of favorable response to anti-PD1 monoclonal antibodies in metastatic skin melanoma*. Diagnostic and Interventional Imaging, 2021.

[25] Haralick, R. M., Shanmugam, K., and Dinstein, I. H., *Textural features for image classification*. IEEE Transactions on Systems, Man, and Cybernetics, 1973. SMC-3(6): p. 610-621.

[26] Kim, T. Y., *et al.*, *3D texture analysis in renal cell carcinoma tissue image grading*. Computational and Mathematical Methods in Medicine, 2014. 2014: p. 536217.

[27] Ojala, T., Pietikäinen, M., and Harwood, D., *A comparative study of texture measures with classification based on featured distributions*. Pattern Recognition, 1996. 29(1): p. 51-59.

[28] Ojala, T., Pietikainen, M., and Maenpaa, T., *Multiresolution gray-scale and rotation invariant texture classification with local binary patterns*. IEEE Transactions on Pattern Analysis and Machine Intelligence, 2002. 24(7): p. 971-987.

[29] Ahonen, T., Hadid, A., and Pietikainen, M., *Face description with local binary patterns: Application to face recognition*. IEEE Transactions on Pattern Analysis and Machine Intelligence, 2006. 28(12): p. 2037-2041.

[30] Shabat, A. M. and Tapamo, J. R., *A comparative study of the use of local directional pattern for texture-based informal settlement classification*. Journal of Applied Research and Technology, 2017. 15(3): p. 250-258.

[31] Nanni, L., Brahnam, S., and Lumini, A., *Local ternary patterns from three orthogonal planes for human action classification*. Expert Systems with Applications, 2011. 38(5): p. 5125-5128.

[32] Fu, Y. and Aldrich, C., *Flotation froth image analysis by use of a dynamic feature extraction algorithm*. IFAC-PapersOnLine, 2016. 49(20): p. 084-089.

[33] Bonomi, M., Pasquini, C., and Boato, G., *Dynamic texture analysis for detecting fake faces in video sequences*. Journal of Visual Communication and Image Representation, 2021. 79: p. 103239.

[34] Arita, Y., *et al.*, *Diagnostic value of texture analysis of apparent diffusion coefficient maps for differentiating fat-poor angiomyolipoma from non-clear-cell renal cell carcinoma*. European Journal of Radiology, 2021. 143: p. 109895.

[35] Moolman, D. W., *et al.*, *Digital image processing as a tool for on-line monitoring of froth in flotation plants*. Minerals Engineering, 1994. 7(9): p. 1149-1164.

[36] Soille, P., *Morphological Texture Analysis: An Introduction, in Morphology of Condensed Matter: Physics and Geometry of Spatially Complex Systems*, K. Mecke and D. Stoyan, Editors. 2002, Springer Berlin Heidelberg: Berlin, Heidelberg. p. 215-237.

[37] Jing, L. and Shan, Z. *A novel image retrieval based on texture primitive. in 2010 5th International Conference on Computer Science & Education*. 2010.

[38] Wang, H. and Na, J. *Texture image segmentation using spectral histogram and skeleton extracting*. in *2009 International Conference on Electronic Computer Technology*. 2009.

[39] Unser, M., *Texture classification and segmentation using wavelet frames*. IEEE Transactions on Image Processing, 1995. 4(11): p. 1549-1560.

[40] Laws, K. I. *Rapid texture identification*. in *Proceedings of the SPIE 0238*. 1980.

[41] Laws, K. I. *Texture energy measures*. in *Proceedings of the Image Understanding Workshop*. 1979.

[42] Jain, A. K. and Farrokhnia, F., *Unsupervised texture segmentation using Gabor filters*. Pattern Recognition, 1991. 24(12): p. 1167-1186.

[43] Cross, G. R. and Jain, A. K., *Markov random field texture models*. IEEE Transactions on Pattern Analysis and Machine Intelligence, 1983. PAMI-5(1): p. 25-39.

[44] Yang, X. and Liu, J., *Maximum entropy random fields for texture analysis*. Pattern Recognition Letters, 2002. 23(1): p. 93-101.

[45] Pentland, A. P., *Fractal-based description of natural scenes*. IEEE Transactions on Pattern Analysis and Machine Intelligence, 1984. PAMI-6(6): p. 661-674.

[46] Kashyap, R. L. and Khotanzad, A., *A model-based method for rotation invariant texture classification*. IEEE Transactions on Pattern Analysis and Machine Intelligence, 1986. PAMI-8(4): p. 472-481.

[47] Xianghua, X. and Mirmehdi, M. *Localising surface defects in random colour textures using multiscale texem analysis in image eigenchannels. in IEEE International Conference on Image Processing 2005*. 2005.

[48] Li, J., *et al.*, *Flotation froth image texture extraction method based on deterministic tourist walks*. Multimedia Tools and Applications, 2017. 76(13): p. 15123-15136.

[49] Bashier, H. K., et al., *Texture classification via extended local graph structure*. Optik, 2016. 127(2): p. 638-643.

[50] Gaetano, R., Scarpa, G., and Sziranyi, T. *Graph-based analysis of textured images for hierarchical segmentation*. in *British Machine Vision Conference, BMVC 2010*. 2010. Aberystwyth, UK.

[51] Jernigan, M. E. and Astous, F. D., *Entropy-based texture analysis in the spatial frequency domain.* IEEE Transactions on Pattern Analysis and Machine Intelligence, 1984. PAMI-6(2): p. 237-243.

[52] Silva, L. E. V., *et al., Two-dimensional sample entropy: Assessing image texture through irregularity.* Biomedical Physics & Engineering Express, 2016. 2(4): p. 045002.

[53] Varma, M. and Zisserman, A., *A statistical approach to texture classification from single images.* International Journal of Computer Vision, 2005. 62(1): p. 61-81.

[54] Bastidas-Rodriguez, M. X., *et al., Deep learning for fractographic classification in metallic materials.* Engineering Failure Analysis, 2020. 113: p. 104532.

[55] Li, J., Chen, L., and Cai, Y., *Dynamic texture segmentation using fourier transform.* Modern Applied Science, 2009. 3(9): p. 29-36.

[56] Leung, T. and Malik, J., *Representing and recognizing the visual appearance of materials using three-dimensional textons.* International Journal of Computer Vision, 2001. 43(1): p. 29-44.

[57] Lowe, D. G., *Distinctive image features from scale-invariant keypoints.* International Journal of Computer Vision, 2004. 60(2): p. 91-110.

[58] Lazebnik, S., Schmid, C., and Ponce, J., *A sparse texture representation using local affine regions.* IEEE Transactions on Pattern Analysis and Machine Intelligence, 2005. 27(8): p. 1265-1278.

[59] Andrearczyk, V. and Whelan, P. F., *Using filter banks in convolutional neural networks for texture classification.* Pattern Recognition Letters, 2016. 84: p. 63-69.

[60] Wu, H., *et al., Deep texture exemplar extraction based on trimmed T-CNN.* IEEE Transactions on Multimedia, 2021. 23: p. 4502-4514.

[61] Lin, T., RoyChowdhury, A., and Maji, S. *Bilinear CNN models for fine-grained visual recognition. in 2015 IEEE International Conference on Computer Vision (ICCV).* 2015.

[62] Dai, X., Ng, J. Y., and Davis, L. S. *FASON: First and second order information fusion network for texture recognition. in 2017 IEEE Conference on Computer Vision and Pattern Recognition (CVPR).* 2017.

[63] Zhang, H., Xue, J., and Dana, K., *Deep TEN: Texture encoding network. in 2017 IEEE Conference on Computer Vision and Pattern Recognition (CVPR).* 2017.

[64] Moolman, D. W., et al., *The interpretation of flotation froth surfaces by using digital image analysis and neural networks.* Chemical Engineering Science, 1995. 50(22): p. 3501-3513.

[65] Moolman, D. W., Aldrich, C., and Van Deventer, J. S. J., *The monitoring of froth surfaces on industrial flotation plants using connectionist image processing techniques*. Minerals Engineering, 1995. 8(1): p. 23-30.

[66] Kistner, M., Jemwa, G. T., and Aldrich, C., *Monitoring of mineral processing systems by using textural image analysis*. Minerals Engineering, 2013. 52: p. 169-177.

[67] Aldrich, C., *et al.*, *Multivariate image analysis of realgar–orpiment flotation froths*. Mineral Processing and Extractive Metallurgy, 2017. 127(3): p. 146-156.

[68] Schmid, C. *Constructing models for content-based image retrieval*. in *Proceedings of the 2001 IEEE Computer Society Conference on Computer Vision and Pattern Recognition. CVPR 2001*. 2001.

[69] Jemwa, G. T. and Aldrich, C., *Estimating size fraction categories of coal particles on conveyor belts using image texture modeling methods*. Expert Systems with Applications, 2012. 39(9): p. 7947-7960.

[70] Krizhevsky, A., Sutskever, I., and Hinton, G.E., *ImageNet classification with deep convolutional neural networks*. Communications of the ACM, 2017. 60(6): p. 84-90.

[71] Simonyan, K. and Zisserman, A., *Very Deep Convolutional Networks for Large-Scale Image Recognition*. Published as a conference paper at ICLR 2015, 2014. arXiv 1409.1556.

[72] Szegedy, C., *et al. Going deeper with convolutions. in 2015 IEEE Conference on Computer Vision and Pattern Recognition (CVPR)*. 2015.

[73] Tian, X. and Chen, C. *Modulation Pattern Recognition Based on Resnet50 Neural Network. in 2019 IEEE 2nd International Conference on Information Communication and Signal Processing (ICICSP)*. 2019.

[74] Kingma, D. P. and Ba, J., *Adam: A Method for Stochastic Optimization*. CoRR, 2015. abs/1412.6980(Published as a conference paper at the 3rd International Conference for Learning Representations, San Diego, 2015).

[75] Marais, C. and Aldrich, C., *Estimation of platinum flotation grades from froth image data*. Minerals Engineering, 2011. 24(5): p. 433-441.

[76] Horn, Z. C., *et al.*, *Performance of convolutional neural networks for feature extraction in froth flotation sensing*. IFAC-PapersOnLine, 2017. 50(2): p. 13-18.

[77] Liu, X. and Aldrich, C., *Deep learning approaches to image texture analysis in material processing*. Metals, 2022. 12(2): p. 355.

12

Internet of Things Based Enabled Convolutional Neural Networks in Healthcare

Joseph Bamidele Awotunde[1,*], Akash Kumar Bhoi[2], Rasheed Gbenga Jimoh[3], Sunday Adeola Ajagbe[4], Femi Emmanuel Ayo[5], and Oluwadare Adepeju Adebisi[6]

[1,3]Department of Computer Science, University of Ilorin, Nigeria
[2]Department of Computer Science and Engineering, Sikkim Manipal Institute of Technology (SMIT), Sikkim Manipal University (SMU), India
[4]Department of Computer Engineering, Ladoke Akintola University of Technology, Nigeria, a Department of Computer Science, University of Zululand, South Africa
[5]Department of Computer Science, McPherson University, Nigeria
[6]Department of Electronic and Electrical Engineering, Ladoke Akintola University of Technology, Nigeria
E-mail: awotunde.jb@unilorin.edu.ng; akashkrbhoi@gmail.com; jimoh_rasheed@unilorin.edu.ng; saajagbe@pgschool.lautech.edu.ng; ayofe@mcu.edu.ng; oadebisi44@pgschool.lautech.edu.ng
*Corresponding Author

Abstract

The Internet of Things (IoT) systems have revolutionized the medical systems with the use of devices and sensors for the collections of data for various uses. Data is generated by these devices in a variety of formats, including text, photos, and videos. Hence, getting accurate and useable data from the massive amounts of data generated by the IoT-based system is a difficult task. The diagnosis of various diseases using data generated by IoT has recently emerged as a potential topic that necessitates sophisticated and effective methodologies. Due to the wide range of disease symptoms and indications, reliable diagnosis is difficult. Existing solutions either rely on handcrafted features or conventional machine learning model. Therefore, this chapter presents the applicability of IoT-based enabled convolutional neural network (CNN) in healthcare diagnosis. The challenges and future prospects of IoT-based enabled CNN are discussed. The chapter proposes an intelligent IoT-based enabled CNN for the diagnosis of patients' health status, and the CNN was used to diagnose the capture data using IoT-based sensors for the disease. As a result, the system takes advantage of the dataset's and CNN's properties, assuring excellent reliability and accuracy. For a case study on healthcare dataset classifications, the suggested system demonstrates real-time health monitoring and tests the system's performance in terms of various metrics. The performance of the proposed system shows better performance when compared with existing methods with an accuracy of 98.4%, specificity of 98.7%, sensitivity of 98.9%, and *F*-score of 98.3%.

Keywords: Internet of Things, convolutional neural network, healthcare systems, diagnosis, deep learning, machine learning.

12.1 Introduction

The new emergence in information technologies has brought about Internet of Things (IoT), and this has really changed the way we think and our behavior globally [1, 2]. This technology has been used in various domains like transportation [3, 4], agriculture [5, 6], smart cities [7, 8], industries [9, 10], and especially in healthcare systems for the monitoring and diagnosis of various diseases [11, 12]. The transformation of IoT in healthcare systems created new innovation called Internet of Medical Things (IoMT) which is used for indoor quality monitoring, elderly person monitoring, disease diagnosis, and treatments, among other [2, 13, 14]. This paradigm shift in healthcare has provided various opportunities with the use of wearable devices and sensors for capturing of physiological signs for enhanced healthcare systems in close relation with mHealth and eHealth medical systems [13]. IoT has brought about various advantages into healthcare systems like availability and accessibility with low cost to medical diagnosis and treatment, thus explaining the increase in the usage of this innovation in recent years. The use of biophysical data for various diseases and monitoring of patients to support healthcare system decisions has helped in various ways to ease the burden of health workers and greatly reduce medical cost.

The new technological innovations like IoT-based systems with artificial intelligence (AI) have been used to process and handle the big data generated by sensors to revolutionize the advancement of medical services [15, 16].

The IoT-based systems used in generating big data can be used to capture data that be processed to give useful clinical information that can be used by specialists for the prediction of patients' health condition [17]. The information examination and information extraction are perplexing cycles that should guarantee further developed security strategies [18]. IoTs convey estimation information to a focal stockpiling area for concentrated dynamic [19]. In the clinical area, such estimation information is typically physiological signs like pulse, heart beat signs, and temperature estimation [20].

The use of AI on generated huge data can offer a few chances for medical service frameworks to accurately predict any diseases from the captured data from the IoT-based system [21]. The use of AI techniques used for the processes of big data can help in the progression of patients' wellbeing worldwide [15, 22]. IoT innovations permit diminishing worldwide expenses for constant sickness anticipation. The constant wellbeing information gathered by these frameworks can be utilized to help patients during self-organization treatments. Cell phones with versatile applications are frequently utilized and incorporated with telemedicine and mHealth through the IoMT [23]. The deep learning (DL) approaches are utilized in wellbeing-based applications to accomplish promising and good exhibitions for a sizeable measure of information. The DL and intellectual calculations have encountered signs of progress as of late and, subsequently, have been utilized to take care of numerous complicated issues.

The huge data captured using IoT-based systems can be processed by DL models since the data can be stored on the cloud database for further processing [11]. The methods can be used for the processing of data to get a proper insight that can be used for the treatment of patients who suffer from any form of diseases. The DL beats the traditional-based models for diagnosis and decision-making. This is significant in a pre-portrayed undertaking, and the norms are acquired from genuine information. DL utilizes the restricted information design to settle on some astute choices in brilliant medical care frameworks that the doctors can make ends. Large information is the center part of the DL procedures' superior [11]. The use of wearable body sensors (WBS) to gather huge data can be used to consolidate the treatment of patient in real time with the use of AI models to process such data. The use of AL techniques will help in getting intelligent results with the captured data and useful for the interpretation of the data. Hence, it helps in getting splendid responses from the huge data for clinical use by the specialists to save lives and time [11].

Therefore, this chapter presents an innovative IoT-based enhanced CNN for patient disease diagnosis. The CNN was used to diagnose the capture data using IoT-based sensors for the disease. A practical case of heart disease is used to test the performance of the proposed systems.

The contributions of the chapter are stated as follows:

(i) The IoT-based enabled CNN applicability in clinical systems is presented, and challenges and future prospects of IoT-based enabled CNN are presented.
(ii) An intelligent IoT-based enabled CNN model for disease diagnosis framework is proposed.
(iii) The performance of the proposed method is presented using a practical case in healthcare system.

12.2 Internet of Things Application in the Healthcare Systems

The IoT-based framework is a new breakthrough that is expected to lead to the discovery of novel pharmaceuticals and clinical treatments. The productivity and nature of healthcare have high potential characteristics like adaptability, flexibility, fondness, cost reduction, and being very fast. This innovation assists us with understanding the particular dangers identified with security and privacy. IoT is primarily to associate the world through connection various devices and sensors. In the healthcare framework, IoT is predominantly used to get to data in real time. IoT is principally interconnected by more devices with the utilization of the Internet. In the healthcare framework, the IoT is fundamentally worked to get to the enormous size of data. It denotes the ability of a matrix of PCs to transmit programs and data. For details, this innovation can be generally refined by various servers as retail organizations, whose necessities can be totally met by combined use.

Here, an ontology-based crisis clinical benefit framework gives the way of gathering, coordinating, and interoperating IoT information. In view of the ontology construction in crisis clinical benefit, choices can be made in IoT with its dynamic interaction, and a choice of decision support system (DSS) can be utilized [24, 25]. Through ontology development, decisions for the medical services system can be made in real time and with ease. In healthcare administration, specialists, patients, and doctors take significant part and they are likewise engaged with a whole overhauling. Specialists need to get to the patient record from anywhere and in real time by putting it away in a conveyed way [26]. Patients additionally need to think about the specialists' accessibility and the situation with the equipment (occupied/free) [27]. To assist patients with getting to specialists' accessibility status, an IoT asset model is required for this openness.

12.2.1 Internet of Things Operation in Healthcare Systems

The applications of IoT-based systems play a critical role in medical services for taking care of patients in real time [13, 28]. The use of a computer-aided diagnosis data gathering approach reduces the danger of human mistakes on data gathering [29]. This would increase diagnosis quality and lessen the chance of human errors, such as the gathering or transmission of erroneous information, which is harmful to individuals' wellbeing [13]. In remote identification, forecast, reconnaissance, recuperation, and treatment, where IoT innovation has a significant influence, telemedicine has, as of late, been comprehensively applied. With interest in planning keen advancements, for example, medical care global positioning frameworks, clinical analysis, forecast and therapy systems, and keen healthcare, IoT has, as of late, been executed in the clinical area. Data gotten from clinical devices, like the use of wearable sensors, CT, and MRI machines, has been used to improve the telemedicine and smart healthcare generally.

Sensors and devices may communicate within a smart environment, and knowledge can be easily exchanged across healthcare networks, thanks to IoT-based technology. However, the new fields arising expediently, they likewise have their difficulties, especially when the objective is medical service frameworks with a muddled issue, troublesome in energy-proficient, protected, adaptable, appropriate, and reliable arrangements. IoT is expected to reach a market size of $300 billion in the medical services framework by 2020, comprising clinical apparatuses, foundations, services, and solutions [30]. This longing for modified e-medical care is additionally prone to be advanced by government drives. Numerous dataset groups and assets are required overall to store huge data, and these have turned into a test. It is a significant issue to get substantial patterns from huge data, like patient analytic data.

These days, an assortment of arising applications for various conditions is being created. Coldhearted frameworks and sensors are most generally utilized for genuine or not-so-distant future applications [31]. Recently, various studies have developed several wearable devices for monitoring patients' physiological parts especially in the remote areas [32, 33]. For example, this has been used to monitoring various illnesses like body temperature, blood sugar, and blood pressure among others for patients' wellbeing. The systems have been used in the areas of elderly activities and food propensities for proper monitoring of these patients like working, sleeping, etc. The cloud warehouse database is used to store the huge data gathered from the IoT-based devices and sensors which can be processed for the prediction of disease in patient body [34]. Patients in hospitals, particularly those in life support, require close surveillance and attentiveness in order to respond to potential crises and preserve lives.

Sensors are used in these activities which are directed to capture biomedical signals, which are then evaluated and stored in the cloud before being given to online caregivers [13]. By analyzing the flow of data collected by the sensors, a group of healthcare professionals interact and thereafter analyze patients according to their specialties. Then, for high-risk individuals, determining the emergency condition (patients requiring urgent or emergency operations, cardiovascular problems, and so on), the process will be simple.

In healthcare IoT systems, context awareness is a key criterion. As it can track down the patient's condition and the climate where the patient was found, it will extraordinarily help the medical service experts to comprehend the varieties that can impact the wellbeing status of these patients. Moreover, the difference in the actual condition of the patient might build the level of its weaknesses to illnesses and be a reason for his/her wellbeing disintegration [11, 13, 14]. The utilization of a few kinds of particular sensor catching different data about the patient's state of being like his strolling, running, dozing, and so on or the climate where the patient is like wet, chilly, warm, and so forth, and the coordinated effort between them to gather the significant data, will give a superior comprehension of the patient's conditions, while they are hospitalized, at home, or anyplace. Besides, it will give assistance in crisis cases to find the patients and know about the sort of crisis mediation that can be taken.

In terms of performance, the integrated cloud and IoT-based application outperform traditional cloud-based apps. Military, banking, and medical systems can all benefit from the cloud-based IoT system. The cloud-based IoT strategy, in particular, would aid in providing therapeutic uses with efficient resources for monitoring and retrieving data from any remote place. Data can be collected in real-time update over a predetermined period of time using the data-centric embedded system. This can be used to monitor a patient in real time for the wellbeing of individuals, especially elderly patients in remote areas; it is useful for disease diagnosis and treatment in most cases before serious health challenges.

Due to the massive data generated by smart devices, in large data processing, DL models can be used to make intelligent decisions. The approach of applying data processing procedures for certain sectors necessitates identifying the data involved, such as its velocity, variety, and volume. Standard data analytics architecture entails the creation of a neural network model, an implementation of complex, and a clustering process model, as well as the deployment of sophisticated methods. The IoT gadgets can be utilized to produce different kinds of information from a few sources and, consequently, portray the components of the information created for appropriate information dealing with. This assistance in taking care of the different qualities of catch information for versatility and speed accordingly helps in tracking down all the models that can give the best outcomes continuously internationally with no difficulties. These are well-known to be amongst the IoT's major problems. In any case, in the most recent advances, these are generally issues that produce an enormous number of potential outcomes. Such data can be gotten to utilizing the most recent medical care applications, and the information is safely put away on the cloud server.

12.2.2 Internet of Things and Patient Monitoring Systems

For many real-world applications, monitoring systems is a significant principle. Many people's health is at risk today all across the world because of a lack of adequate healthcare monitoring [35]. Almost every day, the elderly, children, and chronically ill people are required to be inspected. The feasibility of a remote monitoring system will assist in avoiding unnecessary hospital visits, as shown in Figure 12.1. Due to their vital condition, their health can frequently go unrecognized until diseases progress to a crisis point. The remote access sensor allows caregivers to perform pre-diagnosis and intervention before problems arise.

As a result, people-centric IoT will be employed for various units of neurocognitive disabilities, allowing them to live more autonomously and with simple existence [36]. As the sensor is connected to the skin at explicit areas, it tends to be utilized for diagnosing the heart condition and the impact of the medication on its exercises. Numerous patients who experience the ill effects of constant sicknesses like cardiopulmonary illness, asthma, and heart disappointment are situated far away from the clinical consideration offices. The ongoing observation of such patients through remote checking

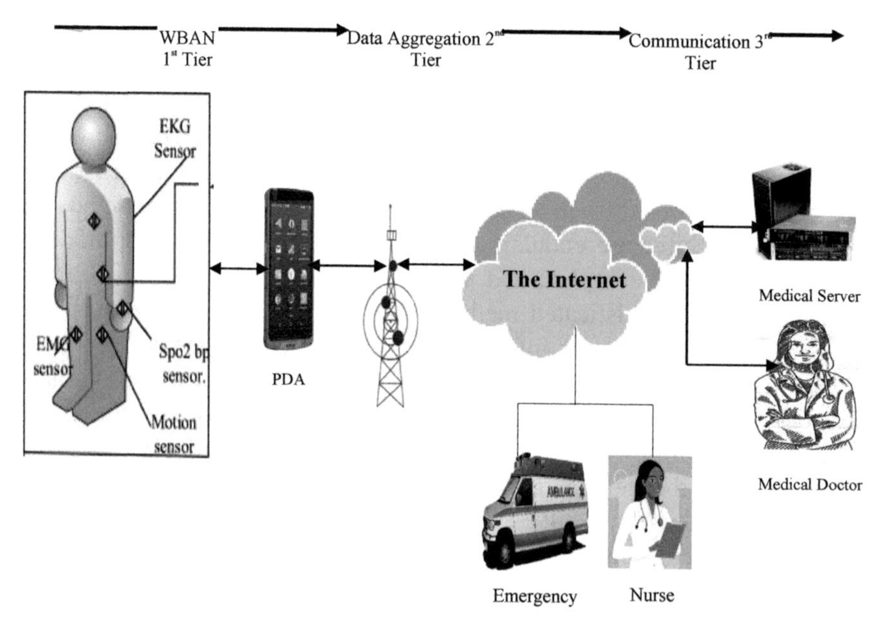

Figure 12.1 Real-time remote monitoring system.

frameworks is the most encouraging application. A portion of the continuous medical care observing frameworks is far off quiet following and checking framework, remote checking of cardiovascular patients, and heartbeat checking framework. The continuous checking framework comprises a distant clinical observing unit and a checking focus. It examines the data from the sensor dependent on ongoing investigation and an admonition sign will arise for crisis and analysis. The signs from the body sensor are taken to the comparing clinical focus through the wireless local area network (WLAN) framework. Accordingly, this ongoing observing framework gives data about patients' ailments, and it might likewise decrease more confusion and give treatment most punctually. Consequently, it gives a precise and constant checking framework in the medical care area. It additionally helps for quicker recognition of info sensors and recoveries a day to day existence.

Due to a lack of accessible access to good structural health monitoring, many health conditions may go misdiagnosed, which is a problem that exists all across the world. On the other side, the IoT enables small, efficient wireless technologies to bring monitoring to patients rather than the other way around. The secure recording of patient health information is possible with these solutions. Before being exchanged via computer networks, the data is evaluated using a network of equipment and complex computations. Medical specialists can then provide appropriate health advice via the Internet.

12.2.3 Internet of Things and Healthcare Data Classification

The advancement of IoT-based systems can work better when compared with the traditional networks methods with well-organized devices and sensors. This has played an important role in society since its inception, spanning everything from traditional hardware to ordinary family protests [37], and has recently attracted serious attention in various fields like transportation, agriculture, education, and, most especially, in healthcare systems. The systems can be used remotely to successfully regulate and check elderly patients without seeing any doctors or specialist be treatment was done upon any sickness. The devices and sensors used can easily communicate with each other with the use of Internet and other wireless connection devices. These devices can even be used to make decisions with the assistance of doctors or specialists. To make IoT more brilliant, a slew of inquiry improvements have been incorporated, including some of the most significant advances in datum mining.

Information mining incorporates discovering novel, captivating, and conceivably supportive models from colossal enlightening assortments and applying estimations to the extraction of stowed away information. Various terms are used for information mining, for example, information exposure (mining) in informational collections (KDD), information extraction, information/plan assessment, information antiquarianism, information burrowing, and information gathering [38]. The objective of any information mining measure is to develop capable judicious or illuminating algorithms that can best explain huge data captured for proper decision-making and, if possible, summarize the data to a meaningful status [39]. Considering a far reaching viewpoint on information mining's helpfulness, information mining is the most well-known method for discovering interesting data from a ton of dataset aside in informational indexes, data conveyance focuses, or different information files.

There are three areas of data mining to really enjoy the full potential of data mining in healthcare systems. The following measures are counted in data mining processes and progression.

(i) Data preparation: The removal of noise from data, synchronization of data captured from various sources, and the extraction of meaningful information from the captured data are the three main sub-steps in data preparation. The data has to be organized in such way that it is going to be meaningful for the users and the processing will become easier.

(ii) The application of computations approach on data is called data mining to discover and evaluate the classes of collected data.

(iii) The interpretation of the gathered results from the mined data to become meaningful decision-making information to the clients and users.

Characterization is significant for the administration of dynamic. The order in mining is called the predefined or allocating and given a definition and meaningful to what is call data mining. To precisely define a class for each case in a data is the main objective of characterization in data mining [40]. For instance, the use of low, medium, or high in credit hazards can be used in a characterization model to define various classes to recognize candidates in data mining in various fields [41].

Various data mining techniques have been used for the processing of huge data in healthcare systems; such techniques are *K*-means, rule-based mining, clusters, progressive grouping, neural organizations, Bayesian organization, and backing vector machines.

The most important component of IoT applications is an effective data analytical method capable of performing tasks such as classification, clustering, regression, and so on. In IoT applications, DL is commonly employed for data analytics. DL and the IoTs have been named as two of the top three strategic technical trends for 2017 [42]. The inadequacy of traditional ML approaches to match the expanding analytical needs of IoT systems is the reason for this zealous promotion of deep learning. Thankfully, advances in ML paradigms are allowing desirable data analytics in IoT applications to enter the picture. Picture recognition, data recovery, discourse recognition, regular language handling, indoor restriction, physiological and mental state recognition, and so on have all shown significant results using DL models, and these are the foundation administrations for IoT applications [43, 44].

DL has cleared a way for enormous forward leaps in the medical care field by finding historic structures like hierarchical computing design (HiCH); this, when combined with notions such as convolutional neural networks, generates CNN, allowing IoT devices to circumvent WBAN inaccurate restrictions [45]. ML classifiers that focus on missing values, decision tree generation, and other AI advancements, such as C4.5, C5.0, KNN, and EM, make the working module/architecture significantly more efficient [46, 47]. There are also several meta approaches that augment ML techniques for improved performance [48].

12.3 Application of Internet of Things Enabled Convolutional Neural Network in Healthcare Systems

Because of its significant level and inescapable checking, a few current advancements, similar to the Internet of Things, are turning out to be more open these days. The IoT likewise permits a methodical and skilled way to deal with patient medical care dependent on distant patient checking and portable wellbeing. Moreover, DL models are utilized in wellbeing-related applications to produce promising and good outcomes for a lot of information. There are breaks in information transmission to the cloud when following the patients' wellbeing status. Since it has shown extraordinary results in different enterprises, DL would assume a basic part in building a more intelligent IoT. Picture recognizable proof, recovery of data, sound ID, computational etymology, indoor situating, psychophysiological condition recognition, etc., are altogether instances of such applications, and these are the administrations that IoT applications depend on. Understanding the potential outcomes of DL for IoT prescient examination becomes urgent at this stage. This is because of the way that DL models are appropriate to handling the exceptionally mind boggling information produced by IoT gadgets. To blend enormous datasets detected from numerous modalities, certain profound learning models have been proposed. DL models have likewise been produced for separating fleeting associations from IoT information. DL models are more powerful in certain fields, and RBMs offer a great deal of possibilities with regard to including extraction and classification.

12.3.1 IoT-Enabled CNN for Improving Healthcare Diagnosis

The sophisticated IoT with unlimited networking possibilities for biomedical data analysis is strengthening the interaction between technology and healthcare society. In the last few years, deep neural networks (DNNs) and the rapid public acceptance of medical wearables have been successfully suddenly transformed. IoT enabled by DNN has allowed for revolutionary medical breakthroughs and unique probability in medical data processing in the healthcare industry [49]. Despite this development, there are still a number of concerns to be resolved in terms of service quality. Applying deep networks to deliver a high level of quality in important aspects in end-to-end reaction time, overhead, and accuracy is the key to prospering in the move from client-oriented to patient-oriented medical data analysis for healthcare society [50].

The profound learning procedures like CNNs, long short-term memory (LSTM), auto-encoders, profound generative models, and profound conviction networks have, as of now, been applied to effectively examine conceivable enormous assortments of information [50]. Utilizations of these techniques in clinical signals and pictures can help clinicians in clinical dynamic [51]. Due to higher time and network bandwidth, these home automation sensors for smart medical systems are still underdeveloped in clinical support infrastructure.

In [45], the authors define DL as a subset of machine learning techniques that have recently been used in a variety of domains. It has been proven to outperform traditional methods in speech signals and visual object detection. Multiple processing layers in deep learning models are capable of learning important aspects of input that is originally raw, all without the need for a domain level competence. Conventional ML models, on the other hand, typically require a significant amount of domain expertise to extract features before performing classifications. CNN is a form of deep neural network (DNN) that is commonly employed with two-dimensional inputs like movies and images. Using millions of images as inputs, they can learn hundreds of thousands of items.

The learning limit of CNN can be changed by changing the breadth and significance of the model. Besides the two-dimensional signs, CNNs are used with one-dimensional signs like electrocardiography (ECG) or sound signs. Commendable designing of CNN used for picture affirmation is made by clubbing various layers of handling units with differentiating occupations. The critical unit of CNN configuration is its convolutional layer which contains learnable channel banks that are started when a specific component is perceived. Max-pooling layer using CNN designing is used to lessen the quantity of limits and license overfitting [50]. Completely related layers overall follow series convolutional and max-pooling layers.

These layers' function is to act as a classifier for newly learned items. A wellbeing contextual analysis on an ECG characterization was also used by the authors in [45] to confirm a proposed design driven by CNN. In this case, the dynamic is aggregated at the edge, resulting in the client receiving notifications as a result of sickness recognition. The response speed and high accessibility are compared to a traditional IoT-based framework in which the cloud server performs all of the estimations, after which the HiCH's accuracy is determined. This provides a dynamic thought at the underlying phase of the observing and working throughout the checking.

12.3.2 IoT-Enabled CNN for Improving Healthcare Monitoring System

Wearable sensors and long-range casual correspondence stages expect a basic part in giving one more system to assemble patient data for powerful clinical benefits checking. Regardless, predictable patient perception using wearable sensors makes a ton of clinical consideration data. Moreover, the customer made clinical consideration data on long-range relational correspondence objections that come in tremendous volumes and are unstructured. The current clinical consideration checking structures are not useful at eliminating critical information from sensors and individual-to-individual correspondence data, and they experience issues taking apart it sufficiently. Also, the ordinary ML approaches are adequately not to manage clinical consideration colossal data for anomaly assumption. Accordingly, an original medical services observing system dependent on the IoT climate and a major information investigation motor utilizing DL models is vital to definitively store and dissect medical services information and to further develop the arrangement exactness.

Artificial intelligence (AI) allows machines to mimic human behavior, and ML algorithms are a subset of AI. In order to classify the given dataset, it includes training and learning components. ML, on the other hand, cannot perform effectively as the dataset grows in size and heterogeneity. As a result, contemporary data analytic research has concentrated on DL approaches, which continuously and reliably learn and classify massive volumes of data. Because of their effectiveness, CNN, auto-encoders, and their combinations are the often used models for monitoring assessment among the other frameworks in DL. Surveillance monitoring is becoming more important to ensure security in all types of buildings, from tiny businesses to major corporations. The IoT is currently used to facilitate faster and easier connectivity. This is because the majority of enterprises turn to the Internet to receive and transfer data. In addition, IoT devices deployed in remote sensing applications create a large amount of data. A Raspberry Pi, a CCTV camera, mobile devices, and other sensors are all part of the IoT circuit.

The checking system includes recording the exercises to watch the unusual exercises alarming individuals. Public spots like shopping centers, air terminals, and different spots where countless individuals meet are checked, which assists in getting people in general. Observing individuals fouling up things is called designated checking. A portion of the current checking devices are "smoke alarms," "entryway counters," and counting the tram travelers. In specific spots, observation checking utilizes electronic cameras, electronic listening gadgets, and building access cards to stay away from misconduct and ill-advised utilization of spots. Additionally, different ongoing applications are utilizing an observation checking framework. Among them are the following: (i) healthcare monitoring and controlling system; (ii) airport surveillance system; (iii) building and industry system; (iv) remote healthcare system; and (v) forest tracking system.

As of late, the utilization of interpersonal interaction in the medical care industry has been quickly expanding. The informal community information can likewise be used to recognize different factors like enthusiastic status and accumulated pressure, which may be converted into the situation with a patient wellbeing. Individuals with diabetes and strange BP share their feelings and encounters with others on informal communication locales [52]. They share important data and inspire each other to battle against diabetes and high/low BP. Likewise, diabetes patients distribute their viewpoints about explicit medications [53]. Another patient sees the assessments of others and reacts to them about similar medications. Along these lines, the medical services observing frameworks for diabetes and unusual BP need interpersonal interaction information to distinguish enthusiastic unsettling influences in patients by utilizing their posts, and to screen drug incidental effects by utilizing drug surveys. Be that as it may, the data on interpersonal interaction locales about persistent feelings and medication encounters are unstructured and unforeseen, and it would be a difficult errand for healthcare observing frameworks to extricate the data and investigate it to screen the patient's psychological wellbeing and to anticipate drug incidental effects. In this manner, there is a need of shrewd methodologies that can consequently separate the most significant elements and diminish the dimensionality of the datasets for better exactness of medical services observing framework.

Both healthcare and person-to-person communication information have significantly expanded in a couple of years, which is called huge information (both unstructured and organized) [54]. The conventional methodologies and ML strategies may not deal with these data very well for the extraction of significant data and for anomaly expectation. Moreover, these data may not help the medical care industry until they are handled keenly progressively.

This requires a major information cloud stage and a high level of profound learning approach, for example, CNN model [55].

12.4 The Challenges of Internet of Things Enabled Convolutional Neural Network in Healthcare Systems

Because IoT is now also employed in real-world applications, QoS is directly tied to data quality, which can be used in decision support applications. The information taken from medical service sensors should be gathered, moved, prepared, broke down, and utilized on schedule; nonetheless, now and again, IoT gadgets cannot offer the required information at its appropriate time. It represents a few difficulties to the nature of IoT medical service frameworks. Since clinical wearable systems work with critical second projects, they require a picky affirmation of QoS. It is a thrilling opening in the field of heterogeneous data collection, checking the patient in its continuous and supporting modified dynamic, according to QoS. As such, beating these hardships, according to QoS, is fundamental.

Today, the expenses of medical care and therapy gadgets observe to be a higher priority than some other time. Thinking about the creator's information, no relative review has been directed by the examination of IoT medical service frameworks' expenses. As a result, we believe cost analysis is an open question in the IoT medical care framework. Even in wealthy countries, the high cost of testing equipment in the IoT medical service framework is a serious challenge. The IoT has not yet made treatment administrations more cost-effective for the general public. The rise in the expense of healthcare devices is a cause for concern for everyone [56].

Since a major volume of unstructured information is created, understanding them is extremely complex. The projects gather the information identified with the patient's wellbeing consistently; so unique stockpiling systems are required contrasted with the typical datasets. Accordingly, one of the difficulties in this field is information stockpiling and the board. These have a major volume of data and incorporate a mind boggling, various, and rich ground of medical care data. Vulnerability in this information is exceptionally high. In this way, the improvement and viability of the legitimacy of the information identified with wellbeing and securing helpful information from them are troublesome. Henceforth, the examination in investigating these large data related with wellbeing in a medical services climate will be a need for settling on better clinical choices and incorporating required strategies.

Medical service program designers experience a few difficulties like security, protection, and dataset innovation; information protection and security are significant things in medical care programs dependent on IoT [57, 58]. Security can be characterized as legitimacy the board and setting some entrance rules to the patient's projects and data. Information security in IoT-based medical care is of high significance Medical service program designers experience a few. Medical service programs are planned and created by getting data from IoT devices. Broad information that are moved and put away consistently can be hacked and taken advantage of in the patient and the doctor's essence. These programmers can create counterfeit identifiers for purchasing drugs all together to take advantage of them. Security and protection and keeping away from data spillage are the primary worries in keeping up with the patient or the client's data, guaranteeing information offering to others with the patient's understanding [56, 58].

12.5 Framework for Internet of Things Enabled CNN in Healthcare Systems

The proposed model has four layers having a wearable device, gateway, IoT-cloud layer with CNN processing model, and the user connectivity called application layer. The captured data can be processed using CNN model and the result is used to monitor and predict the cancer patients; thus, the specialist and caregivers can handle the treatment using related information in the dedicated server of the IoT-cloud database. The captured data from the patients can be sent to the cloud database through the Wi-Fi wireless connectivity gateway for the processing of the collected data in the cloud network. The CNN-based model was used to detect humans with cancer to produce more precise calculation and better accuracy. The system modeled contains the wearable devices connected to IoT cloud sensor nodes. The cancer patient is detected with the captured data using various devices stored in the cloud database.

The data is sent to the data model and applies the CNN to detect and monitor the cancer patient with the use of the captured data stored in the cloud database by collecting the stored physiological signs and performing the task with the model. The user application layer is made up of I users who are human beings (cancer's and non-cancer's) in a smart environment. The data server sends the information to the clinical assessment whenever an emergency alarm is recognized. A signal processing unit is activated with

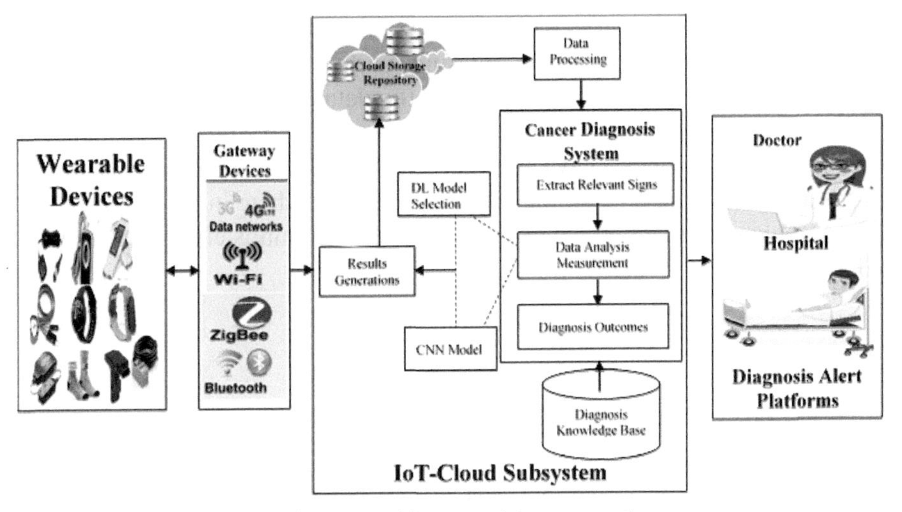

Figure 12.2 The architecture of the proposed system.

IoT-based assistance in the data model to control parallel computation and cloud base for permanent access with the acquired data. A period successive investigation of the gathered examined information is acted in the equal handling, and, thus, the information is thought to be of the downrange of inspecting with 125 Hz. Figure 12.2 displays the proposed system framework.

12.5.1 The Practical Application of the Proposed Framework

12.5.1.1 Dataset

The dataset "Wisconsin Diagnostic Breast Cancer (WDBC)" was made by Dr. William Wolberg at the University of Wisconsin and is accessible at the UCI AI store [59]. It was utilized as a dataset for execution of the proposed study for planning AI-based framework for the conclusion of bosom disease. The dataset has a size of 569 subjects with 32 qualities and 30 elements being genuine worth components. The objective yield mark analysis has two classes to address the threatening or harmless subject. The class dispersion is 357 harmless and 212 threatening subjects. Subsequently, the dataset is a 569 $*$ 32 component framework.

12.5.1.2 Dataset pre-processing

Prior to applying the AI calculations for grouping issues, information handling is important. The handled information [60] diminished the calculation season of the classifier and expanded the characterization execution of the classifier. Techniques, for example, missing worth location, standard scalar, and min–max scalar are broadly applied to the dataset pre-processing. Standard scalar guarantees that each element has mean 0 and difference 1; consequently, all provisions have a similar coefficient. Min–max scalar moves the information so that all components are gone somewhere in the range of 0 and 1. The element that has a vacant worth in the column is eliminated from the dataset.

$$Y = \frac{Y - \min(Y)}{\max(Y) - \min(Y)}. \tag{12.1}$$

12.5.2 Convolutional Neural Network Classification

A CNN model to group the malignant growth dataset was acquainted in this section with help in disease finding. The engineering of the proposed CNN model comprised essentially of three convolutional layers and a completely associated layer. Each convolutional layer was trailed by a batch normalization (BN) layer that standardized the yields of the convolutional layer, a rectified linear unit (ReLU) layer that applied an initiation capacity to its feedback esteems, and a maximum pooling layer that directed a down-examining activity. The proposed CNN embraced a normal pooling layer before the completely associated layer to diminish the elements of the component esteems contribution to the completely associated layer. Following the work by [61], a dropout pace of 0.5 was utilized between the normal pooling layer and the completely associated layer to keep away from overfitting and increment the exhibition. The proposed model additionally attempted spatial dropout between every maximum pooling layer and its after convolutional layer however tracked down that such dropouts brought about execution corruption [62]. Thus, the model did not matter spatial dropout. As info, the organization takes the component upside of a malignancy dataset, and it yields the likelihood that the example has a place with a specific class (e.g., the likelihood that the comparing patient has bosom disease). The occasions were taken care of into the model layer by layer. The contribution to each layer is the yield upside of the past layer. The layers perform explicit changes on the information esteems and afterward pass the handled qualities to the following layer.

The deep learning employed in this chapter is called CNN for the classification of various physiological symptoms and sighs of persons collected using wearable devices and sensors [63]. To obtain the initial value of features, the dataset segmentation is done semantically, and the trained data labeled is collected as

$$\{(x_k,\ y_k)\,|k = 1,\ 2,\ \ldots,\ N\} \tag{12.2}$$

on the dataset. From eqn (12.2), x_k represents the center class of the target size from the instances, and $y_k = (y_{k1},\ y_{k2},\ \ldots,\ y_{ky}) \in \mathbb{R}^4$ represents the label of the target instance. If x_k belongs to the *i*th class if label. From eqn (12.2), x_k denotes the center patches of the target pixel size of 128×128, and $y_k = y_{k1},\ldots,y_k\ 4 \in R^4$ denotes the label of the target pixels. If the label $y_k = 1$, else $y_k = 0$. The principle objective of the underlying assessment of CNN is addressed as a capacity f^* and then it is viewed as that the fix *X* has four names and the capacity for preparing information as

$$f^* = \underset{f}{\operatorname{argmin}} \sum_{k=1}^{N} \|(fX_k) - y_k\| \ . \tag{12.3}$$

In view of the CNN design, the capacity f is composed as

$$f(X) = f^{(1)}\left(f^{(0)}\ldots\left(f^{(0)}(X)\right)\right) \tag{12.4}$$

where $f^{(j)}$ addresses the jth layer on CNN. Likewise, $f^{(j)}$ perhaps any of the layers, for example, convolutional, pooling, or completely associated. In such an event, assuming we are utilizing $f^{(j)}$ the inclination and weight in f, the capacity is dictated by θ the boundaries as

$$\theta = (W_1,\ b_1,\ldots). \tag{12.5}$$

Eqn (12.4) can be expressed as eqn (12.6), using eqn (12.5) as a starting point.

$$\theta^* = \min_{\theta} \frac{1}{N} \sum_{k=1}^{N} L(K_k,\ y_k;\ \theta) \tag{12.6}$$

where eqn (12.6) portrays, rather than deciding f^*, the arrangement of θ^* boundaries can be determined. *L* shows the misfortune work for cross-entropy used to assess the blunder happened between label *y* and $p_i\ (x,\ \theta)$ yield acquired from each mark (class) as

$$L(x_k,\ y_k;\ \theta) = -\sum_{i=1}^{4} Y_{ki}\log p_i(x_k;\ \theta)\ . \tag{12.7}$$

It returns the least-squares error for all training patches, as referred to in [64].

12.5.3 Performance Evaluation Metrics

The classifier was evaluated by computing various performance metrics using the following.

The evaluations of the proposed classifier were assessed by processing the rates of sensitivity (SE), specificity (SP), and characterization precision; the separate definitions are as follows:

a. Accuracy in %

$$\text{accuracy}\,(T) = \frac{\sum_{i=1}^{|T|} \text{assess}\,(t_i)}{|T|} \times 100 \tag{12.8}$$

where $\text{assess}\,(t_i) = \begin{cases} 1, \text{if classif } (t) = t.c \\ 0, \text{ otherwise} \end{cases}$.

Here, T is the initial activities, $t \in T$, $t.c$ is the object's class, and class (t) provides the algorithm's categorization of t.

b. Sensitivity (SE) in %

$$\text{SE} = \frac{\text{TP}}{\text{TP} + \text{FN}} \times 100. \tag{12.9}$$

c. Specificity (SP) in %

$$\text{SE} = \frac{\text{TN}}{\text{TN} + \text{FP}} \times 100 \tag{12.10}$$

where TP, TN, FP, and FN signify the following:

True positives: no cancer is classified as such;
True negatives: correctly labels cancer patients as not cancer;
False positives: no cancer is classified as cancer;
False negatives: incorrectly labels cancer patients as cancer.

12.6 Results and Discussion

The planned technique has been implemented in R programming language software and the outcomes are evaluated to determine its efficacy. The R program is deployed on a system with an Intel Core i7-7th Gen CPU running at 3.40 GHz and 64 GB of RAM. To examine the effectiveness, the set is generated in matrix form and obtained during cancer screening. During the evaluation, the system used the screen's input data together with other factors. To avoid false alerts, each class and module has a threshold level specified to it.

Table 12.1 Classification matrix for the proposed model

Real/predicted	Malignant	Benign	Total
Exist	198	7	205
Non-exist	13	351	364
Total	205	364	569

Table 12.2 Proposed method evaluation

Dataset	Accuracy	Specificity	Sensitivity	PPV	NPV	F-score	AUC
WDBC	98.4%	98.7%	98.9%	97.6%	97.5%	98.3%	0.992

12.6.1 Performance Evaluation of the Proposed Model

Table 12.1 shows the classification matrix for the proposed model employed in this study to classify the dataset.

Table 12.2 summarizes the suggested model's accuracy, specificity, AUC, sensitivity, positive/negative predictive value, and F-score characteristics. The proposed model has an accuracy of 98.4%, specificity of 98.7%, sensitivity of 98.9%, positive predictive value (PPV) of 97.6%, negative predictive value (NPV) of 97.5%, F-score of 98.3%, and AUC of 0.992 values, respectively. Figure 12.3 shows the performance evaluation of the proposed classifier.

12.6.2 Performance Comparison of the Proposed Model

The proposed model was compared with existing methods that used the same dataset to evaluate the performance of their model after implementing the technique. Table 12.3 shows the results of the comparison with other existing models. The various current models used DL techniques like dense convolutional network (DCN), multiple-weight SVM (MWSVM), genetic grey based neural network (G2NN), and fuzzy genetic grey-wolf based CNN model (FG2CNN) in classifying the cancer dataset.

The proposed model was compared with recent works that used the same dataset especially those that used DL methods for cancer classification. The model with higher classification results used feature selection which the proposed model failed to use on the dataset; feature selection helps the highest model to classify the dataset better than other DL techniques without feature extraction. The highest model has an accuracy of 99.7%, specificity of 98.4, sensitivity of 98.7%, and F-score of 98.5%, respectively. The proposed came second with an accuracy of 98.4%, specificity of 98.7%, sensitivity of 98.9%, and F-score of 98.3%, respectively. The least performance classifier has an

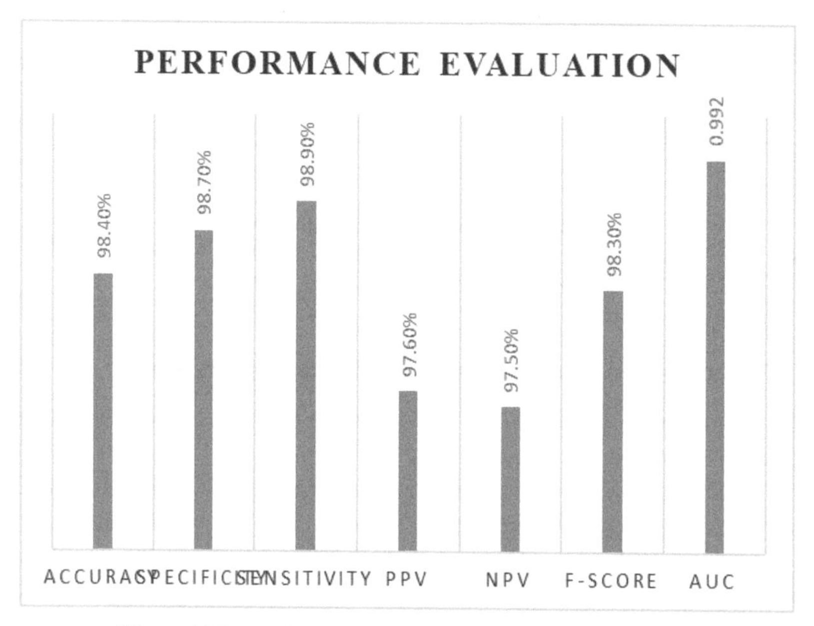

Figure 12.3 Performance evaluation of the proposed model.

Table 12.3 Performance comparison of the model with other existing methods using the same dataset

Models	Accuracy	Specificity	Sensitivity	*F*-score
MWSVM [65]	98.5	96.2	97.9	98.8
G^2NN [65]	98.9	97.2	97.8	97.4
Fuzzy + G^2CNN [65]	99.7	98.4	98.7	98.5
DCN [66]	94.9	94.5	96.9	96.0
KNN [67]	96.5	95.7	95.9	96.2
Proposed model	98.9	98.7	98.4	97.9

accuracy of 94.9%, specificity of 94.5, sensitivity of 96.9%, and *F*-score of 96.0%, respectively. The results show that the proposed did not perform badly and yielded a better classification using various metrics with other DL models.

12.7 Conclusion and Future Directions

One of the dangerous illnesses among women in the world today is breast cancer. Researchers have tried to use ML models for the classification of breast cancer, but their accuracy and efficiency still remains questionable.

Hence, DL techniques have been proposed for high accuracy and efficiency especially when huge amount of data is involving. DL models became popular in various fields for the classification of huge data especially in healthcare systems. The accuracy and efficiency of CNN classifier has been used in classification of various illnesses. Therefore, this chapter proposed an intelligent IoT-based enabled CNN for the classification of breast cancer for better accuracy and efficiency and, thus, will help medical doctors to detect and treat cancer patient in real time and remotely. The performance of the proposed system using various metrics shows that the CNN performance better compare with existing methods with an accuracy of 98.4%, specificity of 98.7%, sensitivity of 98.9%, and F-score of 98.3%, respectively. Future work will consider the use of feature selection to select revenant features by removing unwanted and unproductive features from the dataset. The security and privacy of the IoT-based systems will also be taken into consideration to enable the users have total trust in the use of the IoT-based systems.

References

[1] Rhayem, A., Mhiri, M. B. A., & Gargouri, F. (2020). Semantic web technologies for the Internet of Things: Systematic literature review. Internet of Things, 11, 100206.

[2] Adeniyi, E. A., Ogundokun, R. O., & Awotunde, J. B. (2021). IoMT-based wearable body sensors network healthcare monitoring system. Studies in Computational Intelligence, 933, pp. 103–121.

[3] Bojan, T. M., Kumar, U. R., & Bojan, V. M. (2014, December). An internet of things based intelligent transportation system. In 2014 IEEE International Conference on Vehicular Electronics and Safety (pp. 174–179). IEEE.

[4] Zhang, H., & Lu, X. (2020). Vehicle communication network in intelligent transportation system based on internet of things. Computer Communications, 160, 799–806.

[5] Pillai, R., & Sivathanu, B. (2020). Adoption of internet of things (IoT) in the agriculture industry deploying the BRT framework. Benchmarking: An International Journal.

[6] Kour, V. P., & Arora, S. (2020). Recent developments of the internet of things in agriculture: A survey. IEEE Access, 8, 129924–129957.

[7] Gomathi, P., Baskar, S., & Shakeel, P. M. (2021). Concurrent service access and management framework for user-centric future internet of things in smart cities. Complex & Intelligent Systems, 7(4), 1723–1732.

[8] Zhang, Y., Xiong, Z., Niyato, D., Wang, P., & Han, Z. (2020). Information trading in internet of things for smart cities: A market-oriented analysis. IEEE Network, 34(1), 122–129.

[9] Abikoye, O. C., Bajeh, A. O., Awotunde, J.B., ... & Salihu, S. A. (2021). Application of Internet of Thing and cyber physical system in Industry 4.0 smart manufacturing. In Advances in Science, Technology and Innovation (pp. 203–217).

[10] Malik, P. K., Sharma, R., Singh, R., Gehlot, A., Satapathy, S. C., Alnumay, W. S., ... & Nayak, J. (2020). Industrial Internet of Things and its applications in industry 4.0: State of the art. Computer Communications.

[11] Awotunde, J. B., Folorunso, S. O., Bhoi, A. K., Adebayo, P. O., & Ijaz, M. F. (2021). Disease diagnosis system for IoT-based wearable body sensors with machine learning algorithm. In Hybrid Artificial Intelligence and IoT in Healthcare (pp. 201–222). Springer, Singapore.

[12] Birje, M. N., & Hanji, S. S. (2020). Internet of Things based distributed healthcare systems: A review. Journal of Data, Information and Management, 2, 149–165.

[13] Awotunde, J. B., Jimoh, R. G., AbdulRaheem, M., Oladipo, I. D., Folorunso, S. O., & Ajamu, G. J. (2022). IoT-based wearable body sensor network for COVID-19 pandemic. Studies in Systems, Decision and Control, 378, pp. 253–275.

[14] Awotunde, J. B., Ogundokun, R. O., & Misra, S. (2021). Cloud and IoMT-based Big Data analytics system during COVID-19 pandemic. Internet of Things, pp. 181–201.

[15] Awotunde, J. B., Jimoh, R. G., Oladipo, I. D., Abdulraheem, M., Jimoh, T. B., & Ajamu, G. J. (2021). Big data and data analytics for an enhanced COVID-19 epidemic management. In Artificial Intelligence for COVID-19 (pp. 11–29). Springer, Cham.

[16] Jahan, T. (2021). Machine learning with IoT and big data in healthcare. In Intelligent Healthcare (pp. 81–98). Springer, Cham.

[17] Marques, G., & Pitarma, R. (2018, November). Smartwatch-based application for enhanced healthy lifestyle in indoor environments. In International Conference on Computational Intelligence in Information System (pp. 168–177). Springer, Cham.

[18] Manogaran, G., Varatharajan, R., Lopez, D., Kumar, P. M., Sundarasekar, R., & Thota, C. (2018). A new architecture of Internet of Things and big data ecosystem for secured smart healthcare monitoring and alerting system. Future Generation Computer Systems, 82, 375–387.

[19] Rath, M. (2018). A methodical analysis of application of emerging ubiquitous computing technology with fog computing and IoT in diversified fields and challenges of cloud computing. International Journal of Information Communication Technologies and Human Development (IJICTHD), 10(2), 15–27.

[20] Dias, D., & Paulo Silva Cunha, J. (2018). Wearable health devices—Vital sign monitoring, systems and technologies. Sensors, 18(8), 2414.

[21] Özdemir, V., & Hekim, N. (2018). Birth of industry 5.0: Making sense of big data with artificial intelligence, "the internet of things" and next-generation technology policy. Omics: A Journal of Integrative Biology, 22(1), 65–76.

[22] Oniani, S., Marques, G., Barnovi, S., Pires, I. M., & Bhoi, A. K. (2021). Artificial intelligence for Internet of Things and enhanced medical systems. In Bio-Inspired Neurocomputing (pp. 43–59). Springer, Singapore.

[23] Marques, M. S. G., & Pitarma, R. (2016). Smartphone application for enhanced indoor health environments. Journal of Information Systems Engineering & Management, 1, 4.

[24] Belciug, S., & Gorunescu, F. (2020). *Intelligent Decision Support Systems–A Journey to Smarter Healthcare*. USA: Springer International Publishing.

[25] Phillips-Wren, G., Pomerol, J. C., Neville, K., & Adam, F. (2020). Supporting Decision Making During a Pandemic: Influence of Stress, Analytics, Experts, and Decision Aids. In The Business of Pandemics (pp. 183–212). Auerbach Publications.

[26] Darshan, K. R., & Anandakumar, K. R. (2015, December). A comprehensive review on usage of Internet of Things (IoT) in healthcare system. In 2015 International Conference on Emerging Research in Electronics, Computer Science and Technology (ICERECT) (pp. 132–136). IEEE.

[27] Kumar, V. (2015). Ontology based public healthcare system in Internet of Things (IoT). Procedia Computer Science, 50, 99–102.

[28] Gill, S. S., Tuli, S., Xu, M., Singh, I., Singh, K. V., Lindsay, D., ... & Garraghan, P. (2019). Transformative effects of IoT, blockchain and artificial intelligence on cloud computing: Evolution, vision, trends and open challenges. Internet of Things, 8, 100118.

[29] Ştefănescu, D., Streba, C., Cârţână, E. T., Săftoiu, A., Gruionu, G., & Gruionu, L. G. (2016). Computer aided diagnosis for confocal

laser endomicroscopy in advanced colorectal adenocarcinoma. PloS one, 11(5), e0154863.

[30] Awotunde, J. B., Jimoh, R. G., Folorunso, S. O., Adeniyi, E. A., Abiodun, K. M., & Banjo, O. O. (2021). Privacy and security concerns in IoT-based healthcare systems. Internet of Things, 105–134.

[31] Aleksandrovna, M. E., Fedorovich, N. A., & Viktorovna, P. T. (2016). Potential of the Internet network in formation of the assortment of the trade organizations. European Science Review, 1–2.

[32] Padikkapparambil, J., Ncube, C., Singh, K. K., & Singh, A. (2020). Internet of Things technologies for elderly health-care applications. In Emergence of Pharmaceutical Industry Growth with Industrial IoT Approach (pp. 217–243). Academic Press.

[33] Bharathi, R., Abirami, T., Dhanasekaran, S., Gupta, D., Khanna, A., Elhoseny, M., & Shankar, K. (2020). Energy efficient clustering with disease diagnosis model for IoT based sustainable healthcare systems. Sustainable Computing: Informatics and Systems, 28, 100453.

[34] Matthews, M., Abdullah, S., Gay, G., & Choudhury, T. (2014). Tracking mental well-being: Balancing rich sensing and patient needs. Computer, 47(4), 36–43.

[35] Chiuchisan, I., Costin, H. N., & Geman, O. (2014, October). Adopting the Internet of Things technologies in health care systems. In 2014 International Conference and Exposition on Electrical and Power Engineering (EPE) (pp. 532–535). IEEE.

[36] Luo, J., Chen, Y., Tang, K., & Luo, J. (2009, December). Remote monitoring information system and its applications based on the Internet of Things. In 2009 International Conference on Future BioMedical Information Engineering (FBIE) (pp. 482–485). IEEE.

[37] Tomar, P., Kaur, G., & Singh, P. (2018). A prototype of IoT-based real time smart street parking system for smart cities. In Internet of Things and Big Data Analytics Toward Next-Generation Intelligence (pp. 243–263). Springer, Cham.

[38] Mining, W. I. D. (2006). Data mining: Concepts and techniques. Morgan Kaufinann, 10, 559–569.

[39] Mukhopadhyay, A., Maulik, U., Bandyopadhyay, S., & Coello, C. A. C. (2013). A survey of multiobjective evolutionary algorithms for data mining: Part I. IEEE Transactions on Evolutionary Computation, 18(1), 4–19.

[40] Kesavaraj, G., & Sukumaran, S. (2013, July). A study on classification techniques in data mining. In 2013 Fourth International Conference on Computing, Communications and Networking Technologies (ICCCNT) (pp. 1–7). IEEE.

[41] Sun, S. (2011). *Analysis and Acceleration of Data Mining Algorithms on High Performance Reconfigurable Computing Platforms*. USA: Iowa State University.

[42] Wang, J., Ma, Y., Zhang, L., Gao, R. X., & Wu, D. (2018). Deep learning for smart manufacturing: Methods and applications. Journal of manufacturing systems, 48, 144–156.

[43] Dang, L. M., Piran, M., Han, D., Min, K., & Moon, H. (2019). A survey on internet of things and cloud computing for healthcare. Electronics, 8(7), 768.

[44] Alam, K. M., Saini, M., & El Saddik, A. (2015). Toward social internet of vehicles: Concept, architecture, and applications. IEEE Access, 3, 343–357.

[45] Azimi, I., Takalo-Mattila, J., Anzanpour, A., Rahmani, A. M., Soininen, J. P., & Liljeberg, P. (2018, September). Empowering healthcare IoT systems with hierarchical edge-based deep learning. In 2018 IEEE/ACM International Conference on Connected Health: Applications, Systems and Engineering Technologies (pp. 63–68).

[46] Sharma, S., Agrawal, J., & Sharma, S. (2013). Classification through machine learning technique: C4. 5 algorithm based on various entropies. International Journal of Computer Applications, 82(16).

[47] Awotunde, J. B., Ajagbe, S. A., Oladipupo, M. A., Awokola, J. A., Afolabi, O. S., Mathew, T. O., & Oguns, Y. J. (2021, October). An improved machine learnings diagnosis technique for COVID-19 pandemic using chest X-ray images. Communications in Computer and Information Science, 1455 CCIS, 319–330.

[48] Mehala, B., Thangaiah, P. R. J., & Vivekanandan, K. (2009). Selecting scalable algorithms to deal with missing values. International Journal of Recent Trends in Engineering, 1(2), 80.

[49] Yin, W., Yang, X., Zhang, L., & Oki, E. (2016). ECG monitoring system integrated with IR-UWB radar based on CNN. IEEE Access, 4, 6344–6351.

[50] Folorunso, S. O., Awotunde, J. B., Adeboye, N. O., & Matiluko, O. E. (2022). Data classification model for COVID-19 pandemic. Studies in Systems, Decision and Control, 378, 93–118.

[51] Al-Turjman, F., Zahmatkesh, H., & Mostarda, L. (2019). Quantifying uncertainty in internet of medical things and big-data services using intelligence and deep learning. IEEE Access, 7, 115749–115759.

[52] Oladipo, I. D., Babatunde, A. O., Awotunde, J. B., & Abdulraheem, M. (2020, November). An improved hybridization in the diagnosis of diabetes mellitus using selected computational intelligence. Communications in Computer and Information Science, 1350, 272–285.

[53] Jimenez, G., Lum, E., & Car, J. (2019). Examining diabetes management apps recommended from a Google search: content analysis. JMIR mHealth and uHealth, 7(1), e11848.

[54] Anshari, M., & Alas, Y. (2015). Smartphones habits, necessities, and big data challenges. The Journal of High Technology Management Research, 26(2), 177–185.

[55] Wang, Y., Sun, Y., Liu, Z., Sarma, S. E., Bronstein, M. M., & Solomon, J. M. (2019). Dynamic graph CNN for learning on point clouds. ACM Transactions on Graphics (tog), 38(5), 1–12.

[56] Aghdam, Z. N., Rahmani, A. M., & Hosseinzadeh, M. (2020). The role of the Internet of Things in healthcare: Future trends and challenges. Computer Methods and Programs in Biomedicine, 105903.

[57] Qi, J., Yang, P., Min, G., Amft, O., Dong, F., & Xu, L. (2017). Advanced internet of things for personalised healthcare systems: A survey. Pervasive and Mobile Computing, 41, 132–149.

[58] Awotunde, J. B., Chakraborty, C., & Adeniyi, A. E. (2021). Intrusion detection in Industrial Internet of Things network-based on deep learning model with rule-based feature selection. Wireless Communications and Mobile Computing.

[59] Wolberg, W. H., Street, N., & Mangasarian, O. L. (1995). *Wisconsin Diagnostic Breast Cancer (WDBC)*. USA: University of California.

[60] Kotsiantis, S. B., Kanellopoulos, D., & Pintelas, P. E. (2006). Data preprocessing for supervised leaning. International Journal of Computer Science, 1(2), 111–117.

[61] Srivastava, N., Hinton, G., Krizhevsky, A., Sutskever, I., & Salakhutdinov, R. (2014). Dropout: A simple way to prevent neural networks from overfitting. The Journal of Machine Learning Research, 15(1), 1929–1958.

[62] Tompson, J., Goroshin, R., Jain, A., LeCun, Y., & Bregler, C. (2015). Efficient object localization using convolutional networks. In IEEE Conference on Computer Vision and Pattern Recognition (pp. 648–656).

[63] LeCun, Y., Bengio, Y., & Hinton, G. (2015). Deep learning. Nature, 521(7553), 436–444.

[64] Tieleman, T. (2014). *Optimizing Neural Networks that Generate Images*. Canada: University of Toronto.

[65] Kannan, A. G. (2021). Fuzzy genetic grey wolf based deep learning model for classification on breast cancer dataset. Turkish Journal of Computer and Mathematics Education (TURCOMAT), 12(11), 4532–4541.

[66] Güldoğan, E., Zeynep, T. U. N. Ç., & Çolak, C. (2020). Classification of breast cancer and determination of related factors with deep learning approach. The Journal of Cognitive Systems, 5(1), 10–14.

[67] Wadhwa, G., & Mathur, M. (2020, November). A convolutional neural network approach for the diagnosis of breast cancer. In 2020 Sixth International Conference on Parallel, Distributed and Grid Computing (PDGC) (pp. 357–361). IEEE.

Index

About the Editors

Mohd Naved is a machine learning consultant and researcher, currently teaching in Amity International Business School (AIBS), Amity University for various degree and research programs in data science, analytics and machine learning. He is actively engaged in academic research on various topics in management as well as on 21st century technologies. He has published 40+ research articles in reputed journals (SCI/Scopus/ABDC indexed). He has 17 patents in AI/ML and is actively engaged in the commercialization of innovative products developed at university level. Interviews with him have been published in various national and international magazines. A former data scientist, he is an alumnus of Delhi University. He holds a PhD from Noida International University.

V. Ajantha Devi is working as Research Head in AP3 Solutions, Chennai, Tamil Nadu, India. She received her Ph.D. from University of Madras in 2015. She has worked as Project Fellow under a UGC Major Research Project. She is a Senior Member of IEEE. She has been certified as a Microsoft Certified Application Developer (MCAD) and Microsoft Certified Technical Specialist (MCTS) from Microsoft Corp. She has more than 35 papers in international journals and conference proceedings to her credit. She has written, co-authored, and edited a number of books in the field of computer science with international and national publishers such as Elsevier, Springer, etc. She has been a member of the Program Committee/Technical Committee/Chair/Review Board for a variety of international conferences. She has five Australian Patents and one Indian Patent to her credit in the areas of artificial intelligence, image processing and medical imaging. Her work in image processing, signal processing, pattern matching, and natural language processing is based on artificial intelligence, machine learning, and deep learning techniques. She has won many Best paper presentation awards as well as a few research-oriented international awards.

Loveleen Gaur is Professor and Program Director of Artificial Intelligence, Business Intelligence and Data Analytics at the Amity International Business School, Amity University, Noida, India. Her research areas cover interdisciplinary fields including but not limited to artificial intelligence, machine learning and IoT. She is an established author and researcher and has filed five patents and two copyrights in AI/IoT. She is a senior IEEE member and series editor with CRC.

Ahmed A. Elngar is Assistant Professor of Computer Science at the Faculty of Computers and Artificial Intelligence, Beni-Suef University, Egypt. Dr. Elngar is the Founder and Head of the Scientific Innovation Research Group (SIRG). He is a Director of the Technological and Informatics Studies Center (TISC), Faculty of Computers and Artificial Intelligence, Beni-Suef University. He has more than 55 scientific research papers published in prestigious international journals and over 25 books covering such diverse topics as data mining, intelligent systems, social networks and smart environment. Dr. Elngar is a collaborative researcher and is a member of the Egyptian Mathematical Society (EMS) and International Rough Set Society (IRSS). His other research areas include internet of things (IoT), network security, intrusion detection, machine learning, data mining, artificial intelligence, big data, authentication, cryptology, healthcare systems, and automation systems. He is an editor and reviewer of many international journals around the world. Dr. Elngar has won several awards including the Young Researcher in Computer Science Engineering at the Global Outreach Education Summit and Awards 2019, January 2019, Delhi, India. Also, he was awarded Best Young Researcher Award at the Global Education and Corporate Leadership Awards (GECL-2018).

9788770227254